MIXED-SPECIES ASSOCIATION OF *CERCOPITHECUS* MONKEYS IN THE KAKAMEGA FOREST, KENYA

Mixed-Species Association of *Cercopithecus* Monkeys in the Kakamega Forest, Kenya

by Marina Cords

UNIVERSITY OF CALIFORNIA PRESS
Berkeley • Los Angeles • London

UNIVERSITY OF CALIFORNIA PUBLICATIONS IN ZOOLOGY

Editorial Board: Peter B. Moyle, James L. Patton, Donald C. Potts, David S. Woodruff

Volume 117
Issue Date: February 1987

UNIVERSITY OF CALIFORNIA PRESS
BERKELEY AND LOS ANGELES, CALIFORNIA

UNIVERSITY OF CALIFORNIA PRESS, LTD.
LONDON, ENGLAND

ISBN 0-520-09717-3
LIBRARY OF CONGRESS CATALOG CARD NUMBER: 86-30822

© 1987 BY THE REGENTS OF THE UNIVERSITY OF CALIFORNIA
PRINTED IN THE UNITED STATES OF AMERICA

Library of Congress Cataloging-in-Publication Data

Cords, Marina.
  Mixed-species association of Cercopithecus monkeys in the Kakamega Forest, Kenya.

  (University of California publications in zoology; v. 117)
  Bibliography: p.
  1. Cercopithecus mitis--Kenya--Kakamega Forest Reserve--Behavior. 2. Cercopithecus mitis--Kenya--Kakamega Forest Reserve--Ecology. 3. Cercopithecus ascanius--Kenya--Kakamega Forest Reserve--Behavior. 4. Cercopithecus ascanius--Kenya--Kakamega Forest Reserve--Ecology. 5. Kakamega Forest Reserve (Kenya) 6. Mammals--Kenya--Kakamega Forest Reserve--Behavior. 7. Mammals--Kenya--Kakamega Forest Reserve--Ecology. I. Title. II. Series.

QL737.P93C625 1987   599.8'2   86-30822
ISBN 0-520-09717-3 (pbk.)

To my parents, Jutta and Helmuth Cords

Contents

List of Figures, *ix*

List of Tables, *x*

Acknowledgments, *xii*

Abstract, *xiii*

1. Introduction — *1*
   Other Vertebrate Taxa, *1*
   Competition, *3*
   Hypotheses, *5*

2. Study Site, Study Animals, and Methods — *11*
   The Kakamega Forest, *11*
   Study Animals and Study Area, *13*
   Methods, *20*

3. Dynamics of Mixed-Species Associations — *23*
   Temporal Patterns of Mixed-Species Associations, *23*
   Who Is Responsible?, *28*
   Behavior During Formation and Dissolution of Mixed-Species Groups, *29*
   Evaluation of a Null Hypothesis, *29*

4. Behavioral Relationships Between Species in Mixed Groups — *32*
   Interactions Between Species, *32*
   Interspecific Spatial Relationships, *35*
   Activity Patterns, *36*

5. Predation — 40
    Predators, *40*
    Monkeys' Response to Predators, *41*
    Mixed-Species Groups and Predation, *43*
    Evaluation of the Predation-Avoidance Hypothesis, *46*

6. Feeding — 48
    Methods and Definitions, *48*
    The Diets of Blue Monkeys and Redtails, *49*
    Diet Composition and Temporal Variation in Mixed-Species Grouping, *54*
    Comparison of Diet in and Not in Mixed-Species Groups, *55*
    Evaluation of Hypotheses, *56*

7. Ranging — 60
    Methods, *60*
    Ranging Patterns of Blue Monkeys and Redtails, *61*
    Ranging and Mixed-Species Groups, *70*
    Some Hypotheses Reconsidered, *74*

8. Discussion and Conclusions — 79
    Predation, *79*
    Food, *81*
    Competition, *82*
    Invertebrate Feeding, *83*
    Demography, *83*
    A Comparative Perspective, *85*
    Conclusions, *89*

Appendixes, *91*

Literature Cited, *101*

List of Figures

Figure 1. Rainfall at the study site, *12*
Figure 2. Study area, focal group home ranges, and vegetation plots, *15*
Figure 3. Diurnal fluctuations in association, *25*
Figure 4. Association lengths, *27*
Figure 5. Diurnal variation in activity, *39*
Figure 6. Diurnal variation in distance traveled, *63*
Figure 7. Typical daily ranging patterns, *65*
Figure 8. Cumulative percentage of areas used for different numbers of days in five-consecutive-day samples, *67*
Figure 9. Home range use over 11 sample months, *68*
Figure 10. Diurnal variation in occupation of different strata, *71*

List of Tables

Table 1. Adult body weights (in pounds), *14*

Table 2. Group composition on last complete count , *16*

Table 3. Frequency, dispersion, and phenology of Kakamega forest trees and shrubs, *18*

Table 4. Monthly variation in the amount of time spent associated, *24*

Table 5. Frequency (per day) of associations, *26*

Table 6. Interspecific social interactions, *33*

Table 7. Relationship between relative body size and the outcome of interspecific agonistic encounters, *33*

Table 8. Species identity of nearest neighbors in feeding trees for different age-sex classes (%), *36*

Table 9. Activity scores of blue monkeys and redtails, all hours combined, as a function of association (% in parentheses), *38*

Table 10. Coincidence of alarm vocalizations by adult males, excluding known false alarms, *42*

Table 11. Distribution of eagle encounters during the day, *44*

Table 12. Does association affect the likelihood of attack? (Data Set I), *45*

Table 13. Composition of the plant diet (% of feeding scores), *49*

Table 14. Percentage overlap in the plant diet of blue monkeys and redtails, by month, *50*

Table 15. Selection ratios for top five plant species (ranked top to bottom in each month) in plant diets, *51*

Table 16. Diurnal fluctuations in items consumed by blue monkeys and redtails, *53*

Table 17. Invertebrate capture methods (% of captures) when alone or in a mixed-species group, 11 months combined, *57*

List of Tables

Table 18. Monthly variation in ranging patterns of blue monkeys and redtails: means, standard deviations, ranges (n in parentheses), *62*

Table 19. Percentage of height records in three height classes, *70*

Table 20. Quadrat use in relation to forest structure and time spent by redtails associated with blue monkeys, *73*

Table 21. Hours spent by redtails covering ground used or not used previously on the same day by blue monkeys, *77*

Table 22. Redtails' response to blues monkeys' reuse of space, *78*

Table 23. Demography and range use in two study areas (means and standard deviations given unless n ≤ 3), *84*

Acknowledgments

I am very grateful to the Government of Kenya for permission to work in the Kakamega Forest, and to the Foresters and Forest Station staff for practical and essential assistance throughout the study. C. Kabuye and G. Mungai at the Kenya National Herbarium, and J. Gutwa at the Forest Station helped me immeasurably in identifying plant species. P. Milimo and E. Isiaho made some of the vegetation censuses. The British Museum (Natural History) kindly allowed me to inspect A.J. Haddow's original field notebooks.

On the other side of the ocean, the University of California Research Expeditions Program, the National Science Foundation (Graduate Fellowship), Sigma Xi, the American Women in Science, and the U.C. Berkeley Zoology Department all provided financial support, and U.R.E.P. also gave me 21 good monkey counters. T. Rowell introduced me to the animals, and spurred enthusiasm throughout the project. She, along with R. Caldwell, R. Colwell and K. Milton, generally provoked my mind and read an earlier version of the manuscript. I am also grateful to C. Janson and A. Richard for their advice and encouragement.

Abstract

Blue and redtail monkeys in the Kakamega Forest, Kenya, spend over half their time together in mixed groups. Most behavioral interactions between members of different species are agonistic, and at least half of these involve direct competition over feeding sites. Blue monkeys are at a competitive advantage because they are bigger.

To determine why mixed groups occur, the behavior of both species in and out of mixed groups was compared, and the incidence of such groups was related to ecological factors thought to influence their occurrence.

Mixed groups are not the result of chance encounters of groups of both species moving randomly and independently. It is also unlikely that they result from each species independently exploiting food sources that are shared by both. The fact that redtails are mainly responsible for their formation and persistence suggests that at least this species actively benefits from participating.

Predator avoidance may be an important benefit of mixed-species association, although predation pressure was not directly measured. The marked alarm responses of both monkey species to eagles suggest several ways in which participation in mixed groups might reduce the risk of being captured.

When in mixed groups, redtails, who are more insectivorous than blues, do not appear to benefit by using the blues as "beaters." Neither species improves its access to particular food types rendered more available by the other species' direct action.

In mixed groups, redtails, but not blue monkeys, increase their use of more open forest types. This shift may be related to antipredator advantages of associating with blue monkeys.

Redtails may use the "local" group of blues as guides to rare but preferred food sources not recently visited. Avoiding areas that the blues have exploited earlier on a given day may benefit redtails if, in so doing, they increase their consumption of ripe fruit, whose availability would be depressed by a passing group of blues.

In Kakamega, temporal variation in the incidence of mixed-species groups is related to patterns of range use, whereas anti-predator benefits seem to be relatively constant in time.

1

INTRODUCTION

Associations between sympatric primate species are widespread. They occur on all three continents where at least two species co-occur, and have been observed in most long-term studies of primates that share their habitat with other primates (review in Struhsaker 1981b). Though data on the fraction of time spent in mixed-species associations are scarce, there are at least some species in some places that spend nearly 100% of their time associated with others (Gautier-Hion et al. 1983, Terborgh 1983); in most cases, it seems that associations are not as long lasting, but are nevertheless conspicuous features of the social life of those populations in which they occur.

In primates, mixed-species associations appear to involve species that are quite gregarious intraspecifically as well; thus there are no reports of mixed-species associations in nocturnal prosimians (Jolly 1966, Charles-Dominique 1977) or orangutans (Rodman 1973). Most associations have been reported from diurnal primates inhabiting tropical forests: this is not surprising because this habitat is the home of most primate species, and the richest primate communities. The most long lasting associations are formed by pairs of species that are ecologically and phylogenetically similar (e.g. *Cercopithecus nictitans*, *C. pogonias*, and *C. cephus*, Gautier-Hion et al. 1983; *Saguinus fuscicollis* and *S. imperator*, Terborgh 1983) but associations are not limited to such species, nor do all such pairs associate often.

Other Vertebrate Taxa

The ubiquity of mixed-species associations has long been noted in birds. A thorough but old review may be found in Rand (1954). The bird literature focuses on three main questions. First, which species participate in mixed flocks and what do they do there? The degree of gregariousness and participation varies among species in a bird community, some participating in large numbers in all flocks, others being occasional visitors. Various classifying schemes were devised (Gannon 1934, Davis 1946, Winterbottom 1949, Rand 1954, Moynihan 1962). Second, there has been a long-standing interest in the mechanisms of cohesion, especially in those mixed flocks that travel together (Gannon 1934, Davis 1946, Moynihan 1962, MacDonald and Henderson 1977, Caldwell 1981); morphology, vocalizations, and behavior of flocking species are considered in order to elucidate the mechanisms whereby particular species participate in flocks to a greater or lesser degree. Finally, every description of mixed-species flocks has been accompanied by some suggestion as to what the function of flocking might be.

Mixed-species groupings of fish and mammals have been the subject of explicit interest for a shorter time. Most reviews of social groupings in fishes mention mixed-species associations only briefly (Breder 1959, Shaw 1970, Radakov 1973) and there is little documentation of their ubiquity. Ehrlich and Ehrlich (1973), Morse (1977) and Wolf (1983) suggest that such groups are common around tropical coral reefs, and they are known in freshwater fishes as well (Morse 1977). Fishes that form mixed schools may be herbivorous (Barlow 1974, Wolf 1983) or piscivorous (Radakov 1973, Itzkowitz 1977), and a variety of taxonomic groups are involved. Ehrlich and Ehrlich (1973), Barlow (1974), Itzkovitz (1974, 1977), McFarland and Hillis (1982), and Wolf (1983) explicitly discuss mixed-species schools of particular tropical reef fishes; whereas Ehrlich and Ehrlich, Barlow, and Wolf concentrate on adaptive significance of mixed schooling, Itzkovitz also classifies species on the basis of their participation in schools and briefly considers prerequisites for attraction between different species. His discussion is thus quite similar in nature to those of birds. The fish and bird literature differ, however, in that ichthyologists have argued more over what a "school" actually is (Shaw 1970 and Radakov 1973 review this dispute): students of both taxa realized quite early (Hindwood 1937, Breder and Halpern 1946) that they should distinguish between groups formed by active attraction of individuals, and those that resulted from independent responses of individuals to extrinsic factors (e.g. common food or water currents) or to limited space (Itzkowitz 1977).

In mammals other than primates, mixed-species associations have been noted, but not explicitly studied, in plains living ungulates (Keast 1963, Lamprey 1963, Sinclair 1985) and in cetaceans (Pilleri and Knuckley 1969). These studies provide few quantitative data on the incidence and dynamics of mixed-species groups, but suggestions as to adaptive significance are made based on the identity of participants and their behavior while in such groups. Elder and Elder (1970) report associations between mammalian species that differ more widely taxonomically: they found bushbuck in association with baboons or vervet monkeys in over a third of their observations, and postulated feeding and antipredator benefits for one or both species. Baboons and impala may occur in mixed groups for similar reasons (Altmann and Altmann 1970). Struhsaker (1981b) reports several cases of forest monkeys dropping fruit down to duikers, bush pigs, and baboons on the forest floor.

Whereas most mixed-species groups involve taxonomic relatives (i.e., same class at least; Wilson 1975, Bertram 1978), there are also several reports of associations between mammals and birds. Most of these involve a bird species capturing insects or fish flushed by their mammalian companions (Pilleri and Knuckley 1969, reviews in Fontaine 1980, Rasa 1983), but birds also may lead mammals to rich plant food sources (Friedmann 1955, Rasa 1983). Most of these reports are anecdotal; only Fontaine and Rasa made more systematic observations of the behavior of mixed-species group members, and so were better able to describe the incidence and probable functions of such groups. A few cases of association by more distantly related species have also been reported (McFarland and Kotchian 1982, Newton 1984).

Competition

A question that naturally arises from observations of mixed-species associations is: what function, if any, do they serve? This is an interesting question because, all else being equal, grouping tends to increase competition among group members. Alexander (1974) has gone so far as to say that increased competition is an "automatic" and "universal" consequence of grouping. This statement is true and competition is a problem only if resources are limiting. In the case of multispecific groups, an added precondition is that the species in question are potential competitors (e.g., not on different trophic levels).

Whereas the questions of developing modern ecology in the 1960s and early 1970s often centered on competition itself or concepts closely related to competition (e.g., limiting similarity, niches and species diversity; MacArthur 1972, Pianka 1976, Wiens 1977, Diamond 1978), more recently the importance and ubiquity of competition have been debated (Schoener 1982, Salt 1983). Criticisms have been leveled on theoretical and empirical grounds by a variety of biologists proposing rather different alternatives (e.g., Wiens 1977, Connell 1980, Lawton and Strong 1981). For this discussion, the chief questions are whether the domain of competition includes species that form mixed-species groups, and, if so, what the costs of increased competition in such groups may be.

How can one tell if two (or more) species are competing? The critical observation is that, all else being equal, in the presence of a competitor, the species under study either contracts its niche or decreases its rate of population growth (Colwell and Fuentes 1975, Pianka 1976, Fleming 1979, Connell 1980): thus, demonstration of competition demands a comparison. How this comparison is structured may vary. Experimental additions or removals (Colwell and Fuentes 1975, Connell 1980) or natural introductions and extinctions (Diamond 1978) in one population over time are probably the two best possibilities; however, they are often impractical to study. Alternatively one could compare two populations, one allopatric and one sympatric with the presumed competitor. As many have pointed out, the problem with this approach is the lack of controls: one must establish that the populations differ in no relevant way other than the absence or presence of the competitor (Pianka 1976, Connell 1980), and this is often difficult. Weaker -- and for some unacceptable -- evidence of competition comes from studies of populations only in sympatry with their competitors. In these studies overlap along some niche dimension(s) is considered, but both overlap and lack of overlap have been interpreted as evidence of competition between species (Terborgh and Diamond 1970, Schoener 1982). If lack of overlap is to be taken as evidence for competition, one must ensure that it does not exist for some other reason: Connell (1980) has made a strong case against invoking the "Ghost of Competition Past," but few studies meet his standards, and many invoke interspecific differences as coexistence mechanisms without question. If overlap is construed as evidence for competition, then one must also independently demonstrate a shortage of resources that are shared and presumably competed for, but this is seldom done. Finally, there is one more way to demonstrate competition that does not involve making a comparison: one can observe interference. Interference may involve aggression or establishment of (interspecific) dominance that correlates with priority of access to resources (Wilson 1975, Morse 1977; but see Coelho et al. 1976 for a different view). The problem with this approach is that it cannot detect exploitative competition, and it may require long periods of close observation of the animals in question.

The relevance of competition to mixed-species groups was appreciated first by Rand (1954), who pointed out the "seeming paradox of social behavior that increases competition, which evolution has been working to reduce." Others were aware of possible competition but did not document it (Keast 1963, Goss-Custard 1970); some discounted it (rather informally) as unimportant in their study animals (Lamprey 1963, Brosset 1969, Willis 1966a, MacDonald and Henderson 1977). Only a few of the more recent studies have actually demonstrated the effects of increased competition in mixed-species groups by looking at niche shifts (Morse 1970, Austin and Smith 1972, Werner and Hall 1976, Hogstad 1978, Morse 1978, Alatalo 1981).

In primates, competition for food between species that form mixed-species associations has been inferred frequently based simply on habitat or diet partitioning (Gartlan and Struhsaker 1972, Struhsaker 1978, Fleagle et al. 1981, Pook and Pook 1982 for insects); in some cases, interference has been observed (Gartlan and Struhsaker 1972, Klein and Klein

1973, Struhsaker 1981b; but see Waser and Case (1981) who conclude that exploitation is more important than interference in the Kibale forest primates, and Skorupa's (1983) reply). In only two primate studies have niche shifts, which lead to a reduction of overlap during the dry (lean) season, been documented (Gautier-Hion 1980, Terborgh 1983). These results, however, do not explicitly address the correspondence of competition and mixed-species group formation. Most studies of mixed-species groups in primates conclude that competition is not actually exacerbated in the mixed group, either because no interspecific aggression was observed (Bernstein 1967, Gautier and Gautier-Hion 1969, Gautier-Hion and Gautier 1974, Pook and Pook 1982) and/or because the food eaten by mixed groups was judged to be in superabundant supply (Gartlan and Struhsaker 1972, Pook and Pook 1982). Gautier-Hion et al. (1983) reject competition as important even in principle because "from a quantitative point of view, nothing is altered by association in the short-term since the total biomass of the community living in the study area is unchanged." However, they overlook the fact that animals have limited time in which to feed and cover a limited area each day: if members of one species avoid or are prevented from feeding when in mixed groups, they lose relative to conspecifics not in such groups (all else being equal). The food may be there, but equally free access is not.

It seems, in sum, that increased competition is a possible consequence of mixed-species grouping in many kinds of animals. Sometimes niche shifts coincident with mixed group formation mitigate that competition; in other cases, mixed groups seem to form at times when competition is unlikely to occur. The role of competition in primate mixed-species groups is difficult to evaluate because the evidence is largely anecdotal.

Hypotheses

Why, then, do animals form mixed-species associations, sometimes in spite of the disadvantages of competition? There is clearly no unitary answer to this question: mixed-species associations differ considerably in their frequency, duration, membership, internal structuring, and predominant activity. It should be expected that different animals could benefit differently at different times (Hindwood 1937, Winterbottom 1949, Rand 1954, Morse 1970, 1977, MacDonald and Henderson 1977, Rubenstein et al 1977, Grieg-Smith 1978).

The simplest explanation -- a null hypothesis -- is that mixed-species associations result from chance meetings of members of different species who are moving independently of one another (Waser 1982). Thus mixed groups are the statistical consequence of a number of individuals (or groups) of different species being confined to a finite area, where, now and then, they will encounter each other.

A second kind of null hypothesis again stipulates movement independent of other mixed-species group members (individuals or groups) per se; however all members respond similarly (yet independently) to a common stimulus, and so are brought together (Waser 1982). In other words, mixed-species group members congregate because each is attracted, either actively or passively, to some (and the same) environmental feature, not to other group members. This kind of mixed association corresponds to what was dubbed an "aggregation" (as opposed to a "flock") by ornithologists, who noted, for example, many instances of frugivorous birds converging on trees in full fruit (Hindwood 1937, Willis 1966a, Diamond and Terborgh 1967, Brosset 1969, Terborgh and Diamond 1970), or of ant-eating birds converging on a column of ants (Willis 1966b, Morse 1970). Struhsaker (1981b) and Waser (1982) have also invoked this explanation for some primate mixed-species groups.

Rejection of the above hypotheses implies that (at least some) members of mixed-species associations actively seek one another out and benefit from so doing. The kind and degree of benefit each species derives is not necessarily the same: some may not even benefit at all. Benefit hypotheses have largely fallen into two classes: individuals in mixed-species groups can (1) decrease the probability of being preyed upon, or (2) increase their foraging success, relative to conspecifics who are not in such groups. Many investigators have cast an opinion as to whether food or predation is the primary factor selecting for mixed-species grouping but most would agree that often neither factor alone provides a complete explanation, and that both factors may be related (Morse 1977).

There are several ways in which mixed-species associations can decrease the likelihood of being preyed upon. First, animals in (mixed) groups have more eyes and ears available to detect predators, hence promoting earlier detection or allowing each animal to spend less time being vigilant. This hypothesis is advocated by Moynihan (1962), Goss-Custard (1970), and Morse (1970, 1978) for birds, Elder and Elder (1970) for antelope, Rasa (1983) for mongoose, and Gartlan and Struhsaker (1972), Rudran (1978), Struhsaker (1981b), Pook and Pook (1982), and Gautier-Hion et al. (1983) for primates. Diamond (1981) gives several other examples. Second, mixed-species groups may deter a predator's attack by confusing the predator. Charnov et al (1976) suggest that a predator may be particularly confused if confronted with prey who have different escape tactics because it would be difficult to predict the escape behavior of any individual it chose to pursue. Confusion of a predator may be visual, as in fish (Breder 1959) and some birds (Miller 1922, Winterbottom 1949, Moynihan 1962), or auditory (Morse 1977). Struhsaker (1981b) suggests confusion of predators may also occur in forest monkeys. Third, the members of a mixed-species flock may cooperate in mobbing a predator. There is some evidence for such behavior in birds (Moynihan 1962); it has not been reported in primates. Fourth, mixed-species groups could be "selfish herds" (Hamilton 1971), such that each individual tries to position another prey item between itself and the predator. Nearly all examples of the "selfish herd" effect come from single-species groups of animals (reviewed in Morse 1977), but it also has been tentatively identified in a few mixed groups (Goss-Custard 1970, Hamilton 1971). Finally, mixed-species groups may dilute the successful predator's effect, in that the more other potential prey items are present, the less likely it is for any one individual to be taken (Bertram 1978).

Mixed-species associations can likewise lead to increased foraging success in several ways. A popular idea, which arose in the bird literature, is that one species increases its prey capture rate by feeding on insects or other prey items flushed from their hiding places by other ("beater") species in the mixed group. There are many reports of this occurring in birds (Swynnerton 1915, Rand 1954, Short 1961, Moynihan 1962, Brosset 1969, Siegfried 1971, 1972), in mixed groups of birds with other animals (Davis 1946, Rand 1954, Pilleri and Knuckley 1969, Fontaine 1980, Diamond 1981, Rasa 1983), in fish (Barlow 1974) and possibly in primates (Klein and Klein 1973, Rudran 1978, Pook and Pook 1982). Terborgh (1983) argues that this effect is likely to be more relevant to birds than to primates because mobile insects, the kind that would be flushed, are best caught by stealth and surprise, unless the predator is able to capture them in the air as they flee. In some cases, one member of a mixed-species group will even steal prey already caught by another group member (Grieg-Smith 1978, Barnard and Stephens 1981, Diamond 1981, Klein, cited in Terborgh 1983, Rasa 1983). Another way in which mixed-species groups could enhance the ability of their members to find food is by having one species serve as a guide to temporarily abundant food for another species. Guiding may involve one species following another while foraging, one simply approaching another, or one imitating another (Krebs 1973). Guiding has been proposed as a benefit to members of mixed-species flocks of birds (Short 1961, Paulsen 1969, Goss-Custard 1970, MacLean 1970, Murton 1971, Rubenstein et al 1977, Caldwell 1981, Barnard and Stephens 1983), monkeys (Moynihan 1970, Gartlan and Struhsaker 1972, Rudran 1978, Pook and Pook 1982, Skorupa 1983), and other mammals (Friedmann 1955). A third way that mixed-species groups can reduce search time is by avoiding duplication of effort and regulating return time to renewing resources (Cody 1971). By staying together, individuals (or groups) of one species can ensure that they do not visit food sources that individuals (or groups) of another species have already picked over. This idea has received some support from studies of birds (Short 1961, Morse 1970, Cody 1971, 1974, MacDonald and Henderson 1977) and primates (Gautier-Hion et al. 1983).

Mixed-species groups can lead to an increase in the number of available foods in two ways. One species may make certain food types available to another species by its own direct action on them. For example, monkeys knock down certain seed pods or fruits from trees which antelope below can then feed on (Lamprey 1963, Elder and Elder 1970). Struhsaker (1981b) reports that redtail monkeys "line up" to feed on the flesh of a fruit whose tough outer skin was previously opened by mangabeys, and squirrel monkeys wait to collect *Scheelea* nuts, whose shell and husk they cannot open, dropped by *Cebus* monkeys (Terborgh 1983; see Balph and Balph 1979 for a similar case in birds). Alternatively, certain foraging areas may be available to members of mixed groups that are not available to individuals or groups of one species. Mixed schools of surgeonfishes can feed on territories of lavender tang, whereas individuals are excluded by the territory holder (Barlow 1974). Similarly, Struhsaker (1981b) suggests that redtail monkeys may benefit by joining red colobus because the latter can supplant chimpanzees from fruiting trees to which redtails alone would not have access. Gautier-Hion et al. (1983) have shown that *Cercopithecus cephus* monkeys spend more time in high forest when in mixed groups than when alone, probably because when in such groups they can use warnings by the other mixed group member species rather than hiding in low, dense foliage as an effective antipredator tactic.

Finally, in conjunction with antipredator benefits such as reduced vigilance levels for each group member, association in larger groups may allow more time for foraging. This hypothesis has been tested only in single-species groups of birds (Caraco 1979, Caraco et al. 1980).

Another benefit of mixed-species grouping proposed by Freeland (1977) for primates is that it functions to reduce the number of fly bites individual animals receive, and thus minimizes their risk of acquiring vector-borne disease. This idea has not been applied to other taxa, and has been criticized on empirical grounds by Struhsaker (1981b) with reference to the population in which Freeland initially proposed it. Also, some of the premises of Freeland's argument are based on studies of animals and conditions different from those in his study, and it may well not be fair to to generalize. This hypothesis is not considered further.

The selective benefit hypotheses that have been proposed are largely the same as those used to explain intraspecific gregariousness (reviewed by Bertram 1978), assuming that predators and diet are common to the different species in a mixed group. The remaining question is why mixed-species groups are better (or at least no worse) than single-species groups in performing these functions. There are several possible answers. First, in general there may be less competition between individuals of different species than between individuals of the same species: the costs of competition, if they are likely consequences of grouping, may be lower in mixed-species groups. Second, the chance of being closely related to other group members is lower for each individual in a mixed-species group than in a single-species group. When one individual stands to gain at the expense of other group members (e.g., dilution of a predator's effect, or food piracy), its net gain will be greater if the other group members are not relatives. Third, sometimes the reasons for interspecific association simply involve complementary skills or behavior of different species (e.g., the ability to make certain foods available, or to detect or confuse predators). Finally, especially in highly social animals, the social organization of a monospecific group would be more complex and possibly more unwieldy than that in an equal sized group of heterospecifics (Gautier-Hion and Gautier 1974); in the latter case, a large group is effectively broken down into subunits, each of which has its own social organization.

I have presented these hypotheses without indicating the quantity or quality of data available to test them. This is because, for the most part, and especially in primates (Struhsaker 1981b; but see Waser 1982), hypotheses have not been rigorously tested, though there is often considerable anecdotal or surmised support. In particular, null hypotheses are seldom formally considered or well developed (e.g., Moynihan 1962, McFarland and Kotchian 1982, Wolf 1983, but see Itzkowitz 1977). It is important to test null hypotheses if one wants to evaluate adaptive significance (Waser 1982): members of mixed-species groups that form by chance encounters may well *benefit* from being in those groups, but in such cases, there is no evidence that there has been *selection* for interspecific gregariousness.

If the null hypotheses are rejected, one must distinguish between the many remaining possible explanations for the prevalence of interspecies association. To evaluate the benefits of such association for members of a given species, a comparison is needed between animals associated and not associated. Ideally, this comparison should be made when conditions are similar, so that the only variable is degree of association. A few studies with this design have been made of birds (Morse 1970, 1978, Austin and Smith 1972, Krebs 1973, Caldwell 1981, Barnard and Stephens 1983) in which feeding behavior was monitored in flocks of different sizes or different species compositions, or both. In primates, only two very recent studies have compared animals in and out of mixed-species groups (Gautier-Hion et al. 1983, Terborgh 1983).

The problems with this comparative approach are practical ones. First, some associations are not variable in their occurrence, so comparisons within study populations are not possible. Second, it is seldom possible to ascertain that all aspects of the animals' biology do not covary systematically with association: Gautier-Hion et al (1983), for example, compared two different groups of moustached monkeys. Suppose associations only and (nearly) always occur when certain environmental conditions are met (e.g., at certain seasons or times of day)? In this case, the ideal comparison is simply not possible. However, the very patterning of interspecific association may be instructive to someone searching for its cause. At the very least, one can identify environmental factors that promote or inhibit interspecific association.

In this study, I explore mixed-species groups of blue (*Cercopithecus mitis*) and redtail (*C. ascanius*) monkeys in the Kakamega Forest, western Kenya. These species belong to a genus in which mixed-species associations of sympatric congeners have been long known (Booth 1956, Gautier and Gautier-Hion 1969, Bourliere et al 1970, Gartlan and Struhsaker 1972, Gautier-Hion and Gautier 1974, Galat-Luong 1979), and are sometimes dramatic in their persistence in time (Gautier-Hion et al 1983). In this particular species pair, formation of conspicuous mixed groups has been reported previously from Uganda by Hayashi (1975), Rudran (1978), and Struhsaker (1981b) working in the Kibale Forest and by Aldrich-Blake (1970) and Marler (1973) in Budongo.

My goals are fourfold: (1) to quantify the incidence of mixed-species groups, (2) to illuminate ecological relationships between the two species, (3) to identify ecological factors that enhance or inhibit mixed-species group formation, and (4) to evaluate quantitatively different hypotheses about why such groups occur. I use natural variation in the occurrence of mixed-species groups to compare the same animals in the same location when they are and are not in mixed groups. Detailed observations of feeding, ranging, antipredator and social behavior in both conditions are provided. Such an approach has not been attempted previously in Old World primates.

2

STUDY SITE, STUDY ANIMALS, AND METHODS

The Kakamega Forest

General Description

The Kakamega Forest officially occupies 238 square kilometers in western Kenya, about 40 km northeast of the town of Kisumu on Lake Victoria (Diamond 1979, Oruko 1979). At most, 48% of this land is under natural forest (Diamond 1979 after 1976 aerial survey), the remainder supporting plantation forest or small-scale agriculture. The actual study area is part of the indigenous forest and is 1,580 m above sea level. The Kakamega Forest Station, located at the southern end of the whole forest and adjacent to the study area, receives 2215 $\pm$ 26 mm of rain per year (this and the following climatic data are from six years of unpublished Kakamega Forest Station records, 1976-1981). Each month contains from 0 to 22 rainy days (mean = 12.2 $\pm$ 2.2). Rain falls seasonally, especially from March or April to July or August (long rains) and again for a month or two around October and November (short rains), but there is considerable variability from year to year in the timing of rains and the distinctness of the seasons (Figure 1). Average monthly maximum temperatures range from 18 to 29 degrees Celsius; average monthly minimum temperatures are between 11 and 21 degrees Celsius.

Flora and Fauna

The indigenous vegetation of the Kakamega Forest resembles that of the lowland Congo basin forests farther west (Lucas 1968), and has been variously classified (Zimmerman 1972, Hamilton 1974), most recently as a "drier type Guineo-Congolian lowland rainforest" (UNESCO/AETFAT map). Kakamega is one of few remaining representatives of this vegetation type in Kenya and it forms the easternmost boundary of the Congo-basin type forests stretching to the west coast of the African continent (Lucas 1968, Hamilton 1981). These forests become increasingly fragmented east of present-day Zaire, and today the

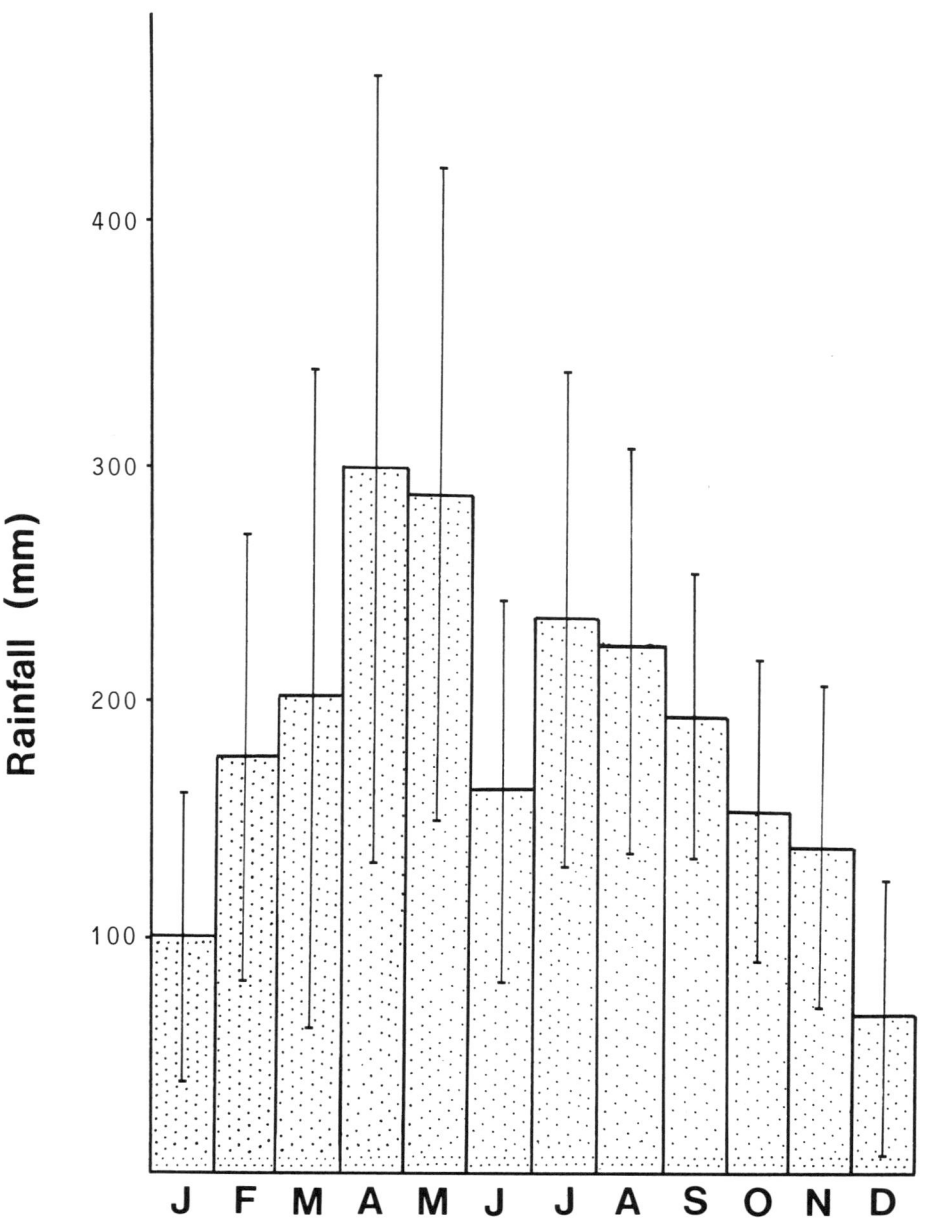

FIGURE 1. Rainfall at the study site. Stippled bars give means for six years, 1976-81 inclusive; thin lines indicate standard deviations.

Kakamega Forest is truly an island. In terms of flora and fauna, the Ugandan and western Kenyan forests also become more depauperate from the west to the east, probably as a consequence of Pleistocene climatic fluctuations that limited lowland forest to a few refugia during periods of extreme aridity and from which it expanded when the climate became more permissive (Hamilton 1974, Livingstone 1975).

The flora and fauna of the Kakamega Forest show many similarities to those of central African forests and thus confer on the forest a unique status in Kenya (for plant species list, see Cords 1984b). The Congolese-lowland character of the animals has been documented most thoroughly in birds (Zimmerman 1972), mammals (Kingdon 1971), butterflies (Carcasson 1964), and snakes (Spawls 1978). Monkeys are by far the most conspicuous mammals in the forest. Blue monkeys, redtails, and black and white colobus (*Colobus guereza*) are regular inhabitants. Baboons (*Papio anubis*) are occasional visitors, and *Cercopithecus neglectus* has been seen outside the study area (pers. obs.).

Study Animals and Study Area

Study Animals

Blue and redtail monkeys belong to the genus *Cercopithecus,* which contains the largest number of African primate species. Most members of this genus live in forested habitat, ranging from the relatively dry eastern coastal forests to gallery, montane bamboo, and lowland rain forest (Wolfheim 1983).

Blue and redtail monkeys co-occur in many forests in east-central Africa, where their geographical ranges largely coincide. Exceptions are high altitude forests, like Mt. Elgon in Kenya, where redtails are absent, and the forests on the edge of Lake Victoria, where there are no blue monkeys. In three places, hybrid individuals have been seen: Gombe Stream (Clutton-Brock, pers. comm.), Budongo (Aldrich-Blake 1968), and Kibale (Struhsaker 1981a, 1984). In Budongo and Kibale, hybrid females are fertile, and are evidently permanent members of groups of redtails (Kibale) or blues (Budongo). Such interspecies mating seems rare, however, and no evidence for its occurrence was seen in Kakamega, even though blues and redtails were frequently in close proximity. Hybridization may be most likely to occur when the population densities of the two species are very different (Struhsaker 1981a, Clutton-Brock, pers. comm.).

Blue monkeys are larger than redtails. Table 1 presents data on body weight from Haddow (1952, and unpublished field notes). Haddow determined age classes based on tooth wear and reproductive state (females). These data clearly have their limitations, but they are the best data available for field specimens; values agree with those reported by Haltenorth and Diller (1980). In my study, age was assumed to be correlated with size, and

TABLE 1

Adult Body Weights (in pounds) *

	BLUE MONKEYS		REDTAILS	
	Weight	n	Weight	n
Adult female	8.8 ± 0.9	8	6.5 ± 1.1	30
Old female	9.6 ± 0.9	6	6.6 ± 1.1	13
Adult male	15.0 ± 1.2	2	9.2 ± 1.2	31
Old male	15.3 ± 0.7	4	9.7 ± 1.5	8

* Data from A.J. Haddow's unpublished field notes, British Museum, Natural History. "Old" individuals were distinguished from "adults" mainly on the basis of tooth wear.

four to five size-age classes were distinguished. In general, blue monkeys of a given size-age class were as big as redtails of the next biggest-oldest size-age class.

The study reported here spanned 21 months from July 1979 to April 1981, with a revisit in June-August 1982. Observations concentrated on the T group of blue monkeys and the L group of redtails. These groups were chosen because the redtail home range completely circumscribed the blue monkey home range (Figure 2), thereby minimizing differences in the foods and habitats available to the focal groups. In both groups, all adults were recognized individually by June 1980; all redtail juveniles and about half of the blue juveniles were individually known by the second year of the study. In addition, four neighboring groups of blues and two neighboring groups of redtails were periodically censused.

Both species live in social groups with females as permanent members. Young males leave their natal groups as large juveniles or subadults. Adult males are passing members of social groups; males' tenures may vary considerably in length (Cords 1986b). When males are not part of social groups, they are usually solitary and cryptic; occasionally, two to five such blue males were seen within 25 m of each other, but redtail extra-group males were never seen in proximity to one another. Males who spend most of their time outside social groups may nevertheless join a group during the mating season and breed with the females (Cords 1984a, Tsingalia and Rowell 1984, Cords et al. 1986).

Table 2 presents data on group size and composition for all groups repeatedly counted. The average group size is larger for blues, but differences are not statistically significant (Mann-Whitney U, $p > 0.10$); from incomplete counts of other groups, however, I have the impression that blue groups are usually larger. Total numbers and composition fluctuated

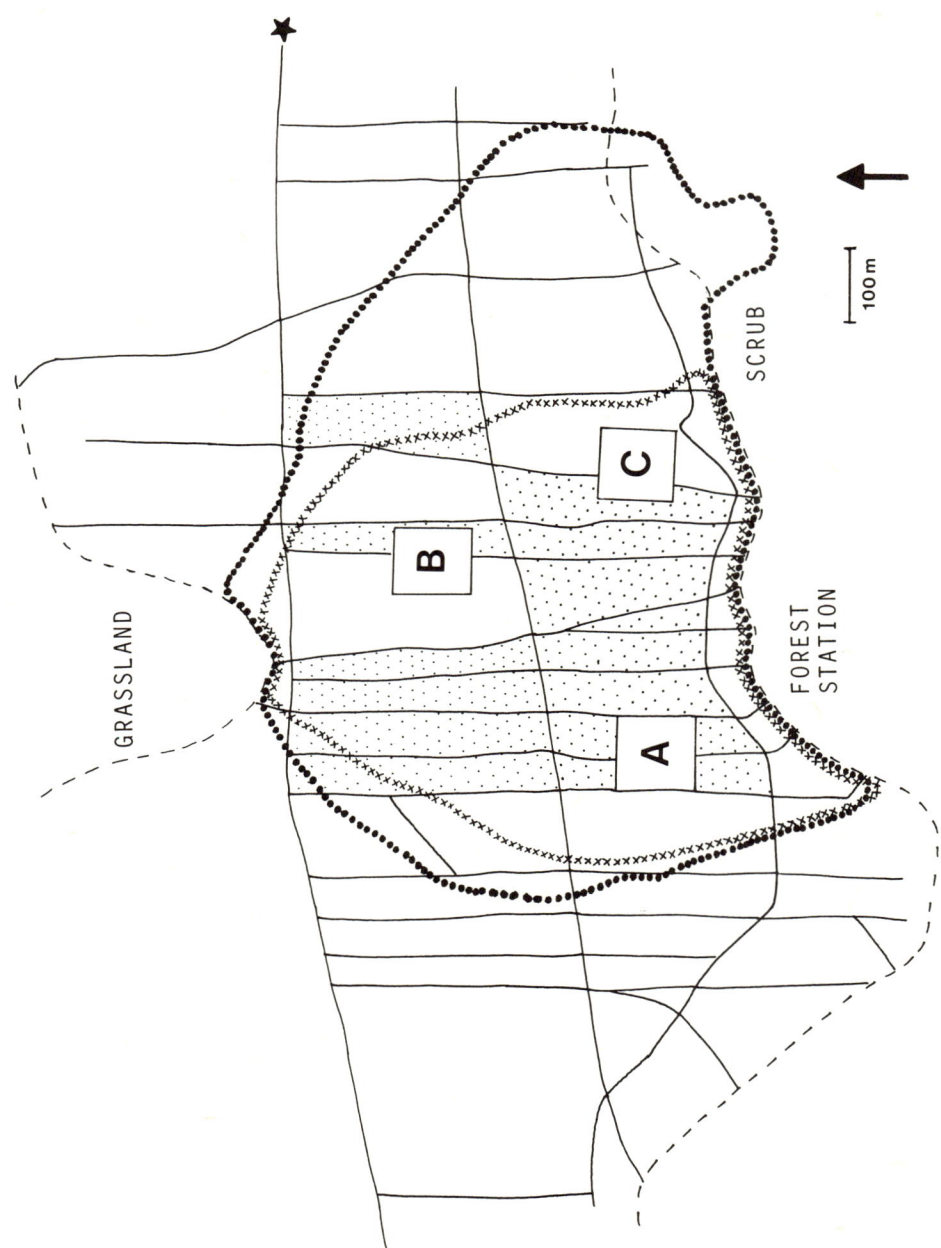

FIGURE 2. Study area, focal group home ranges, and vegetation plots. Stippled area designates 13 forest blocks used for overall census; lettered quadrats were used for detailed spatial analysis. Heavy dots show redtail home range boundary, and small x's show blue monkey home range boundary.

TABLE 2
Group Composition on Last Complete Count *

(a) BLUE MONKEYS

Group	Date	Total	AM	AF	"Big"	LJ	MJ	SJ	TJ/in	Unidentified
T	4/1/81	45	1	17	-	3	9	6	9	-
F	4/8/81	35	1	18	1	1	3	2	9	-
S	3/19/81	34	1	13	4	2	4	5	5	-
R	8/3/80	28	1	6	3	1	7	4	3	3
G	3/18/81	21	1	6	3	1	4	2	4	-

(b) REDTAILS

Group	Date	Total	AM	AF	"Big"	LJ	MJ	SJ	TJ/in	Unidentified
L	3/7/81	25	1	1	9	1	5	3	5	-
O	3/17/81	34	1	0	11	4	6	7	4	1
N	7/10/82	20	1	0	5	3	5	2	1	3

* Age-sex classes are adult male, adult female, large, medium, and small juveniles, and tiny juvenile/infants. "Big" animals are either AF or LJ, because the presence of nipples could not be confirmed.

within 15% of the mean as a result of birth of infants, maturation of juveniles, and transient adult male membership. Nevertheless, between December 1979 and August 1983, there was little change in overall numbers in each species and it appeared that the populations were stable.

Groups of either species defend territories against neighboring groups of conspecifics (see also Struhsaker and Leland 1979). Between November 1979 and April 1981, territorial encounters were witnessed 70 times in blue monkeys (0.39 per observation day) and 24 times (0.16 per day) in redtails. Mostly adult females and medium and large juveniles participated in intertroop encounters; adult males, especially redtails, were seldom involved.

Study Area

The two focal troops and their immediate neighbors, who comprised the study population, lived in an area of 1.14 square kilometers located between the Kakamega Forest Station and a 400 hectare grassland to the north (Figure 2). Zimmerman (1972) gives a general description of this area, in which he undertook extensive studies of birds in the 1960s. The forest department maintains a grid of trails running north-south at 50-100 m intervals, and east-west at about 300 m intervals. These trails greatly improve visibility because one is able to move freely along them; in most places, there is a fairly dense ground layer of vegetation that would otherwise prevent easy access. Karr (1976) estimated 60% cover by this vegetation layer. Part of the study area (north of the trail marked with a star on Figure 2) is included in a Nature Reserve where human interference with the flora and fauna is forbidden, but not entirely controlled. The remainder of the study area was partially felled in the 1940s, when some very large trees were removed. Subsequently, there was some enrichment planting of indigenous timber trees, notably *Olea welwitschii*, *Vitex fischeri*, and *Khaya*, and some Southeast Asian species (*Bischoffia javanica*, *Acrocarpus fraxinifolius*), mainly along one of the north-south trails. This human influence is probably partly responsible for the relatively (to most tropical moist forests) dense undergrowth, but such undergrowth is not uncommon in some parts of the Kakamega Forest that were never felled. Indeed, these areas resemble the study area in lacking a well-defined middle canopy layer. The major structural difference between the study area and the pristine forest lies in the height of the canopy and degree of canopy cover, both of which are lower in the study area. Canopy height in the study area was measured with a Keuffel and Esser clinometer in six randomly chosen locations and averaged 32.5 ± 6.1 m. A few emergents reached 45-50 m. Zimmerman (1972) estimated cover to be 75% in 1965.

Because there had been no previous systematic study of the Kakamega Forest vegetation, various descriptive analyses of the vegetation in the study area were made, mainly in July-August 1980 and June 1982. First, all trees with a diameter at breast height (DBH) equal to or greater than 20 cm were mapped in all blocks of forest used by the focal monkey groups (blocks are defined by the paths on their edges). The accuracy of the mapping of the trees was influenced by the size of the block: area and stem density were negatively correlated ($r_s = -0.59$, $p < 0.01$, $n = 20$). Therefore the analysis was limited to 13 blocks (shaded in Figure 2), each of whose area did not exceed 2.35 ha, or in which visibility was improved by small access trails. This sample showed no correlation between block area and stem density. The total sample included 1,955 trees with DBH > 20 cm representing 62 species in a combined area of 21.78 ha. Species diversity, calculated as $H = - \Sigma p_i \ln p_i$ (Pielou 1966), is 3.455. This high diversity means that few species are common: in the sampled area, 60% of the species occurred at densities less than 1 tree per hectare, 35% occurred 1-5 times per hectare, and only 5% occurred at densities higher than 5 per hectare (Table 3).

TABLE 3

Frequency, Dispersion and Phenology of Kakamega Forest Trees and Shrubs *

	Freq	Rel Freq	Density (per ha)	Dispersion	Phenology Species Fruiting Season	Inter-Individual Synchrony
Funtumia latifolia	231	11.81	10.6	C-1, R-5		broad
Croton megalocarpus	151	7.72	6.9			
Antiaris toxicaria	145	7.42	6.7	R-6	4.5	broad
Celtis durandii	122	6.24	5.6	C-5, R-1	4-4.5	broad
Celtis africana	114	5.83	5.2	R-6	3 #	v.broad
Albizia gummifera	84	4.30	3.9	C-6	1-1.5	tight
Bosqueia phoberos	80	4.09	3.7	C-6	2-3.5 #	broad
Croton sylvaticus	74	3.79	3.4	C-4, R-2	4	broad
Olea welwitschii	74	3.79	3.4		2-3	moderate
Cordia abyssinica	63	3.22	2.9	C-5, R-1	3.5	moderate
Fagara macrophylla	52	2.66	2.4	C-5, R-1	3-4	broad
Ficus mallatocarpa	48	2.46	2.2	C-6	all mos	no
Teclea nobilis	45	2.30	2.1	C-6	7	broad
Trema guineense	42	2.15	1.9		most mos	no
Prunus africana	42	2.15	1.9		most mos	no
Aningeria altissima	40	2.05	1.8			
Polyscias spp.	39	1.99	1.8			
Ficus exasperata	35	1.79	1.6	C-5, R-1	4	moderate
Diospyros abyssinica	32	1.64	1.5			
Markhamia platycalyx	30	1.53	1.4	C-6		
Blighia unijugata	28	1.43	1.3			
Strombosia scheffleri	27	1.38	1.2			
Ficus thonningi	23	1.18	1.1		all mos	no
Premna angolensis	22	1.13	1.0		1.5-2	moderate
Trichilia emetica	20	1.02	0.92			
Bequaertiodendron oblanceolatum	20	1.02	0.92	C-2, R-4	1	tight
Casearia battiscombei	20	1.02	0.92			
Chaetacme aristata	19	0.97	0.87	C-5, R-1		
Margaritaria discoidea	18	0.92	0.83	C-3, R-3	5-6 #	broad
Croton macrostachyus	17	0.87	0.78		3-3.5	broad
Bischoffia javanica	16	0.82	0.73		6 #	v.broad
Chrysophyllum albidum	15	0.77	0.69			
Harungana madagascariensis	15	0.77	0.69		3.5	moderate
Chlorophora excelsa	15	0.77	0.69			
Khaya spp.	14	0.72	0.64			
Vitex fischeri	10	0.51	0.46			
Bridelia micrantha	9	0.46	0.41			
Manilkara butugi	9	0.46	0.41		4-5	v.broad
Morus lactea	9	0.46	0.41		4	broad

TABLE 3 (continued)

Frequency, Dispersion and Phenology of Kakamega Forest Trees and Shrubs *

	Freq	Rel Freq	Density (per ha)	Dispersion	Phenology Species Fruiting Season	Phenology Inter-Individual Synchrony
Craibia brownii	8	0.41	0.37			
Strychnos spp.	7	0.36	0.32			
Ficus ?dawei	6	0.31	0.28		1-1.5	tight
Spathodea nilotica	6	0.31	0.28		1-2	tight
Ficus brachylepis	6	0.31	0.28		2-3 #	moderate
Albizia grandibracteata	5	0.26	0.23			
Fagaropsis angolensis	5	0.26	0.23			
Maesopsis eminii	5	0.26	0.23		1-2	tight
Tipuana tipu	5	0.26	0.23			
Drypetes gerrardi	4	0.20	0.18			
Apodytes dimidiata	4	0.20	0.18			
Kigelia moosa	4	0.20	0.18			
Szygium guineense	3	0.15	0.14			
Cassipourea ruwensorensis	3	0.15	0.14			
Sapium ellipticum	3	0.15	0.14			
Acrocarpus fraxinifolius	2	0.10	0.09			
Bersama abyssinica	2	0.10	0.09			
Ficus cyathistipula	2	0.10	0.09			
Rothmania urcelliformis	2	0.10	0.09			
Acacia abyssinica	1	0.05	0.09			
Alangium chineense	1	0.05	0.05			
Ensete ventricosum	1	0.05	0.05			
Ehretia cymosa	1	0.05	0.05			
Aulacocalyx diervilleoides				C-5, U-1	2	moderate

* Frequency and density data from stems with DBH > 20 cm in 13 forest blocks. Dispersion data from trees with DBH > 15 cm in plots A, B, C (Fig. 2). Number refers to number of scales, varying from 10x20 m to 40x80 m, in which dispersion was clumped (C), random (R), or uniform (U). Phenology data from ad lib. data on selected species (Cords 1984b).

Some stragglers

To quantify the spatial distribution of individual species, a more detailed analysis was made of the trees and lianas (with GBH ≥ 15 cm) located in three plots selected to be representative of the study area (labeled A, B, C on Figure 2; Cords 1984b). The dispersion of individuals belonging to 17 species, chosen because they were important food sources for the monkeys and were represented by over 23 sample specimens each, was assessed as clumped, random, or uniform by comparing the variance to mean number of individuals per quadrat, using six quadrat sizes (Kershaw 1964). Results are summarized in Table 3, and show that individuals of most species are significantly clumped at more than one spatial scale. Spatial patterning of this sort renders the forest a nonhomogeneous environment. As Milton (1980) has pointed out, species that are very rare may also contribute to this patchiness when each tree is considered a patch of food.

The effects of spatial patterning on homogeneity are intensified by temporal patterning. Data on phenology were collected on an ad lib. basis from October 1979 to April 1981 by noting the phenological activity of 40 species at least once a month, including records of marked individual trees (Cords 1984b). Few species show a tight synchrony of fruit production; a more common pattern involves a broad overall fruiting seasonality for a species, composed of only partly overlapping periods of fruit production in different individuals (Table 3).

Interspecies differences contribute to the complexity of food distribution because species may differ in (i) timing of phytophase, (ii) number of certain phytophases per year, (iii) length of phytophases, and (iv) timing of different phytophases relative to one another.

Methods

Observations

Systematic observations of the focal groups of monkeys were made from November 1979 to April 1981. Each observation day was devoted either to redtails or to blue monkeys, even though the other species might be associating with the study group. Observations were made continuously from 0700 to 1900 (essentially dawn to dusk), except when a group could not be located first thing in the morning, or was lost in the evening (rare). On most days there was a 50 minute break sometime during the day, usually when the monkeys were low in the trees (therefore difficult to see), or inactive, or both. The animals seldom moved much, if at all, during this interval. Between 1 March 1980 and 28 February 1981, the period to which many subsequent analyses are confined, 1,128 hours were spent following the L redtails and 1,021 hours following the T blue monkeys. In the entire study, over 4,000 hours were spent in contact with one or both of these focal groups. These totals do not include hours logged by a field assistant who monitored the groups during my absence in July.

During each month from March 1980 to February 1981 (except July), redtails were followed an average of 8.6 ± 2.2 days and blues 7.9 ± 1.2 days per month. Each month, first the redtail and then the blue monkey group was followed for 5-6 consecutive days; then another 2-3 days were spent on each group, beginning with the redtails. Most observations were made in the first 20 days of each month.

The exact protocols involved in collecting specific kinds of data will be described when those data are presented. The main problem was to generate a behavioral sample representative of the group. A predetermined schedule of focal animal samples offers one solution to this problem, but they were impracticable under visibility conditions at Kakamega. For this reason, I circulated through the group, trying to observe as many individuals as possible, none for prolonged periods unless they were engaged in a rarely seen activity.

I was an active searcher rather than a sit-and-wait observer. Sit-and-wait observers of forest monkeys concentrate on large fruiting trees or a few places where visibility is relatively good; therefore they are likely to collect biased data on foraging and ranging, and they may sample group members nonrandomly. Active search should reduce these biases. However, it is unlikely that data collected actively are entirely unbiased: certain individuals simply are more visible than others (e.g., because of differential habituation) and certain places allow better visibility than others (e.g., because of vegetation structure or proximity to paths).

A rough estimate of the degree to which my data are representative can be made by comparing the proportion of feeding records of each age-sex class to its average proportional composition in each focal group over the year. Feeding records were instantaneous samples collected throughout the day and only those in which the individual was identified as to age-sex class are included here. Average proportional group compositions were calculated excluding infants because they were not seen to feed from plants. In general, feeding scores of different age-sex classes are proportional to their representation in the group: differences between observed and expected percentage contribution to the total number of feeding scores by each age-sex class do not exceed 7%. Observed and expected values are statistically different (G test, $p < 0.05$), however, because of the very large sample sizes. If animals identified only as "big" were included, these differences would be reduced because most "big" monkeys are probably adult females, who are otherwise somewhat underrepresented. Deviation may also be due to bias, to different feeding patterns among age-sex classes (e.g., few long versus many short feeding bouts), or to different absolute amounts of time spent feeding by age-sex classes.

One can also check for relative representation of known individuals within age-sex classes under the null hypothesis of equal sampling rates. Here, only feeding scores with an individually recognized actor are included, and the analysis is limited to adult females, because too few juveniles were easily recognizable. Thirteen of sixteen adult female blue monkeys and six of ten adult female redtails were scored as frequently as expected (G test, $p > 0.05$, Cords 1986a). For four of the seven remaining females, deviations from the expected were in the direction predicted by the unusual ease or difficulty with which they could be identified; the other three females were overrepresented (by 18-20%) for unknown reasons.

Statistics

Most statistical tests are nonparametric because the data do not conform to the requirements of parametric testing. Unless otherwise mentioned, tests of significance are two-tailed and alpha = 0.05. In the absence of significant results, p-values are not always given. Correlations are measured with Spearman rank coefficients. All G values were adjusted with William's correction factor (Sokal and Rohlf 1981).

3

DYNAMICS OF MIXED-SPECIES ASSOCIATIONS

The foundation for much of the analysis that follows is the natural variation over time in the occurrence of mixed-species associations in the two focal groups. A prerequisite first step is the description of such associations: What criteria define an association? When and how do mixed-species group form? The data on incidence of mixed-species groups are taken from days during the year when there were ten or (usually) more hours of observation of a focal group, unless indicated otherwise. Time spent in mixed-species groups was measured to the nearest half hour.

A group of redtails and a group of blue monkeys were considered "associated" if a group member, other than an adult male, of one species was 20 meters or less from a similar group member of the other species. Adult males were excluded because they could be far from the main body of their groups; i.e., if an imaginary line were wrapped around a group of females and juveniles, adult males often would be found outside the resulting polygon.

Temporal Patterns of Mixed-Species Associations

Annual Trends
Over the course of the year March 1980-February 1981, the L redtails spent 73.9% of their time in mixed groups with blue monkeys (n = 94 days), and the T blues spent 47.2% of their time with redtails (n = 84 days). The two groups differ from one another because of the relative sizes of their home ranges. The L redtails overlapped parts of four blue home ranges, and so had four groups available for association; the T blues had mainly only the L redtails available, and only when the L redtails were in that part of their range coincident with the T range. Occasionally the T blues associated with the O redtails, whose range overlapped T's slightly. The amount of time the L redtail group associated with each group

of blues was correlated with the degree to which the two home ranges overlapped (r_s = 1.00, n = 5, p < 0.05).

Intermonthly Variation

In both species, there is variation in the proportion of time spent in mixed-species groups from month to month (Table 4). Intermonthly variation is less marked in the redtails than in the blue monkeys. The amount of time spent in mixed groups (per month) by the redtails and the blues is not correlated. This again reflects the fact that the L redtails had several blue monkey groups with whom they could associate, while the T blues did not.

TABLE 4

Monthly Variation in the Amount of Time Spent Associated

	Blue Monkeys		Redtails	
	% Time Associated	n *	% Time Associated	n *
March	7.8	194	69.0	235
April	50.9	216	84.5	242
May	71.5	216	80.7	243
June	69.2	100	75.0	122
August	27.4	158	76.8	163
September	64.3	186	62.0	217
October	52.1	192	87.5	194
November	65.5	192	71.4	228
December	41.7	212	62.0	206
January	63.5	192	75.0	214
February	29.9	184	67.0	192

* Number of half hours in 94 days (redtails) and 89 days (blues)

Intermonthly variation does not follow an obvious seasonal pattern: in neither species is the monthly proportion of time spent in mixed-species groups correlated with rainfall, number of rainy days, or proportion of observation time when it was rainy, cloudy, or sunny.

Diurnal Variation

Figure 3 shows how the tendency to be associated fluctuates with hour of the day; data come from 98 days (for each species) when associations were monitored for at least five

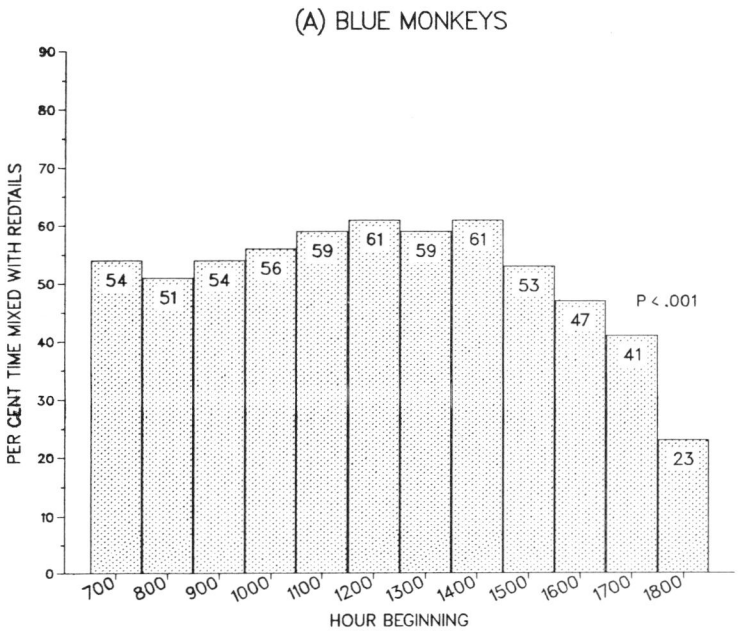
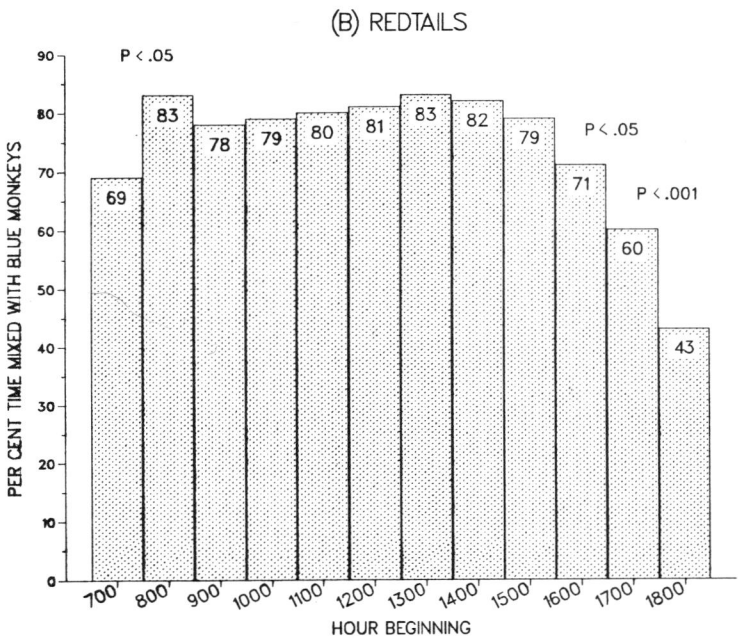

FIGURE 3. Diurnal fluctuations in association.

consecutive hours. Half days were used only when they could be paired with other half days to give coverage over all (or most) of the 12 hours of the day. The hypothesis that association is equally likely at all hours of the day can be rejected for blue monkeys except during the period from 0700 to 1700, and for redtails except from 0800 to 1700 (G tests, $p < 0.05$). Thus both species show little change in association tendency during most of the day; associations occur relatively less frequently only during the last 2 hours of the day, and for redtails, also early in the morning.

Frequency and Length of Association

Table 5 presents data on association frequency per day. A new association begins if the focal group has spent an hour or more alone, or if the identity of the group with which it associates changes. Redtails took part in 1-7 associations per day ($\bar{x} = 2.49 \pm 1.21$), whereas blues participated in 0-3 associations per day ($\bar{x} = 1.22 \pm 0.70$). The difference between species is again not unexpected given different home range sizes.

TABLE 5

Frequency (per day) of Associations

Number of Associations Per Day	Blues: Number of Days	Redtails: Number of Days
0	13	0
1	49	19
2	29	38
3	2	20
4	0	13
5	0	5
6	0	0
7	1	0
	n = 96 days	n = 93 days

Associations can last up an entire day but most are shorter (Figure 4). The average duration of L redtail associations was 4.2 ± 6.8 hours (n = 239); T blue monkey associations averaged 5.1 ± 3.7 hours (n = 113). Differences between species are significant (Mann-Whitney U test, $p < 0.001$). Many of the half-hour associations involved two groups sideswiping one another; in those lasting more than half an hour, the two groups actually traveled together.

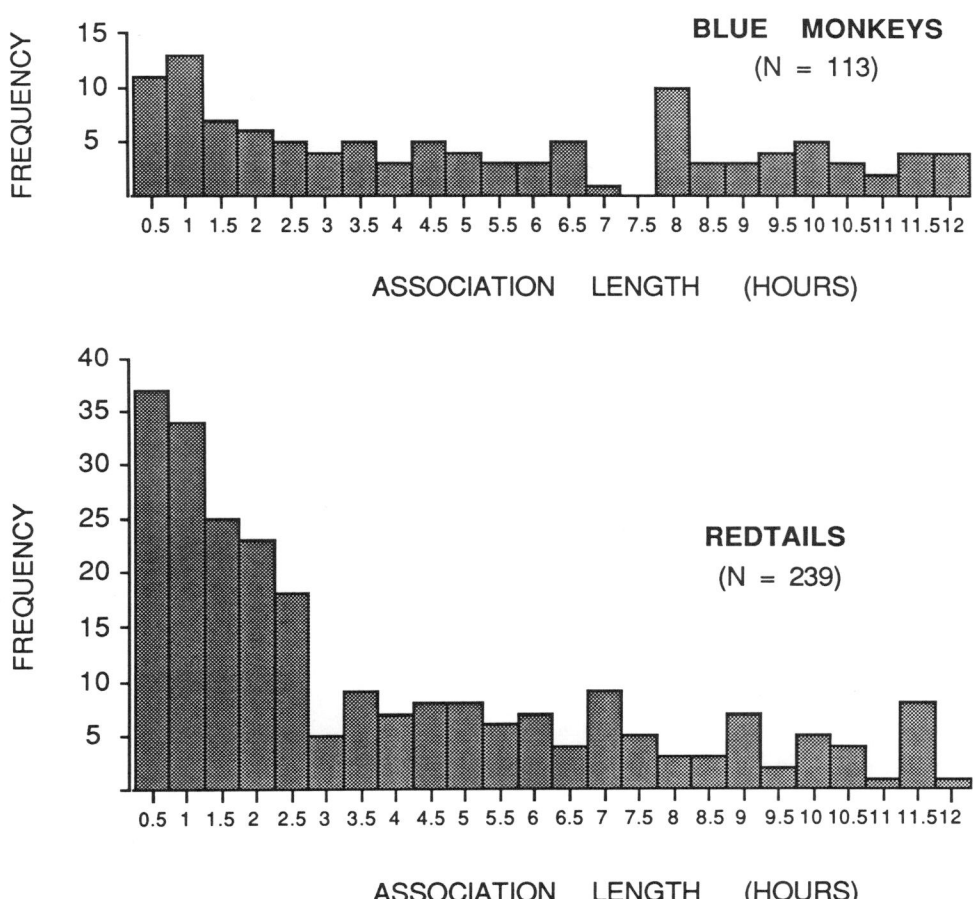

FIGURE 4. Association lengths.

Given the previous results, it is not unexpected that redtails spend more time in association per day than do blues (Mann-Whitney U test, p < 0.001). The L redtails spend an average of 8.9 ± 2.5 hours per day (n = 89) with blue monkeys; the T blues spend 6.0 ± 4.0 hours per day (n = 84) with redtails.

Who Is Responsible?

The question of which species is responsible for mixed-species associations is an important one for understanding why they occur. Operational definitions of responsibility are based on the relative movement of the two groups: the one responsible either (1) approaches more often, (2) leaves less often, or (3) approaches more than it leaves, relative to the other species. The first two criteria address respectively responsibility for forming and disbanding associations. The third criterion, developed in the context of developmental study of dyadic relationships (Hinde and Atkinson 1970), combines reducing and not-increasing proximity as two equally important facets of responsibility for maintenance of an association (or relationship). It is insensitive to absolute frequencies of approaching and leaving which may depend on the relative mobility of different groups, and is determined only by their relative frequencies within one group.

It was necessary to define "approach" and "leave" in terms of groups of monkeys. In the case of individuals, this is usually easy because one animal (being approached or left) is motionless. Because groups of monkeys are seldom motionless, however, one must evaluate relative speed: group A can be said to approach group B if A moves (toward B) faster than B moves. When associations formed in the field, the relative speeds of the two groups were assessed; this was possible because the presence of the nonfocal group could often be detected before associations had formed. Sometimes it seemed that the two groups moved at equal rates; in such cases, a group that abruptly turned or backtracked, and subsequently associated with another group, was determined to have approached the latter group. Such group movements are unusual for both species, and it seemed that by turning or backtracking, a group was "going out of its way" when it subsequently joined or left the other species.

The above description is complicated by the fact that blue monkey groups (in their smaller home ranges) ran into territorial boundaries where redtails did not. Associations formed and dissolved when redtails crossed these boundaries, joining a blue group on the other side, or leaving one behind, or both. Because blues often slow down or tarry at home range edges (they usually go there because of a rich food source) before setting off in another direction, approaches and leaves occurring in these circumstances would be biased toward redtails. For this reason, formations and dissolutions of mixed-species groups in groups in these circumstances were not included in the following calculations.

Redtails approach blues more often than blues approach redtails (71/83 approaches by redtails); if approaching is taken as an indication of responsibility, redtails should be considered more responsible than blues for the formation of mixed-species associations. However, redtails also leave blues more often than blues leave redtails (65/93 leaves by redtails); thus if not leaving is taken as a way of maintaining an association, blue monkeys are responsible. These apparently contradictory conclusions arise simply because redtails generally move faster than blue monkeys (see Ranging).

The dilemma can be resolved by considering approaching and not-leaving together as aspects of responsibility for proximity. Their summed percentages (% approaches - % leaves) will be greater than zero for the responsible species. By this formulation, redtails are considered responsible (% A - % L = 16), because they leave (relatively) less often than they approach (G test, $p < 0.02$).

Behavior during Formation and Dissolution of Mixed-Species Groups

Gautier and Gautier-Hion (1983) have described distinctive vocalization patterns of adult male *Cercopithecus* in Gabon during formation, but not dissolution, of polyspecific groups. A thorough analysis of vocalizations in Kakamega is not yet complete: however, it is clear that adult male loud calls (described in Marler 1973) do not regularly accompany either association formation or dissolution. In only 6 of 22 formations (the sample included the first two per observation month) and 5 of 22 dissolutions did male loud calls occur up to half an hour before or after the point of joining or splitting of the two groups. In every case, both blue and redtail males made loud calls, whereas Gautier and Gautier-Hion seldom heard vocalizations from all species.

No other remarkable behaviors occurred as associations formed or disbanded. Rather, it usually seemed that the two groups simply flowed together gradually, without overtly acknowledging one another. Gartlan and Struhsaker (1972) similarly noted no specific behavior patterns during polyspecific group formation and dissolution in *Cercopithecus* in Cameroon.

Evaluation of a Null Hypothesis

Most discussions of mixed-species associations in primates seem to reject the null hypothesis that such associations are due to chance (Waser 1982). By chance we mean that groups of different species move independently of one another in a common area where they will occasionally encounter one another. The reasons for rejecting chance as an explanation are seldom explicit: mixed-species associations appear to be so common or so conspicuous that chance is not even considered as a contributing factor (e.g., Rudran 1978, Pook and Pook 1982).

Three arguments against the null hypothesis for the Kakamega *Cercopithecus* can be derived from data presented here. First, given that all areas of the study area are accessible to both species so they need not move along specific limited routes, it simply seems unreasonable to consider two groups of monkeys moving together for up to 12 hours (i.e., all day) as independent units: they clearly are not. Of course, most associations do not last this long, but even those of average length (4-5 hours) would seem to be unlikely consequences of independent group trajectories.

Second, if groups of different species run into one another by chance, one would not expect to be able to assign responsibility to one species. The faster species would be expected to approach and leave more than the slower species, but there is no reason why the percentages of approaches and leaves should differ within a species. The fact that they do significantly differ speaks against the null hypothesis.

A third more formal argument relies on models developed by Waser (1982, unpublished manuscript) which predict the incidence of mixed-species groups under the null hypothesis. Monkey groups are analogized to perfect gas molecules moving in a two-dimensional space. The first model predicts the collision rate between groups of different species based on rates of movement, group spread, and group density. The assumption that rates of movement follow two-dimensional Maxwell-Boltzmann distributions appears to hold for the Kakamega monkeys (Cords 1984b).

Mean observed encounter rates for both species exceed the expectation under the null hypothesis (1.08 encounters per day expected for blues, 2.03 for redtails). It is not yet known how one can evaluate these deviations statistically (Waser 1982), but a qualitative conclusion is possible: these species encounter each other slightly more frequently than expected by chance. This conclusion is strengthened because the model overestimates encounter rate by ignoring the fact that monkey groups are not perfectly elastic: when they meet and stick together, they are not available for further "collisions." The fact that observed encounter frequencies exceed even inflated expectations further supports the conclusion that chance is an insufficient explanation for the occurrence of mixed-species groups.

Encounter rates might, however, be even less than those expected by chance and associations could still be biologically significant if they lasted a long time: for example, *C. nictitans* and *C. pogonias* in Gabon spend 97% of their time in mixed groups, and consequently the rate of encounters between troops must be extremely low (Gautier and Gautier-Hion 1983). In his second model, Waser (unpublished manuscript) predicts expected encounter lengths under the null hypothesis, assuming now that molecules pass through rather than bounce off one another. Observed association lengths at Kakamega exceed the predicted value of 1.58 hours by more than a factor of two, and a deviation of this magnitude is unlikely to have occurred by chance ($p < 0.05$, Waser, unpublished manuscript). Thus it seems that mixed groups of blue monkeys and redtails in the Kakamega Forest form more frequently and last significantly longer than would be expected if they were simply random encounters of independently moving units.

This conclusion also holds when expected and observed values are calculated for each month separately. In all months, observed average encounter lengths are at least twice the values expected under the null hypothesis. In April, May, and October, the three months when redtails spent most time with blues, average association lengths of both species exceed expectations by the largest margins; encounter rates for these months are consequently slightly less than expected.

The null hypothesis rejected here, however, is not the only one we should consider (Waser 1982). As we shall see, the diets of blue and redtail monkeys in Kakamega are very similar. Groups of different species might tend to converge on common feeding sites, though their movements are independent of one another per se. This second null hypothesis will be discussed later.

4

BEHAVIORAL RELATIONSHIPS BETWEEN SPECIES IN MIXED GROUPS

Interactions Between Species

Social Behavior

Overt social interactions between blue monkeys and redtails in mixed-species groups are rare, especially relative to interactions occurring within species. Between August 1979 and April 1981, a total of 431 interactions between members of different species in mixed-species groups were observed. Interactions are broken down into dyads: thus, if two blue monkeys simultaneously threatened one redtail, two separate interactions were scored. Table 6 shows that most interactions were agonistic. Friendly interactions such as grooming and play occurred relatively infrequently. Within species, on the other hand, friendly interactions composed the largest fraction of those observed.

Of the 295 interactions involving overt aggression, at least 49% occurred in trees or lianas where one or both interacting animals were feeding; these interactions appeared to result from competition over food or feeding sites. In nearly all cases, these were fairly large feeding trees which could accommodate at least two or three monkeys simultaneously. The recipient of the aggression always left the immediate vicinity of the aggressor, and frequently moved out of the feeding tree altogether, while the aggressor continued or began feeding. No cases of direct piracy of food items were observed.

In the remaining agonistic and approach-retreat interactions, the reason for conflict was not known. In about half the cases, the events leading up to the confrontation were not observed. At other times, however, particularly when one animal simply avoided another without having been threatened or chased, the functional significance of the interaction was simply not obvious.

It appears that blue monkeys are generally dominant to redtails, but that this pattern is a consequence of relative body size rather than species identity per se (Table 7). Of 330 agonistic interactions in which the identity of the "winner" could be established, blue

TABLE 6

Interspecific Social Interactions

	N	%
Agonistic and approach-retreat		
(a) aggression	295)	78.3
(b) avoid	47)	
Play	56	12.8
Solicit grooming	10	3.7
Grooming	9	2.1
Approach to within 1 meter	9	2.1
Present	2	0.5
Other	3	0.5

monkeys "won" in 63% of the cases. In 196 of these cases it was possible to judge the body sizes of the contestants; blues were bigger than redtails in most instances, and were seldom smaller in size. When redtails "won" agonistic interactions, they were usually as big as or bigger than their blue monkey opponents. In many (18/31) of the instances in which a smaller redtail defeated a larger blue monkey, the redtail was a juvenile who was aided either by a conspecific adult female as big as or bigger than the blue monkey opponent, or by another juvenile of the same or greater size as the redtail juvenile. When aiding behavior occurred in interspecific confrontations, juveniles were nearly always the aided individuals. In general, however, aiding was rare (it involved 48 of the 330, or 15% of, contestants who "won") and interspecific agonism was mainly limited to dyads.

Whereas interspecific agonistic interactions involved individuals of all age-sex classes in both species, friendly interactions were limited mainly to young animals. All interspecific play bouts involved medium and small juveniles. In 32 of the 56 bouts (57%), play

TABLE 7

Relationship Between Relative Body Size and the Outcome of Interspecific Agonistic Interactions

Size relationship *	Blue wins	Redtail wins
blue > redtail	119 (61%)	31 (27%)
blue = redtail	62 (31%)	42 (36.5%)
blue < redtail	15 (8%)	42 (36.5%)

* Only cases in which the relative sizes could be judged are included.

partners were equal in size; in the remaining 43% of bouts, blue juveniles were bigger than redtails. Given that there are more juveniles in a blue monkey group than in a redtail group, and that most blue juveniles are bigger than redtail juveniles, these results indicate that heterospecific play partners are not randomly chosen: more dyads are equally matched for size than would be expected by chance.

Most interspecific grooming bouts also involved juveniles. Adult female redtails solicited grooming five times, but only once was an adult female groomed. Both solicitation for grooming and grooming itself were very asymmetrically distributed between the two species: only redtails were observed to solicit grooming from another species, and in seven of nine grooming bouts observed, blue monkeys groomed redtails.

Interspecific solicitation for grooming was observed on six additional occasions outside of a mixed-species group when a solitary adult male redtail traveled with the T blue monkey group and solicited grooming (very persistently) from blue monkey juveniles. In three cases, such solicitations were successful.

In sum, behavioral interactions between members of the two *Cercopithecus* species appear to be primarily competitive, and direct interference in the use of food or feeding sites was conspicuous. This pattern contrasts with intraspecific social interactions that include a greater proportion of friendly behaviors. Relative body size was a good predictor of the outcome of agonistic interactions between members of different species. Friendly interactions were rarer between individuals of disparate body sizes, and consequently occurred most frequently between young animals.

Vocalizations

The two species also interacted vocally. The contact calls of females and juveniles (croaks or grunts, Marler 1973) seemed to elicit responses only by conspecifics. The loud calls of males, however -- the redtail "pop" and blue monkey "pyow" (Marler 1973) -- were usually given at the same time. Hundreds of calling bouts involving both species were heard, and with one exception, the blue monkey male always started them. After the blue male's first or second pyow, the redtail male would begin to insert pops between every 2 or 3 pyows (see also Marler 1973) and he was heard to pop as many as 10 times in a given calling sequence. More typically, the redtail male would pop 2-4 times. In single-species groups, redtail males popped only once.

Blue monkey pyows were usually given 4-5 times per day, and as noted before, their timing did not appear to be closely related to the formation or dissolution of mixed-species groups. Rather, one came to expect a pyow before 0830 hrs and one after 1800 hrs, and 1-3 more sometime during the middle of the day. If the redtail group was with or near (within about 200 meters) a blue monkey group when its male pyowed, the redtail male would pop almost invariably. On 10-15 occasions, the redtail male did not answer the pyows of the accompanying blue male. These cases were so exceptional that a special effort was made to explain the redtail's failure to respond; however, spatial relations of the two groups were in no way unusual relative to times when both males vocalized, and no unusual external stimulus was apparent or absent. If redtail groups were farther than about 200 meters from blue monkey groups, pops were less likely to occur in response to blue monkey pyows.

Pops were seldom given on their own (i.e. not associated with blue monkey pyows), unless as an alarm response. This was especially true in mixed-species groups: not once did a redtail male in a mixed group pop without being accompanied by blue monkey pyows.

Interspecific vocal exchanges by adult males have been noted in several other *Cercopithecus* communities, where again, the loud calls of the larger *Cercopithecus* species typically precede the "pop" or "hack" of the smaller one (Aldrich-Blake 1970, Struhsaker 1970, Marler 1973, Gautier and Gautier-Hion 1983). It has been suggested that the species vocalizing first and more frequently has a lower threshold for giving loud calls because it is territorial, while the second species is not; however, the second species is selected to respond to the first in order to maintain group cohesion in the face of group-dispersing stimuli detected initially by the first species (Struhsaker 1970). A second explanation is similar, except that the response thresholds of the two species are related to their roles in mixed groups: the "leader" species vocalizes more often and vocalizes first more often and thus is able to coordinate movement and cohesion among all members of a mixed group (Gautier and Gautier-Hion 1983). In this *Cercopithecus* community, the latter explanation seems more applicable, because redtails are territorial and because, as we shall see, blue monkeys may determine where the group travels. The significance of the vocal exchanges, however, cannot be evaluated without more detailed data on group responses to loud calls, which have not been carefully documented in any population.

Interspecific Spatial Relationships

Even when the two groups were spatially fully overlapping, members of each species were likely to be nearer to a conspecific than to members of the other species. Table 8 shows that the likelihood of different age-sex classes having each species as their nearest neighbor in feeding trees is variable: for both blues and redtails, adult females with clinging infants were most likely to have conspecifics as nearest neighbors, whereas large and medium juveniles were most likely to have members of the other species as nearest neighbors. If

TABLE 8

Species Identity of Nearest Neighbors in Feeding Trees for Different Age-sex Classes (%)

(a) BLUE MONKEYS	AF	AF+in	AM	LJ	MJ	SJ
Blue neighbor	94.9	97.8	94.9	91.0	93.1	92.0
Redtail neighbor	5.1	2.2	5.1	9.0	6.9	8.0
n	1125	138	136	279	1277	853

(b) REDTAILS	AF	AF+in	AM	LJ	MJ	SJ
Redtail neighbor	89.1	96.6	89.9	85.2	87.1	91.7
Blue neighbor	10.9	3.4	10.1	14.8	12.9	8.3
n	1688	176	385	256	1367	895

nearest neighbors in other (nonfeeding) contexts were included in this analysis, the tendency for conspecifics to be neighbors would appear even stronger, because animals at rest or engaged in social activities such as grooming or play were clearly more often in proximity to conspecifics than to members of different species. The general spatial pattern, then, is one of small subgroups of redtails (two or more) interspersed among small subgroups of blue monkeys.

Spatial patterning on a larger scale was usually not apparent. Members of neither species were predictably concentrated at either the front or back of the mixed group, though a van of redtails sometimes ran ahead to large fruiting trees that most blues would reach later. In progressions, the first animal to cross through a particular tree or set of trees was a blue monkey more than 50% of the time; however because blue monkeys were twice as common as redtails in mixed groups, this pattern would be expected by chance. The most obvious spatial relationship between the two species in a mixed group was that the redtails were usually more concentrated on the outside of a curved path. The redtails as a group thus moved faster than the blues even when the two species were moving together.

Activity Patterns

Scan samples made once an hour 13 times per day were used to look for changes in activity patterns in each species as a function of time of day and being associated. On each scan, the first activity lasting at least 5 seconds of the first four (or fewer, if fewer were seen) animals seen within a 2 minute period was recorded. Activities were divided into five categories: (1) Feed and forage, including manipulation of potential food or inspection of microhabitats for invertebrate prey; (2) locomote, including any directional movement such as walking, running or jumping not immediately associated with foraging; (3) rest, indicating inactivity, either sitting awake or sleeping; (4) groom or be groomed; and (5) other, including play and aggression. Scores were made on 48 days for each species, yielding 2,273 records for for blues and 2,129 for redtails.

The data from scan samples cannot be interpreted as an activity budget because samples as short as 5 seconds and as infrequent as once an hour do not accurately reflect a time budget, and the relative visibility of different activities may be a confounding variable (Altmann 1974). The main purpose of the activity scans, however, was to evaluate differences, in which case any biases can be assumed constant. In particular, activity scores were used to detect diurnal changes, and differences that might occur between the monkeys in monospecific and mixed-species groups.

The different activities are not equally likely to occur at different hours of the day (Figure 5, G test, $p < 0.001$ for both species). Feeding and foraging activity occurs most commonly in the morning and evening; resting is more common in the middle of the day. This trend is apparent in spite of the fact that the monkeys are at lower elevations in the middle of the day, where inactive animals are particularly difficult to observe.

The activity profile of redtails in mixed groups does not differ from that of redtails when alone (G test). Blues, on the other hand, show a greater proportion of feeding and locomotion scores and a smaller proportion of resting scores when in mixed groups, relative to when they are alone (Table 9). When each hour of the day is considered separately, this same pattern is significant only from 1200 to 1300 and from 1500 to 1600. Other indications of the blue monkeys' increased activity when with redtails are discussed in the section on Ranging.

TABLE 9

Activity Scores of Blue monkeys and Redtails, All Hours Combined, as a Function of Association (% in parentheses)

(a) BLUE MONKEYS	Alone		Mixed	
Feed, forage	527	(43.8)	529	(49.4)
Locomote	158	(13.1)	169	(15.8)
Rest, inactive	482	(40.1)	339	(31.7)
Groom	19	(1.6)	13	(1.2)
Other	17	(1.4)	20	(1.9)
n	1203		1070	

$G_{adj} = 18.85$, 4 d.f., $p < 0.001$

(b) REDTAILS	Alone		Mixed	
Feed, forage	239	(37.5)	581	(38.9)
Locomote	257	(40.3)	552	(37.0)
Rest, inactive	129	(20.3)	319	(21.4)
Groom	3	(0.5)	23	(1.8)
Other	9	(1.4)	17	(1.1)
n	637		1492	

G test, n.s.

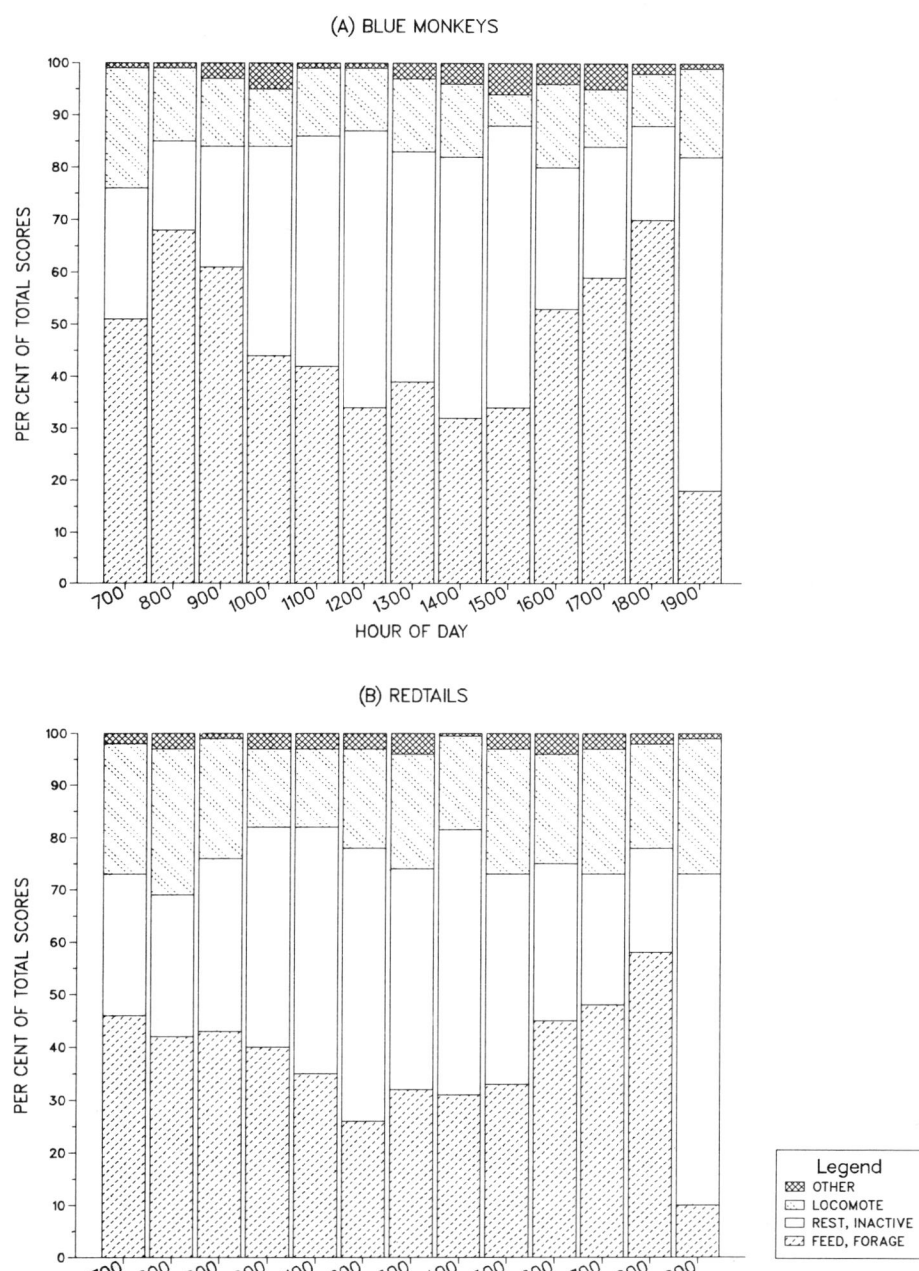

FIGURE 5. Diurnal variation in activity.

5

PREDATION

Members of different species may associate in mixed-species groups to reduce the risk of being preyed upon. After a description of the predators on *Cercopithecus* monkeys in Kakamega, and a discussion of the monkeys' response to predator attacks, the mechanisms by which mixed-species grouping could reduce risk are considered. The predation-avoidance hypothesis is then evaluated by comparing the incidence of predator attacks with the incidence of mixed-species grouping.

Predators

In the Kakamega Forest, monkeys are preyed upon mainly by the crowned hawk eagle (*Stephanoetus coronatus*). Large cats, such as leopards (*Panthera pardus*) and golden cats (*Felis aurata*), are now extremely rare. (I once glimpsed a large cat at dusk, but was unable to identify it.) People occasionally hunt monkeys in the forest, though there was no evidence of poaching from in or near the study area during my study.

The crowned hawk eagle and its habits are described by Brown (1970, Brown et al. 1982). It is distributed throughout Africa where there are forested or heavily wooded zones. Pairs are resident year-round in home ranges of 10-146 km^2; ranges are smaller in forests than in woodland-grassland mosaic. Each pair produces young biennially. In Kakamega, a juvenile was seen in March 1980.

Before the attacks that I observed, the bird typically perched high in the canopy at some distance (where it was apparently not perceived) from the monkey group. In a few cases, an eagle was seen or heard displaying overhead up to 30 minutes before attacking. To attack, the bird swooped down through the trees into the middle of the monkey group. It seemed to rely on surprise and speed. This description largely coincides with that of Brown et al. (1982). Members of pairs have been observed to hunt together (Daneel 1979); in Kakamega, the simultaneous presence of two adult birds was confirmed on two occasions.

The diet of the crowned hawk eagle comprises a variety of medium-sized mammals weighing up to 18-20 kg. It is widely reputed to feed mainly on monkeys, but in parts of central Kenya where it has been most thoroughly studied, antelope constitute by far the largest fraction of its prey intake (Brown et al 1982). There are no data available on the diet of the crowned hawk eagle in western Kenyan forests: however, the importance of monkey in the diet may be substantial, especially since antelope populations have been hunted to very low levels in recent years by man. (Nevertheless, the one time I saw an eagle nearly succeed in catching a prey item, it was a red duiker, *Cephalophus harveyi*.)

The issue at hand, however, is not how important monkeys are to eagles, but how important eagles are to monkeys. At present, the answer must be speculative: I never observed an eagle make a successful attack on a monkey group. No recognized individuals ever disappeared after an attack. When complete group counts were made shortly before and after an attack, there were no missing individuals. However, an eagle was once observed to catch a "medium-sized" colobus (Rowell, pers. comm.). Other sources of mortality for the Kakamega monkeys are also undocumented. Several individuals disappeared during the study (most during breaks in observations). Most of these animals showed no signs of disease or moribundity, which suggests that predation may well limit population growth.

Monkeys' Response to Predators

The marked response of monkeys to their predators supports the view that predation is a major source of mortality and therefore an important selective force on the prey. In Kakamega, blues and redtails respond similarly to crowned hawk eagles: one or a few animals, probably the first to see the bird, utter a single short coughlike growl, whereupon most group members stop their activities and look up. Almost immediately most individuals drop from more exposed branches in the canopy down into the dense foliage of understory shrubs or liana tangles where they may remain from 10 to about 60 minutes. During and after this "dive down," many monkeys chirp and adult males may give loud calls (see below). The chirp (Marler 1973) is a general alarm call given by adult females and juveniles to a variety of stimuli including civets, palm civets, mongooses, domestic dogs, and people. The cough-growl on the other hand, again limited to adult females and juveniles, was almost always heard in response to a bird. It was frequently misdirected, most often at smaller hawks (e.g. *Accipiter melanoleuca*, *Polybroides typus*, n = 33) that do not prey on monkeys, or at hornbills (*Bycanistes subcylindricus*) when they soar through the trees (n = 14). False alarms were also given in response to turacos (*Corythaeola cristatus*, n = 5), herons, storks, unidentified falconiform and passerine birds, and flying squirrels. False alarms were usually accompanied by starts from individuals near the caller, but rarely did any individuals dive. Between July 1979 and April 1981, 129 presumed false alarms were witnessed; in 74 of these, the mistaken animal was seen.

The calls of females and juveniles, especially cough-growls, could not be attributed to individuals or even species in most cases. In the six cases (all false alarms) in which the first cough-growler in a mixed-species group was seen, it seemed to be the monkey nearest the bird's line of flight into the group; four times this was a blue monkey and twice it was a redtail. There was no evidence that one species was more alert than the other.

The adult male alarm call, or ka-train (Marler 1973), was different enough between species to allow identification, and because there were fewer adult males than other group members, the individual caller could also be identified. Cough-growls always preceded ka-trains except when males called at the commotion in a neighboring group. Ka-trains were given only twice during the 129 false alarms, both times by redtails when in single-species groups. Ka-trains were always given when the eagle flew into a group (n = 21 times), and their presence, along with the alarm calling and diving of most or all other group members, was the basis for inferring the eagle's presence in or near a group on 67 other occasions. Ka-trains were also given when there were alarm calls and responses in neighboring groups (n = 18). On 21 occasions, however, no reason was detected for a male's calling and other group members did not appear to be alarmed.

Blue monkey males ka-trained more often than redtail males. When both species vocalized, blue monkey males gave the first alarm call more often than redtails (Table 10), but differences were significant only when blue monkeys ka-trained and redtails popped (Binomial test, $p < 0.0002$). Also, blue males, but not redtails, were seen to charge and displace eagles perched in the group, ka-training all the while (n = 3). No other group members were observed mobbing eagles. Males of both species often made their species-typical loud calls (pyows and pops, Marler 1973) between and after ka-trains. Only redtails made such calls without ka-training (themselves) in alarm situations.

TABLE 10

Coincidence of Alarm Vocalizations by Adult Males, Excluding Known False Alarms

Blue Vocalization	Redtail Vocalization	N of Cases	First Alarm by Blue	First Alarm by Redtail	First Caller Not Identified
ka-train	ka-train	18	8	5	5
ka-train	pop	31	25	5	1
ka-train	(none)	11			
(none)	ka-train	3			

Mixed-species Groups and Predation

The monkeys behave as if the crowned hawk eagle poses a serious threat, both by their alarm response and by their sensitivity to eagle-like stimuli. Mixed-species grouping might reduce the risk of being preyed upon to individual monkeys in several ways. The greater number and spread of eyes in such groups could increase the group's chances of detecting the eagle early. Because the monkeys' only recourse is flight and concealment, early detection and subsequent early response could be important. Furthermore, the rain of monkeys diving down after an alarm call may confuse the eagle as it makes its swooping attack. Mixed-species groups would enhance this confusion effect because there are more falling bodies. Finally there is the effect of dilution: for any individual, the chances of being a victim are reduced the more other prey items are available. Dilution seems to be an automatic benefit for members of larger groups as long as predator success rates do not increase proportionately with group size.

To test whether mixed-species groups do inhibit predation by eagles, one should compare the success rate of eagle attacks on single- versus mixed-species groups. Unfortunately I did not witness a single attack that was known to be successful. This fact in itself could be taken as evidence consistent with the predation-avoidance hypothesis; however, it cannot be used to explain variation in the incidence of mixed-species groups over time.

An approximate test of the hypothesis involves examining the relationship between being in mixed groups and being attacked. If mixed-species groups deter successful predation attempts, monkeys may spend more time in mixed groups when predators are likely to attack. This hypothesis was tested using two data sets: (I.) cases in which the eagle's presence in or near the group was confirmed or strongly suspected, based on the monkeys' very strong alarm response (n = 99 from July 1979 to April 1981); and (II.) cases in which an eagle was actually observed to attack the group (n = 21). Except in 18 cases in the first data set when the focus of alarm appeared to be in a neighboring group, the full complement of alarm responses by the monkeys occurred; in those 18 cases, all types of alarm responses were given, but they were not expressed by as many individuals. In most cases, other species, especially hornbills and colobus, called alarms concurrently. Normal activity did not resume within 10 minutes of the attack. The reason for using the (second) restricted data set is that the monkeys' response to eagle overflights, when the bird is not about to attack them, and true attacks is indistinguishable. In Kakamega, at least three full-scale responses to overflights were witnessed. In the Kibale Forest, Uganda, such responses were seen repeatedly in the vicinity of an eagle nest whenever birds flew to or from it (Skorupa, pers. comm.). If the eagle was not seen, it was therefore impossible to distinguish between attacks and overflights. The first data set includes some (at least three) overflights, whereas the second includes none.

Diurnal Patterns

Table 11 shows how eagle encounters are distributed during the day for the two species in both data sets. Only data from days with ten or more observation hours are included. One cannot reject the hypothesis that encounters are equally likely throughout the day (G tests, p > 0.05; for Data set II, data were combined into three 4-hour time blocks). Nevertheless, there is some variation in eagle encounter rates during the day; this variation is incorporated into the expected values in Table 12. These expected values are calculated under the (null) assumption that the probability of being in a mixed-species group (at a certain time of day) and the probability of encountering an eagle (at that same time of day) are independent. Thus the total number of encounters for a given time period is multiplied by the likelihood of being in mixed groups at that time period to generate an expected number of encounters. while in mixed groups for that time. Time periods were combined for statistical reasons. Some expected values are still less than five, which is generally considered to be a "safe" lower limit. When expected values are less than five, G-test rejects the null hypothesis too easily (Sokal and Rohlf 1981): this bias is conservative for the data in Table 12, however, because the null hypothesis is still not rejected.

TABLE 11

Distribution of Eagle Encounters During the Day

	(a) Data set I		(b) Data set II	
Time	Blues	Redtails	Blues	Redtails
0700-0900	11	11	4	1
0900-1100	7	8	1	1
1100-1300	2	10	0	2
1300-1500	11	8	2	3
1500-1700	8	9	3	1
1700-1900	5	8	0	3

Using data set II, the sample size is so small that it is possible statistically to test only a crude null hypothesis, i.e., that the fraction of attacks occurring when the monkeys are in mixed groups equals the fraction of time spent in such groups, calculated over all hours. For the blue monkeys, this hypothesis cannot be rejected; for the redtails, differences

TABLE 12
Does Association Affect the Likelihood of Attack? (Data Set I)

	Time	Observed Attacks Mixed	(Expected)	Observed Attacks Alone	(Expected)
Blues:	0700-0900	3	(6.1)	8	(4.9)
	0900-1300	4	(5.0)	5	(4.0)
	1300-1500	8	(6.1)	3	(4.9)
	1500-1900	10	(7.8)	4	(6.2)

$G_{adj} = 3.33, 4\ d.f., n.s.$

	Time	Observed Attacks Mixed	(Expected)	Observed Attacks Alone	(Expected)
Redtails:	0700-1300	26	(25.2)	3	(3.7)
	1300-1900	21	(21.8)	4	(3.2)

$G_{adj} = 0.29, 2\ d.f., n.s.$

between observed and expected values are significant (Binomial test, all one-tailed p = 0.036 since we predict mixed-species groups when attacks are frequent), with fewer (0/11) attacks than expected when the monkeys are alone. If the data are broken down into three 4-hour time blocks, however, neither species was attacked more often when in mixed groups than expected (Binomial tests).

Another way of examining the relationship between likelihood of attack and being in mixed groups is by calculating correlation coefficients (Freeland 1977, Waser 1982). For both blues and redtails, neither the number of eagle encounters nor the number of actual attacks was correlated with the fraction of time spent in mixed groups in each of the 12 hours of the day (Pearson's r, all one-tailed p > 0.147). In other words, mixed-species associations were not more likely to form at times of day when eagle encounters or attacks occurred more frequently.

Monthly Patterns
Sample sizes are too small to test monthly patterns except by correlation analysis. For both species, neither the number of eagle encounters nor the number of attacks was correlated

with the fraction of time spent in mixed groups in each of the 11 sample months from March 1980 to February 1981 (Pearson's r, all one-tailed $p > 0.089$). The monkeys did not associate more (or less) in months when eagle encounters or attacks were more common.

Evaluation of the Predation-Avoidance Hypothesis

Predation by crowned hawk eagles appears to be an important factor influencing the behavior of blue monkeys and redtails in Kakamega. Each species has marked vocal and behavioral alarm responses. Each pays attention to the other's alarm response, as well as to those of other primate (i.e., colobus) and avian (e.g., hornbill) species. In Kakamega and Gabon, female and juvenile guenons are the first members of the group to give alarm calls (Gautier and Gautier-Hion 1983). There is no evidence from Kakamega that females or juveniles of one species are intrinsically more likely to call than those of the other species, though sample sizes are small. In Gabon, *C. cephus* females or juveniles are most likely to give the first alarm when predators are terrestrial, but the identity of the first alarm caller for aerial predators was determined only twice. Adult males of different species, on the other hand, are not equally likely to make their loud alarm calls: in Kakamega, blue males call more often and usually call first. This general pattern also holds in Gabon (Gautier and Gautier-Hion 1983); however, there *C. pogonias*, which does not have an ecological or taxonomic equivalent at Kakamega, is most likely to give an alarm call first at aerial predators. These observations raise the possibility that benefits of increased early predator detection accrue unequally to the species participating in mixed groups. Because blue monkey groups in Kakamega are larger, redtails may gain relatively more by associating with blues than vice versa, assuming that the predator's preference for each prey species is the same. (This assumption may be unwarranted, however; there are no relevant data.) The apparently greater readiness of male alarm calls by blues may further accentuate the difference between species, though it seems that the earlier alarms of females and juveniles are more important than males' calls in stimulating early flight and concealment.

It is difficult, however, to demonstrate directly the importance of predation as a contributing factor to the formation and incidence of mixed-species groups because the critical test, a comparison of predator success rates on monkeys alone or in mixed groups, is impossible with the existing data. The fact that no attacks on mixed-species groups were known to be successful provides only weak evidence that association with other species provides antipredator benefits. This is because there were also no successful attacks observed on single-species groups, and because the lack of observations of successful predation can be explained in at least two other ways. (1) An observer often is unlikely to be able to tell if an attack is successful because of poor visibility conditions in the forest.

(2) If attacks really were not successful, some mechanism other than mixed-species grouping might be responsible.

The approach taken here, and by others studying mixed-species groups of primates (e.g., Freeland 1977, Waser 1982), is also of arguable value (Struhsaker 1981b). The prediction that prey should congregate more often when predatory attempts are more likely to occur rests on the assumption that present behavior responds to present ecological pressures, and ignores potential flexibility in the behavior of the predator. If the predator is less likely to succeed when attacking a mixed-species group, should it not prefer to attack more vulnerable (i.e., monospecific) groups? Is it right to expect that the response of prey to predators has some sort of logical, or biological, primacy relative to the response of predators to prey at any given time?

In light of these questions, it is difficult to interpret the kind of test of the predation hypothesis that has been attempted here. The null hypothesis cannot be rejected (except in its crudest form for redtails), but that does not mean that predator pressure is irrelevant in explaining the occurrence of mixed-species groups. Indeed, the importance of predation as a factor contributing to the formation of mixed-species groups of African monkeys has been emphasized by Struhsaker (1981b) and Gautier-Hion et al. (1983). Their cases rest largely on descriptive, unquantified data, presumably because they also lack data on predator success rates in single- and mixed-species groups. Neither study examined the relationship between predation pressure and variation over time in the occurrence of mixed groups (n.b. there was very little such variation in the West African study). As in the present study, such relationships have not been found when they were sought (Freeland 1977, Waser 1982). These authors consequently relegate predation to a relatively unimportant role in explaining mixed-species association.

In these and the present cases, rejection of the predation-avoidance hypothesis seems premature, however, not only because it is hard to make biologically sensible predictions. Mixed-species association may afford antipredator benefits other than reduction of risk. Specifically, the cost of maintaining an equal level of vigilance may be reduced for the individual in a larger group if vigilance is shared, and a difference in predator success rates would not be expected. No studies of primate mixed-species associations have tested this hypothesis, e.g., by comparing individual scanning rates in single- and mixed-species groups. In the present study as well, such data are not available.

The role of predation in determining mixed-species association thus remains unsubstantiated, but potentially important. It is clear that other factors must be involved in determining the timing of mixed-species association in the Kakamega *Cercopithecus* over the course of the day or year.

6

FEEDING

Food may be an important determinant of the incidence of mixed-species groups, both because of foraging benefits such groups provide to their participants, and because of its role in interspecific competition. To understand the potential relevance of mixed-species association to feeding, it is necessary first to consider the feeding ecology of participant species separately. Feeding strategies of blue and redtail monkeys at Kakamega have been described previously (Cords 1984b, 1986a) and are briefly summarized and compared here. Then the relation between diet and mixed-species association is explored in two ways: first, by examining the relationship between variation in time spent in mixed groups and dietary fluctuations, and second, by comparing the diet of monkeys when in mixed groups and when alone. These results are used to evaluate specific hypotheses about foraging benefits resulting from mixed-species grouping.

Methods and Definitions

Feeding observations were scored throughout the day whenever possible, subject to rules described below. Analysis is restricted to data collected March 1980-February 1981 (July excluded). Diet was analyzed using feeding frequency scores, which facilitate comparison with other populations of *C. mitis* and *C. ascanius* and are easy to use in the field. A feeding score represents a particular monkey feeding on a particular item from a particular plant species (Rudran 1978). A new feeding score was recorded for the same monkey, item, and plant only if at least 30 minutes have elapsed, or if the identity of any of the three parameters changed. For invertebrate feeding, the substrate and the speed and nature of the motor pattern used for prey capture were also considered. It was seldom possible to identify the prey. I assumed that a monkey's foraging was successful only if it was seen to ingest or chew something after a catch. Unsuccessful foraging was not scored. A total of 10,167 and 9,009 feeding records were made for blue monkeys and redtails respectively.

Three units of diet composition are considered: (1) items, regardless of the species from which they come, (2) plant species, regardless of the item which was consumed, and (3) plant species-specific items. Dietary overlap is measured by summing shared percentages of specific dietary components (Holmes and Pitelka 1968). This overlap measure can vary from 0 (no overlap) to 100 (complete overlap). The degree to which the monkeys concentrated on certain foods was measured by dietary diversity. Diversity was calculated using the Shannon-Wiener equation for entropy, $H = - \Sigma\, p_i \ln p_i$, in which p_i is the proportion of the ith food item, plant species, or species-specific item in the diet (Pielou 1966).

The Diets of Blue Monkeys and Redtails

Plant Parts

The plant diets of blue and redtail monkeys at Kakamega are in general very similar (Cords 1984b, 1986a, Appendix 1 and 2). Interspecific overlap of the plant diets for the year as a whole is 77.0% for plant species, but less (70.4%) for species-specific items, because the monkeys may eat different items from the same plants. Over two thirds of feeding scores for each species represented fruit, but blue monkeys are less frugivorous than redtails. Blues eat more leaves, seeds, and blossoms, and less gum, than redtails (Table 13).

TABLE 13

Composition of the Plant Diet (% of Feeding Scores)

Item	Blue Monkey Diet	Redtail Diet
Fruit	65.6	81.7
Leaves -- young	20.0	9.1
-- mature	2.7	0.5
Blossoms	4.4	2.7
Gum	2.3	3.8
Seeds	3.0	0.6
Stems and shoots	1.3	1.0
Other	0.7	0.6
n	8454	6755

Both species show a bimodal annual distribution of fruit and leaf consumption (expressed as a percentage of total plant feeding scores), with fruit consumption peaks in the middle of the rainy and dry seasons. Fruit and leaf consumption are inversely correlated. In addition, the importance of particular plant species-specific items waxes and wanes through the months (Cords 1984b), largely in response to the changing availability of certain foods.

Because of this seasonal variation in the plant diet of each species, comparisons between them are more meaningful when made for shorter periods than the year as a whole. In each month, both blues and redtails concentrate on a few plant species and species-specific items: 75-80% of their feeding scores were made on less than 25% of the species used, and about 57% of plant feeding observations are accounted for by the five most frequently used species-specific items. Many of these most popular foods (5-8 per month) are shared by the two species. Monthly overlaps for plant species use are large (Table 14a) and positively correlated with the proportion of fruit in the diet of each species ($r_s = 0.69$, $p = 0.019$ for both), suggesting that much overlap involves fruit. If monthly overlaps are calculated separately for fruit and leaves, it is indeed clear that overlap in species used for fruit always exceeds overlap in species used as sources of leaves, both within and between months. Monthly overlap of species-specific items is lower than overlap of plant species (Table 14b), because the blues and redtails differ quantitatively in their consumption of items from shared plant species. Overlap in species-specific item use is positively correlated with the proportion of fruit in the diet of both blues and redtails across the 11 months.

Although many plant food sources are used by both monkey species, they do not necessarily prefer these food sources equally. Preference can be measured by calculating the ratio of use of a species to its availability. Lacking quantitative information on

TABLE 14

Percentage Overlap in the Plant Diet of Blue Monkeys and Redtails, by Month

	(a) Plant Species Overlap	(b) Species-specific Item Overlap
March	56.4	53.1
April	72.1	69.6
May	64.9	59.6
June	59.2	56.7
August	64.5	57.5
September	62.6	56.5
October	70.6	66.5
November	68.1	65.1
December	66.2	58.8
January	75.8	73.8
February	69.6	62.9

availability of different phytophases over time, I am forced to use a (rather crude) static measure derived from species density and average crown dimensions (Struhsaker 1975, Rudran 1978). Specifically, the selection ratio is defined as SR = 100(% use/cover index), with cover index = density x canopy size. Density estimates are taken from the enumeration of forest block contents described on page 17, and include only blocks for which area and stem density were uncorrelated. Density was calculated separately for the blue and redtail home ranges. Canopy size was measured on ten individuals of the most frequently used tree species, by estimating crown length (i.e., major axis parallel to ground), width, and depth. Length and width were averaged and added to depth to get an index of canopy size. (This measure of canopy size was linearly related to measures based on the product of the first two or three dimensions.) Selection ratios for the top five tree species from each month for both monkey species are given in Table 15.

TABLE 15

Selection Ratios for Top Five Plant Species in Plant Diets
(Ranked Top to Bottom in Each Month) *

	BLUES		REDTAILS	
Mar	T. guineense	94.1	F. exasperata	75.8
	F. exasperata	55.5	H. madagascariensis	211.7
	A. toxicaria	6.8	T. guineense	60.8
	M. lactea	101.5	P. guineense	-.-
	F. mallatocarpa	17.7	A. toxicaria	6.9
Apr	B. oblanceolatum	263.1	F. exasperata	111.0
	F. exasperata	103.1	B. oblanceolatum	189.7
	T. guineense	83.2	M. lactea	156.5
	F. mallatocarpa	71.2	F. mallatocarpa	29.9
	M. lactea	71.2	F. thonningi	43.6
May	F. exasperata	125.1	F. exasperata	108.5
	P. africana	31.6	C. macrostachyus	136.1
	C. africana	10.7	P. africana	23.6
	B. javanica	66.1	B. phoberos	8.2
	B. phoberos	5.6	P. guineense	-.-
June	B. javanica	199.7	C. sylvaticus	51.4
	C. sylvaticus	19.5	B. phoberos	15.2
	T. nobilis	45.3	B. javanica	1.5
	F. exasperata	26.6	T. nobilis	40.3
	F. macrophylla	15.0	F. exasperata	29.5

TABLE 15 (continued)

Selection Ratios for Top Five Plant Species in Plant Diets
(Ranked Top to Bottom in Each Month) *

	BLUES		REDTAILS	
Aug	T. nobilis	85.8	B. javanica	141.6
	T. guineense	92.4	T. guineense	59.7
	B. javanica	141.5	T. nobilis	52.6
	M. butugi	191.1	C. africana	8.9
	C. africana	5.6	M. discoidea	68.1
Sep	T. nobilis	111.9	C. sylvaticus	34.5
	T. guineense	70.6	T. nobilis	78.9
	B. javanica	112.4	T. guineense	36.7
	F. mallatocarpa	30.3	C. africana	7.5
	C. africana	8.6	P. africana	18.0
Oct	T. nobilis	108.7	T. nobilis	148.2
	C. durandii	18.2	C. durandii	19.1
	F. thonningi	60.6	C. sylvaticus	18.9
	T. guineense	48.8	C. abyssinica	18.3
	C. abyssinica	25.1	P. africana	13.6
Nov	F. thonningi	77.9	F. thonningi	84.1
	T. nobilis	55.4	C. durandii	18.7
	T. guineense	55.7	F. macrophylla	33.9
	F. macrophylla	24.1	T. nobilis	59.3
	C. durandii	12.5	P. africana	25.4
Dec	A. gummifera	23.0	F. macrophylla	49.5
	F. macrophylla	28.9	B. phoberos	15.7
	B. phoberos	11.7	C. durandii	14.0
	F. thonningi	57.6	F. mallatocarpa	28.0
	A. toxicaria	7.6	P. africana	16.8
Jan	B. phoberos	41.2	B. phoberos	43.5
	H. madagascariensis	124.0	H. madagascariensis	176.4
	F. mallatocarpa	30.7	C. oliveri	-.-
	M. lactea	115.2	A. toxicaria	7.7
	A. toxicaria	7.0	F. macrophylla	21.0
Feb	H. madagascariensis	296.7	H. madagascariensis	308.3
	F. mallatocarpa	39.3	B. phoberos	9.0
	A. toxicaria	9.4	A. toxicaria	7.8
	B. unijugata	72.7	M. lactea	115.6
	C. africana	6.2	F. mallatocarpa	18.2

* SR's are not calculated for climbers, which were underrepresented in the plant census

Both species showed considerable day-to-day variation in the consumption of major dietary items (Cords 1984b). Relative to its mean, fruit consumption was less variable than insect and leaf consumption. Diurnal changes were also apparent (Table 16, Cords 1984b). In the morning, fruits (and seeds, for blue monkeys) are eaten relatively frequently. Leaf and invertebrate consumption peaks in the middle of the day, and falls again toward evening when fruits (and seeds) again become relatively important.

TABLE 16

Diurnal Fluctuations in Items Consumed by Blue Monkeys and Redtails Expressed as a Percentage of Feeding Records for Each Hour, and Excluding Records in Which the Item was Unidentified

BLUES	Hour Beginning											
	07	08	09	10	11	12	13	14	15	16	17	18
Fruit	63	54	55	45	50	43	54	49	55	52	62	64
Leaves	11	11	16	18	20	29	19	23	21	23	19	17
Insects	16	22	21	26	22	20	18	19	14	13	9	7
Seeds	4	4	2	2	1	1	1	2	3	3	3	4
Other	6	9	6	9	7	7	8	7	7	9	7	8
n	990	1132	1055	858	829	667	448	596	761	739	1032	1034

REDTAILS	Hour Beginning											
	07	08	09	10	11	12	13	14	15	16	17	18
Fruit	70	60	53	55	50	55	51	50	60	68	73	72
Leaves	4	4	5	8	11	7	11	14	7	7	6	10
Insects	21	30	38	31	33	31	32	29	28	19	15	10
Other	5	6	4	6	6	7	6	7	5	6	6	8
n	863	986	913	825	687	457	412	435	569	665	1075	1086

Animal Prey

Animal prey composed 16.8% of blue and 25.0% of redtail feeding scores during the year. The relative frequency of animal prey scores varied from month to month, however: for blue monkeys, it was negatively related to the proportion of scores representing other protein sources (seeds, young leaves), but in redtails it was not (Cords 1986a).

The two species were similarly specialized on a few plant species as sources of insect prey: they each captured about 40% of their prey from about 12% of the species used for prey capture. The identity of these plant species showed minor differences (Cords 1984b). Differences in the profile of substrates used for invertebrate capture were significantly different but small in an absolute sense (Cords 1986a).

A significant difference between the species was found in the motor patterns used for prey capture, even though all patterns were used by both (Cords 1986a). Blues ingest more prey directly from the substrate, while redtails use more fast picks, swipes, and pounces. These differences suggest that redtails capture more mobile prey than blue monkeys, a conclusion supported by the observation of more immobile cocoons and sapsuckers in the identified prey of blue monkeys relative to redtails.

Diet Composition and Temporal Variation in Mixed-Species Grouping

Intermonthly Variation

For both blues and redtails, the proportion of time spent in mixed-species groups was not correlated over the 11 months with the intake of any of the most important classes of items in the plant diet (i.e., fruit, leaves, blossoms, gum, or seeds), nor with the intake of any items in the total diet. Furthermore, the amount of time spent in mixed groups was not related to dietary overlap, calculated for either the plant or total diet, whether alone, in mixed groups, or all together. Overall dietary diversity was also not related to the amount of time spent in association by either species. If diversity was calculated separately for when the monkeys were in mixed groups and when they were not, however, the amount of time spent in mixed groups by redtails (but not blues) is higher in months when the diversity of species or species-specific item use for the plant diet when alone is low ($r_s = -0.66$, $p = 0.026$ species, $r_s = -0.65$, $p = 0.029$ species-specific items), i.e., when they concentrate on relatively few food sources when alone. The amount of time redtails spend associated with blue monkeys (and vice versa) is not related to the diversity of the diet when mixed.

The above relationships are rather crude in that they do not take into account the nature of the tree species or species-specific items used. Struhsaker (1981b) has suggested that mixed groups of blues and redtails in the Kibale Forest occur most often when the monkeys are eating low density but preferred foods, which are shared. This same pattern exists for redtails in Kakamega: the proportion of feeding scores from plant species whose density is less than 2 per hectare and for which the selection ratio (per month) exceeds 100 is correlated with the percent of time redtails spend mixed with blues over the 11 months ($r_s = $

0.75, p < 0.02). The plant species that meet these criteria are, for the most part, also heavily used by the blues, especially in those months when the redtails spend much time mixed with blues. This relationship does not hold up for the blue monkeys, however, i.e., when densities, preferences and contribution to the diet are calculated from their perspective.

Day-to-Day Variation

On a day-to-day basis, the likelihood of blues or redtails being in a mixed group is not related to the fraction of the day's total diet composed of fruit, young leaves or invertebrates (n = 83 days for blues, 84 days for redtails, r_s). When each month is analyzed separately, these results persist for most diet components in most months.

Diurnal Variation

For all 11 months combined, both blues and redtails spend more time in mixed-species groups at hours of the day when the proportion of fruit in the diet is low (blue $r_s = -0.739$, $p < 0.02$; reds, $r_s = -0.746$, $p < 0.02$) and the proportion of invertebrates is high (blues, $r_s = 0.634$, $p < 0.05$; reds, $r_s = 0.763$, $p < 0.02$). Association and leaf consumption are not related. When each month is considered separately, however, these correlations persist in only about half of the 11 months. Thus the relationship between being in mixed groups and diet composition in terms of items is not clear cut on a diurnal time scale, as it also is not from month to month or day to day.

Comparison of Diet in and Not in Mixed-Species Groups

Items

There is no systematic difference between the diet composition, in terms of items consumed in the total diet, of either species when in mixed groups and when alone. For redtails, the difference in percentage consumption of items when mixed and not mixed with blues is significant (Chi-square, $p < 0.05$) in only 3 of 11 months (October-December). In those three months, invertebrate consumption is higher when in mixed groups, but other dietary components are sometimes higher and sometimes lower. For blue monkeys, differences in items consumed are significant in 5 of the 11 months (April-June, August, December), but there are no consistent trends for particular items across all, or even most, of these 5 months.

If one combines all 11 months and factors out hour of the day, the results are no clearer.

The relative consumption of different items differs significantly in 4 of 12 hours for redtails and 6 of 12 hours for blues, but the nature of those changes is not consistent in all or most hours. In many cases, sample sizes were too small to analyze statistically changes in items consumed when the monkeys were in mixed groups, factoring out time of day and month simultaneously. Considering only the three major diet constituents (fruit, leaves, and invertebrates) for those cases when sample sizes were sufficiently large to perform statistical tests (Chi-square), it is clear that in almost all cases, the diets of the monkeys in and not in mixed groups do not differ significantly in terms of items consumed.

Plant Diet

The same statistical tests cannot be performed to test differences in plant species or species-specific items used by monkeys in and out of mixed-species groups without lumping categories in a biologically meaningless way to increase expected values. Generally, species or species-specific items that are frequently used when the monkeys are alone are also popular when they are in mixed groups. For these species or species-specific items, differences in consumption in the two situations are quantitative rather than qualitative. Relatively rarely used species and species-specific items, however, are more likely to be included in the diet in one situation but not the other.

For the blue monkeys, diversity of use of plant species-specific items when in mixed groups did not differ significantly from when the group was alone (Wilcoxon Matched Pairs Signed Ranks Test, n = 11 months). For the redtails, however, diversity in use of species-specific items increased when they were in mixed groups (Wilcoxon, $p = 0.003$). Thus redtails concentrate on few foods less when with blue monkeys than when alone.

Invertebrate Diet

The motor patterns and microhabitats used for prey capture were compared for blues and for redtails in and out of mixed-species groups. For both species, neither microhabitats nor capture methods used in these two situations differed significantly for all 11 months combined (Table 17) or for each month analyzed separately.

Evaluation of Hypotheses

A Null Hypothesis

The data presented in this section allow us to evaluate the (null) hypothesis that mixed-species associations result from groups of different species exploiting common resources, for which they search independently. Two of the arguments used against the simpler null hypothesis discussed previously (page 30) are also relevant here. First, under

TABLE 17

Invertebrate Capture Methods (% of Captures) When Alone or in a Mixed-species Group, 11 Months Combined

	BLUES Alone	BLUES Mixed	REDTAILS Alone	REDTAILS Mixed
Slow, bring substrate to mouth, ingest direct	10.7	8.5	8.0	6.9
Slow, substrate not held, ingest direct	4.8	3.5	6.2	5.8
Slow, pick with 1 or 2 hands	5.7	5.7	7.0	6.2
Uncurl leaves, ingest direct	4.4	2.9	1.7	2.0
Medium, bring substrate to mouth, ingest direct	41.5	39.3	26.6	29.3
Medium, substrate not held, ingest direct	3.4	4.4	6.5	6.8
Medium pick with 1 or 2 hands	21.7	27.0	28.4	30.3
Medium grab, swipe or clap	1.3	1.9	2.6	3.3
Fast ingest direct	1.1	0.8	0.7	0.8
Fast pick, grab or swipe	3.4	3.3	6.5	5.8
Fast pounce or clap	2.0	2.8	5.6	2.8
n	704	792	465	1414
	$\chi^2 = 13.81$		$\chi^2 = 11.46$	
	10 d.f.		10 d.f.	
	$p = 0.181$		$p = 0.323$	

this null hypothesis, there is again no reason to expect one species to be responsible for mixed-species groups, in the sense of leaving them less often than it forms them. Second, associations last 4-5 hours on average, during which time the monkeys are usually feeding or foraging, and moving as they do so. The two groups usually do not spend their time together at a single, or even a few, shared rich food sources, but rather progress through a series of concentrated feeding sites (e.g., fruiting trees) or areas together. Between such feeding sites, they feed on invertebrates, leaves, and gum, items that are usually less localized in space than fruit, and are eaten by individuals in smaller quantities at a time. If the null hypothesis is correct, i.e., groups of both species travel independently of one another per se, then we must ascribe the common route they take from feeding site to feeding site over the course of several hours to independent choices of where to go next made by both species, choices that just happen to be identical. Such an explanation might apply to short associations when only a few feeding sites are used, but seems unreasonable in the context of longer associations during which many feeding sites or areas are used in succession.

Nevertheless, this argument is qualitative and intuitive, and therefore perhaps not entirely satisfactory. Waser and Case (1981) developed a quantitative model which allows one to predict the expected proportion of time that monkey groups should be sighted in association with other species, based on (i) the rates at which each species finds feeding sites, (ii) the rates at which one species leaves a feeding site if the other is not present, and (iii) the probability of one species displacing another from a feeding site. The forest is modeled as an array of feeding sites whose occupation by no, one, or more species varies over time. This model does not describe the *Cercopithecus* in Kakamega very well, however, because the groups are large and dispersed, making it impossible to operationally defione group feeding sites, and therefore to measure in the field the parameters on which the prediction is based.

Other predictions under the null hypothesis are possible however. If mixed-species groups result from (independent) attraction to shared resources, then the diets of the monkeys when in mixed groups should be more similar than their diets when alone. There is no significant difference in dietary overlap in plant foods when in mixed groups or alone, when overlap is based either on plant species (Wilcoxon Matched Pairs Signed Ranks Test, $n = 11$ months, $p = 0.091$) or species-specific items (Wilcoxon, $p = 0.859$). However, overlap when mixed is probably underestimated for two reasons: (1) The two focal groups were not observed simultaneously when in mixed groups. (2) The redtails were sometimes mixed with another group of blues outside the home range of the T blues. Therefore differences between species when in mixed groups may be due to the fact that they were exploiting different parts of the forest.

One would also expect that in months when the plant diets of the two species were more similar, there would be a greater chance of meeting at common feeding sites; however, the amount of time spent in mixed groups is not correlated with overlap in the plant diet over the 11 months. When only rare and preferred plant species are considered, however, we saw that redtail participation in mixed groups was related to the fraction of such species in the diet. Such evidence is consistent with the null hypothesis under consideration here, but it is also consistent with other more complicated hypotheses, e.g., that redtails use blue monkeys as guides to food sources that are patchy in space and time (see Ranging). Therefore, these observations do not help in distinguishing the null hypothesis from alternatives.

Foraging Benefit Hypotheses

A number of hypotheses postulating foraging benefits in mixed-species groups were described in the introduction, and will now be reexamined in light of the data on feeding. Do redtails, who are lighter and more insectivorous than blue monkeys, increase their capture rate of mobile insects flushed by the movement of the blues? No specific examples of such behavior were observed. Furthermore, the redtails in mixed-species groups do not increase the fraction of invertebrates captured using motor patterns specifically suited for mobile prey relative to when they are alone. The beater hypothesis thus seems an inappropriate explanation of mixed-species associations in Kakamega *Cercopithecus*.

One species may also benefit if, by associating with the other, its access to certain foods or foraging areas is improved. No such relationship between the two species was directly observed. We have seen that neither species changed its consumption of particular items when in mixed-species groups. Blues appeared to have easier access to only two foods that were also eaten by redtails: (1) *Funtumia latifolia* seeds, which are encased in a hard fruit (about 5 inches long), requiring some strength to break open (even juvenile blue monkeys ate fewer of these seeds than adults, and appeared to have trouble getting at them when they tried); and (2) invertebrates, probably mostly beetles, taken from dead wood, either under bark that was pulled off, or from rotting stumps, whose fibers were torn apart. The redtails could not have usurped the remains of the blues' efforts at exploiting these foods had they wanted to: *Funtumia* seeds and fruits were dropped to the forest floor, and a monkey uncovering insects in dead wood appeared to eat all of them before moving on. The use of different foraging areas is discussed in the next section.

Another possible foraging benefit of mixed-species association is an increase in foraging time allowed by shared vigilance for predators in larger groups, each of whose members can spend less time watchful. This hypothesis cannot be evaluated with my data, because time spent being vigilant was not measured. In addition, the premise that increased foraging time would be advantageous remains untested.

Two other foraging benefit hypotheses, that one species guides another to ephemeral but rich food sources and that both avoid duplication of effort by foraging together, rely on dietary similarity to be applicable. The data on feeding show that the diets of blues and redtails in Kakamega do overlap substantially. To evaluate these hypotheses, however, ranging patterns must be taken into account.

7

RANGING

Examination of patterns of range use allows one to evaluate hypotheses that mixed-species association improves the food-locating efficiency of participating animals. In this section, patterns of range use by blue and redtail monkeys are compared, and are then related to association in mixed-species groups so that particular hypotheses can be evaluated.

Methods

The positions and movements of group members were recorded on a 1:600 map of the study area so that every part of the forest in which I observed at least one group member was indicated each day. To determine distance moved, an "estimated center of mass" (ECM) of the group was indicated on the field map every hour from 0700 to 1900. Because group members were often spread over several hundred meters (see page 70), this measure required subjective judgment. Because I moved around within the group frequently, however, and roughly kept track of individuals' locations over time, locating the ECM was usually not difficult. The distance between subsequent ECM's along the group's path of travel (not always a straight line) was subsequently measured to the nearest 5 meters to give hourly rates of movement; these were summed to obtain the daily path length.

To measure use of space in two dimensions, a grid representing 50 m x 50 m quadrats was laid over the field maps, and all quadrats entered were checked off. Feeding trees were located in all of the quadrats. Those that were entered n times during the day, when each entrance was separated from the previous one by at least one hour, were given a score of n. Quadrats that contained n successive ECM's were also given a score of n. Thus scores reflect very roughly the amount of time that a monkey group spent in each quadrat during a day: each score represents a fraction of an hour, separated from any other score by at least one hour. In some analyses, all non-zero scores were recoded to "1," effectively giving a "used" or "not used" score for each quadrat on each day. These scores of quadrat use are called 0-1 scores.

To measure use of different canopy strata, the height of the group as a whole was assigned to one of three categories 13 times a day, on every hour from 0700 to 1900. Group height was determined as an average over all group members whose locations were known. When about 50% of visible animals fell into one category and 50% fell into another, the group was assigned half a score in each category. (Group members were never seen equally divided between all three categories.) Categories were as follows: (1) low, ground level to 7 meters; (2) medium, 7-20 meters; and (3) high, greater than 20 meters. These categories were chosen because generally they reflected the forest's structural character: "high" represents the tallest canopy trees, whereas "medium" represents the middle story, including crowns of understory trees and many (but not all) liana tangles. The "low" level includes very small trees and shrubs, as well as the lower trunks and stems of the vegetation whose leaves form levels 2 and 3.

Finally, group spread was estimated as the maximum distance between the farthest outlying individuals excluding adult males. Group spread is difficult, if not impossible, for one person to measure accurately. I made estimates only when I knew the locations of many individuals, and immediately after rapidly circling the edge of the group 1.5 times, which took about 5 minutes. In most cases, an individual at the starting point of the walk-around was still in or near the same tree when I had completed one circumference of the group. Still, estimates should be regarded as rather rough, and because I may have overlooked individuals on the group's edge, especially inactive ones, they are probably conservative. Measurements were taken on different days, and at different times during the day. Measurement was not attempted in parts of the forest where visibility was particularly bad, or if the groups were moving in a rapid directed way (which they seldom did).

Ranging Patterns of Blue Monkeys and Redtails

Distance moved

The daily path length of blue monkeys is shorter than that of redtails (Mann Whitney U Test, $p < 0.0001$, Table 18, only days with 13 ECM's recorded). Although the mean daily path length varies from month to month in both species (Table 18), on average redtails move farther per day than blues in every month, and differences between species are significant except in April and August (Mann Whitney U Test). This pattern persists when the two groups move together because redtails take the outside of a curved path, and so can remain with blues even though the blues move at a slower rate. Neither species moves at an equal rate throughout the day (Figure 6). The groups move less during early and middle

TABLE 18
Monthly Variation in Ranging Patterns of Blue Monkeys and Redtails: Means, Standard Deviations, Ranges (n in Parentheses)

	Daily Path Length (m)		No. 1/4-ha Quads Entered per Day		5-day CV of Quadrat Use	
	BLUE	RED	BLUE	RED	BLUE	RED
Mar	1061 ± 115 835-1220 (7)	1649 ± 397 1240-2285 (8)	50 ± 12 28-64 (7)	52 ± 9 36-66 (8)	47.5	57.3
Apr	1179 ± 333 755-1750 (9)	1517 ± 429 995-2385 (12)	41 ± 13 22-58 (9)	49 ± 13 24-81 (14)	52.9	65.3
May	1274 ± 214 960-1675 (9)	1632 ± 148 1405-1940 (9)	50 ± 10 31-63 (10)	66 ± 12 46-83 (11)	55.8	56.3
June	979 ± 123 775-1105 (5)	1504 ± 216 1215-1750 (5)	44 ± 9 33-54 (5)	59 ± 9 47-68 (5)	61.2	48.3
Aug	1218 ± 268 940-1635 (6)	1412 ± 200 1135-1655 (6)	54 ± 9 46-65 (6)	57 ± 10 38-64 (6)	56.1	47.9
Sept	1125 ± 104 970-1305 (7)	1482 ± 281 920-1675 (6)	57 ± 7 50-71 (7)	61 ± 9 45-68 (6)	54.6	63.2
Oct	1155 ± 144 960-1355 (8)	1694 ± 97 1560-1810 (7)	56 ± 12 38-68 (8)	64 ± 7 53-71 (7)	45.2	45.4
Nov	1292 ± 111 1110-1425 (9)	1636 ± 261 1300-2085 (8)	59 ± 6 51-68 (9)	61 ± 11 46-84 (8)	45.0	52.2
Dec	1121 ± 255 785-1520 (8)	1736 ± 182 1470-1990 (8)	52 ± 14 35-71 (9)	63 ± 12 46-79 (8)	50.3	55.7
Jan	1054 ± 199 820-1430 (7)	1394 ± 227 1055-1740 (9)	51 ± 9 38-61 (7)	50 ± 10 38-68 (9)	60.2	56.3
Feb	824 ± 222 600-1075 (5)	1211 ± 259 965-1680 (6)	42 ± 10 28-53 (7)	51 ± 9 42-63 (7)	56.1	53.5
YEAR	1136 ± 228 600-1750 (80)	1543 ± 296 920-2385 (84)	51 ± 11 22-71 (84)	57 ± 12 24-84 (89)	--	--

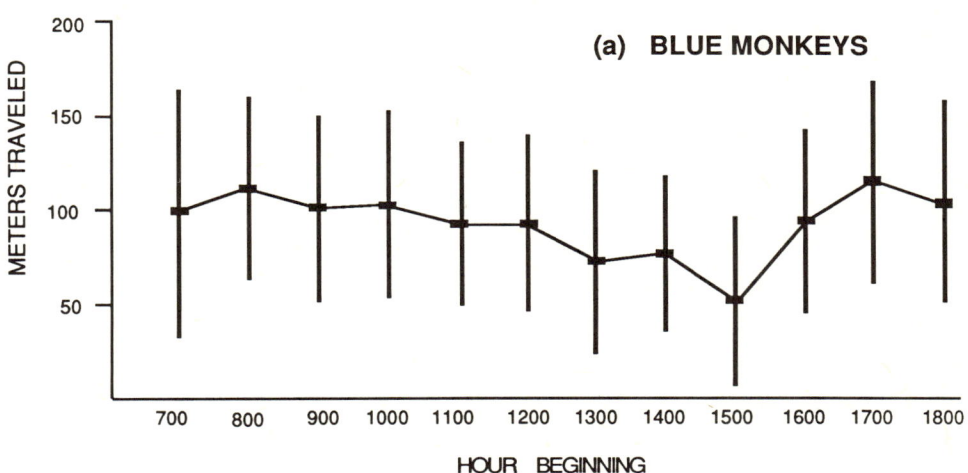

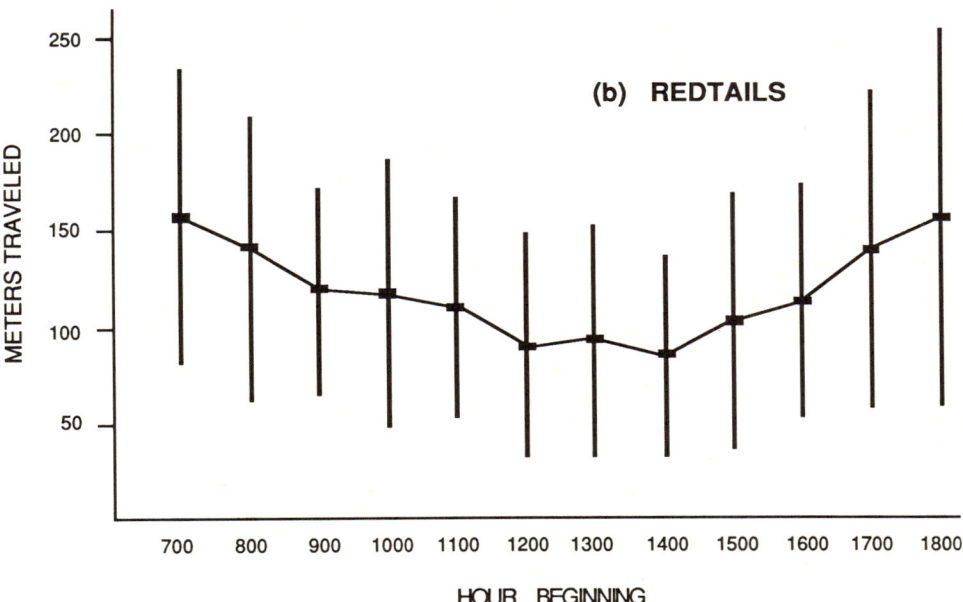

FIGURE 6. Diurnal variation in distance traveled. Squares and thin lines indicate means and standard deviations respectively.

afternoon, a time when group members are more often resting. Redtails move farther than blues in each hour of the day except from 1200 to 1300 when there is no significant difference between the two species. Both species move little overnight (1900-0630).

Quadrat Use

Over the course of the year, the T troop of blue monkeys entered 151 quarter-hectare quadrats, corresponding to a home range of 37.75 hectares. The L redtails entered 239 quadrats, and had a home range of 59.75 hectares. These figures are derived from 84 (blues) and 89 (redtails) days of observations with 11-12 hours of position mapping, but even if one included all other sightings of the focal groups, the home range sizes would not change.

The number of quadrats entered in the first five consecutive days of each month varies from 93 to 120 in blues ($\bar{x} = 105.5 \pm 7.8$, n = 10 months; June was excluded because only four consecutive days of observation were made), and from 119 to 179 in redtails ($\bar{x} = 149.2 \pm 15.3$, n = 10 months, June excluded). Redtails consistently used more quadrats per month than blue monkeys (Mann Whitney U Test, $p < 0.002$), as might be expected based on the larger home range and longer daily path length of the redtails.

The redtails also entered significantly more quadrats per day than the blue monkey group (Table 18, Mann Whitney U Test, $p = 0.003$). In both species, the number of quadrats used per day was correlated with the daily path length ($r_s = 0.731$, $p < 0.001$, n = 80 for blues; $r_s = 0.721$, $p < 0.001$, n = 84 for redtails). The number of quadrats entered per day was variable both between and within months (Table 18).

Each species tends to move about its range without recrossing its path on a given day. In their smaller home ranges, the blues made large circuits covering on average 34% ($\pm 7.5\%$, n = 84) of the home range (Figure 7a). Redtails sometimes moved in a similar pattern (Figure 7b), but, unlike blues, the redtails might also sweep across their range in a more or less straight line from one end to the other (Figure 7c). The redtails covered a smaller fraction of their home range ($24 \pm 5\%$, n = 89) in a day than did blues. This was also true if the area covered in the first five consecutive days of a month were compared.

These patterns are reflected in the number of quadrats which were used two or more times, but not consecutively, in a given day. For redtails, a mean of 3.1 ± 2.9 (n = 89 days) such quadrats were used per day, while for blues, 4.0 ± 3.4 (n = 84 days) such quadrats were used. Expressed as a percentage, redtails recrossed $6.0 \pm 5.8\%$, and blues $8.9 \pm 8.3\%$ of the quadrats entered on a given day: the difference between species is significant when all the data are combined (Mann Whitney U Test, $p = 0.036$). A similar

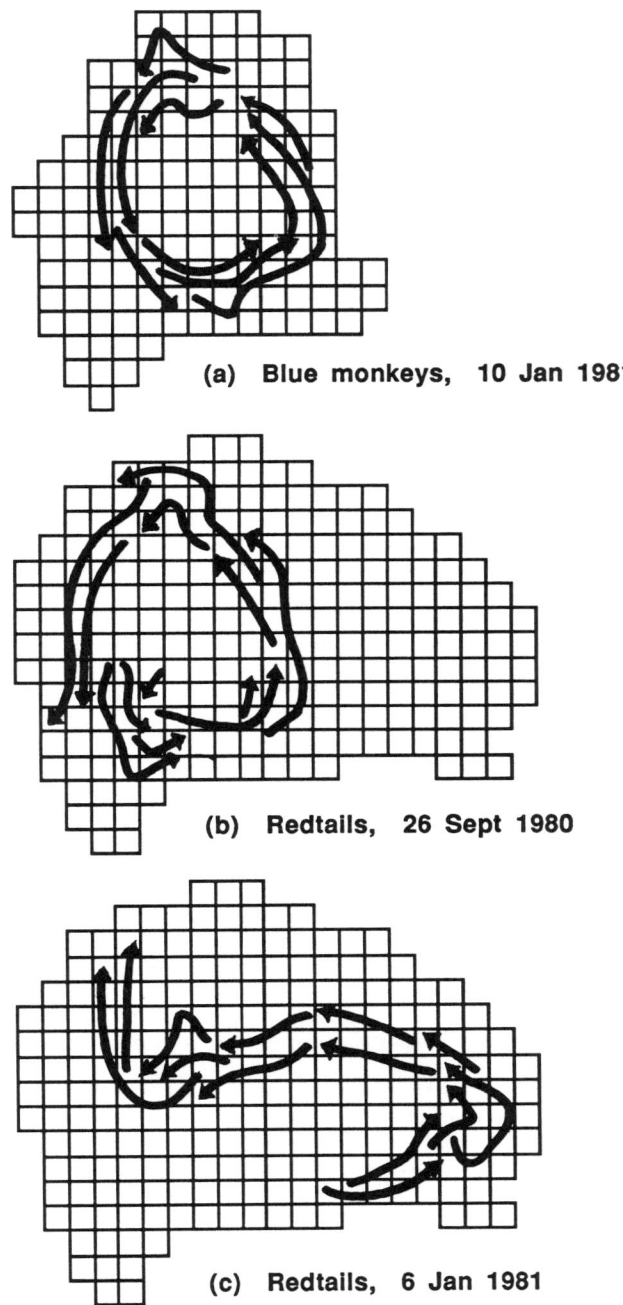

FIGURE 7. Typical daily ranging patterns.

trend is apparent when the two species are compared one month at a time, but the differences are significant only in April.

A rather different pattern emerges if quadrat use on consecutive days is considered: a given quadrat is much more likely to be used on two consecutive days than twice on the same day. 25.9 ± 9.8 quadrats (or 49.1 ± 14.5 % of those entered, n = 52 days) used by blues on one day were entered on the previous day as well. Redtails entered 18.0 ± 10.6 (or 32.2 ± 16.7 %, n = 62 days) quadrats that had been entered the day before.

Patterns of repeated use of quadrats over five consecutive days are shown in Figure 8. Few quadrats were used in five out of five days. An average of about 25% of the quadrats entered in a five-day period by blues, and about 50% of the quadrats entered in a five-day period by redtails, were used on one day only. Figure 9 shows which quadrats were used on different numbers of days for the 11 months combined. Fifty percent of the quadrats used accounted for 77% (redtails) and 83% (blues) of the quadrat-use scores over the entire year.

The above results are calculated using only 0-1 scores. A quadrat may also be assigned values greater than 1 on a given day, thus incorporating information on time spent in the quadrat into its daily score. Using these data, an index of the evenness with which quadrats are used can be calculated. The coefficient of variation (CV) of scores in quadrats is insensitive to the number of quadrats used and to differences in the sum of values in all quadrats (Rasmussen 1980). The CV equals zero if each quadrat is used equally often, and increases with unevenness of use. Table 18 gives coefficients of variation in quadrat use over five-consecutive-day samples each month (except June, when CV's are calculated based on four-day samples). Differences between the two species' five-day CV's are not significant (Wilcoxon Matched Pairs Signed Ranks Test).

On a daily basis, however, the CV of quadrat use by blues ($\bar{x} = 26.3 \pm 28.0$, n = 84 days) is greater than that of redtails ($\bar{x} = 21.4 \pm 22.8$, n = 89 days, Mann Whitney U Test, $p < 0.004$), reflecting the fact that blues move more slowly and are more likely to recross their path than redtails. In both species the daily CV of quadrat use is negatively correlated with the daily path length and number of quadrats entered, and positively related to the proportion of quadrats used per day that were recrossed (r_s, all $p < 0.001$, n = 84 days for blues, 84-89 days for redtails).

Relation to Diet

For blue monkeys, daily path length is positively correlated with the proportion of fruit in the daily diet ($r_s = 0.271$, $p = 0.015$, n = 80 days). Redtail path length is negatively related

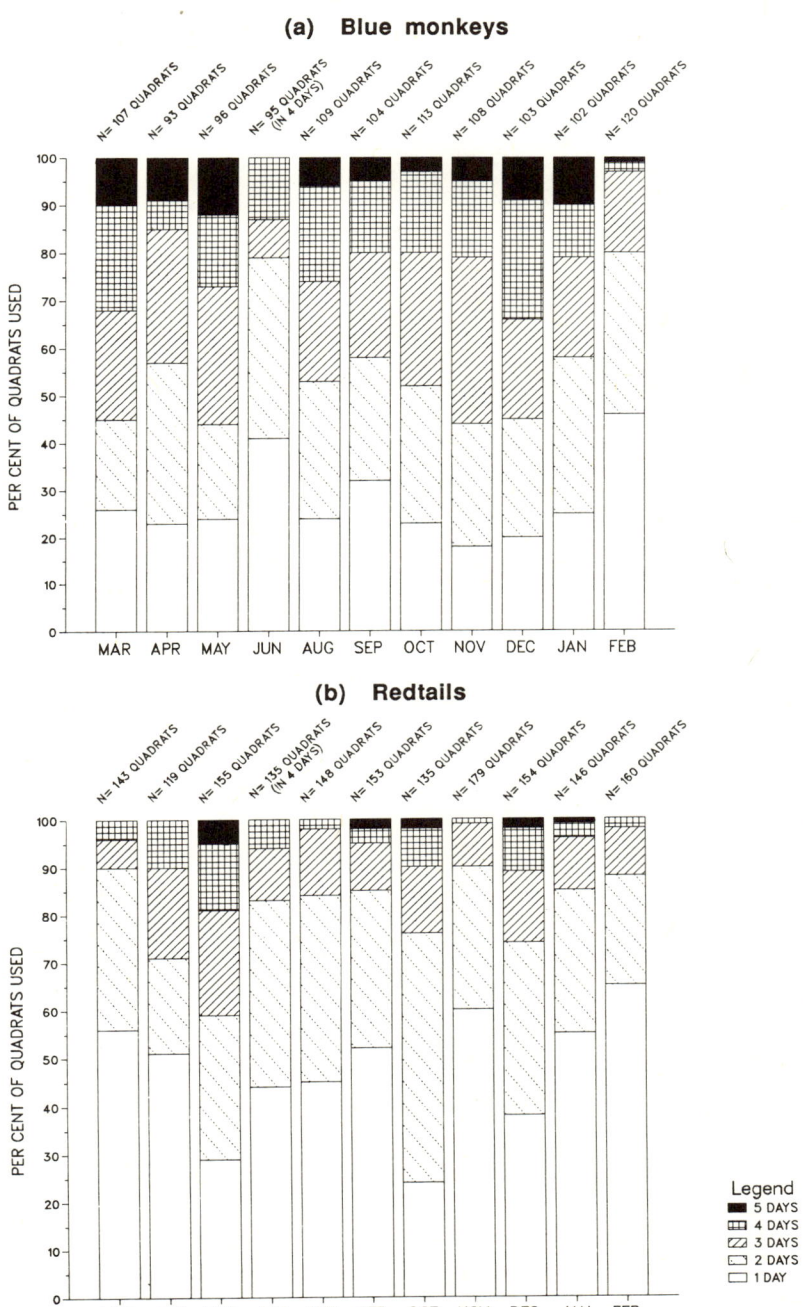

FIGURE 8. Cumulative percentage of areas used for different numbers of days in five-consecutive-day samples. In June, only four consecutive days of observation were made.

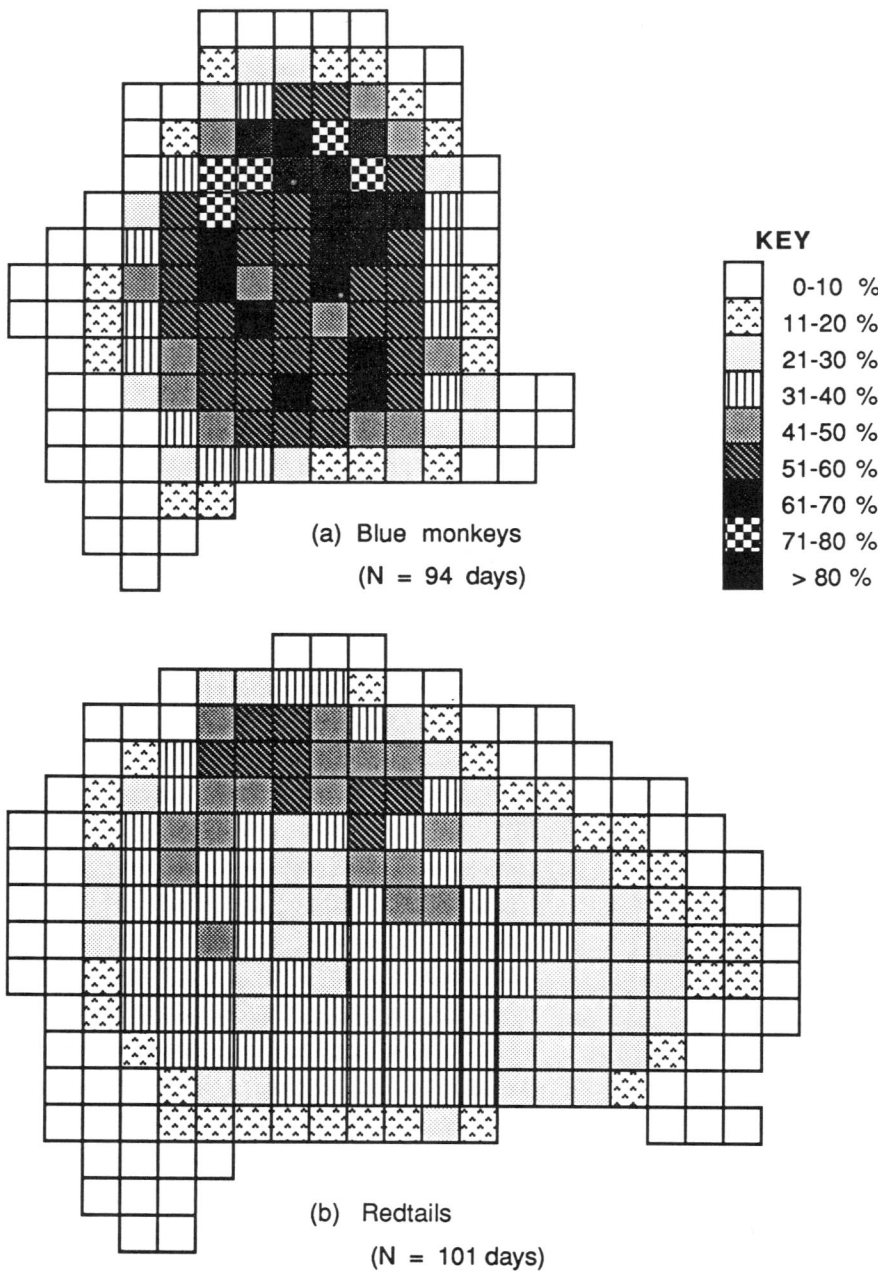

FIGURE 9. Home range use over 11 sample months: percentage of days in which each 50 m x 50 m quadrat was entered.

to the consumption of young leaves (r_s = -0.227, p = 0.041, n = 82 days). In both species, however, the number of quadrats used each day is not related to daily consumption of fruit, leaves, or invertebrates.

For blues, redundancy of quadrat use, as measured by the CV of scores in quadrats, is negatively correlated with the proportion of fruit in the monthly diet (r_s = -0.629, p = 0.010); in other words, over a period of five days, blue monkeys concentrate their activities in relatively few quadrats more in months when the proportion of fruit in the diet is high. The daily CV of quadrat use is negatively correlated with the consumption of young leaves (r_s = -0.357, p < 0.001, n = 83 days). For redtails, neither monthly nor daily CV of quadrat use is related to dietary proportions of fruit, leaves or invertebrates.

The above relationships are grossly simplified, because the spatial distribution of the actual plant species used are not explicitly considered. The monkeys' ranging patterns certainly appeared to be related to the distribution of the specific foods eaten, but the nature of those relationships will not be discussed here. Rudran (1978) gives some evidence for these specific relationships for blue monkeys in the Kibale Forest, Uganda.

Habitat Partitioning

The home ranges of both blue monkeys and redtails were mapped with respect to three structural characteristics: (1) foliage density in the highest ("A") canopy layer, > 20 meters up, (2) foliage density in the layer between 7 and 20 meters ("B"), and (3) abundance of climbers. In each case, three values were recognized: high, medium, and low. A grid of quarter-hectare quadrats was subsequently laid over the field maps, and for each characteristic, a quadrat was assigned the value corresponding to the majority of its area. For each structural characteristic, the number of times the three types of quadrats were used was compared (using a Chi-square One Sample Test) to the representation of those quadrat types in the group's home range, to determine if certain types of quadrats were preferred.

For the 11 months combined, blue monkeys used more quadrats with high foliage densities in the A and B canopy layers, and fewer quadrats with medium or low foliage density than would be expected if quadrats were used in proportion to their abundance (p < 0.001). In addition, they favored quadrats with relatively few climbers (p < 0.036). Redtails showed no preference for particular foliage densities in the top canopy layer, but like the blue monkeys, redtails preferred quadrats with dense foliage in the B canopy layer (p < 0.001), and quadrats with relatively few climbers (p < 0.05).

Height

Both the blue monkeys and the redtails spend the majority of their time between 7 and 20 meters from the forest floor. The two species differ in the degree to which they concentrate their activities at this level (Table 19): blue monkeys are found at the lowest and highest levels more often than redtails, who are more consistently intermediate.

TABLE 19

Percentage of Height Records in Three Height Classes

	BLUE	REDTAIL
> 20 meters	16.5	11.7
7-20 meters	48.1	56.5
< 7 meters	35.4	31.8
n	1113	1162

G_{adj} = 19.22, 2 d.f., $p < 0.001$

Both species come down toward the ground in the middle of the day, but return to higher canopy levels in the evening (Figure 10). This diurnal cycle corresponds roughly with fluctuations in activity: at midday, the monkeys are more likely to be resting or foraging for insects. In the early morning and evening, they are traveling through or feeding on fruit in taller trees.

Group Spread

The blue monkey group's spread was measured 35 times, and the mean was 108.7 ± 31.4 meters. Group spread ranged from 60 to 190 meters in these scheduled measurements; at other times, the group was occasionally seen spread over at least 290 meters. The redtail group was more compact: group spread ranged from 20 to 85 meters with a mean of 56.1 ± 17.3 meters in 31 measurements.

Ranging and Mixed-Species Groups

Distance Traveled

Blue monkeys move farther per hour when they are associated with redtails than when they are alone (Mann Whitney U Test, $p < 0.0001$), even though associations are least likely to occur at times of day when they move farthest (i.e., morning and evening). For redtails, there is no significant difference in the hourly rates of movement when the group is

Mixed-species Association of Cercopithecus Monkeys

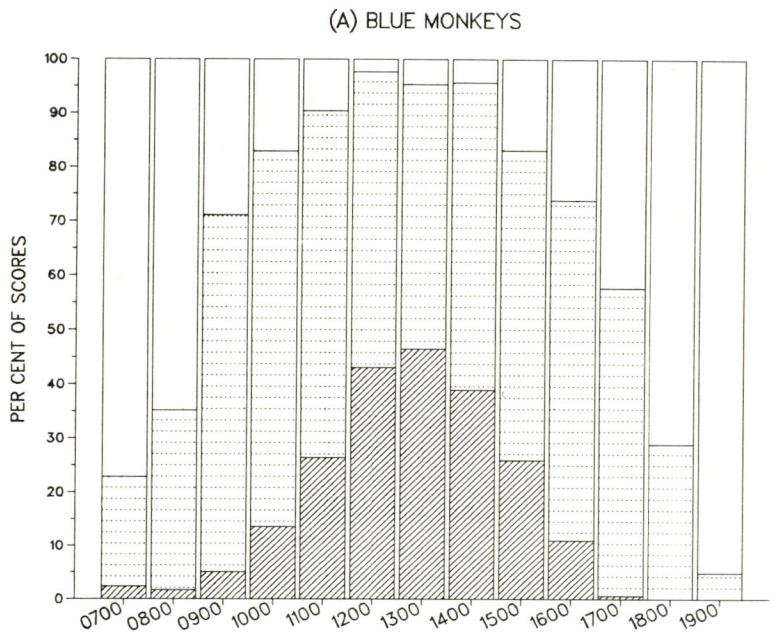

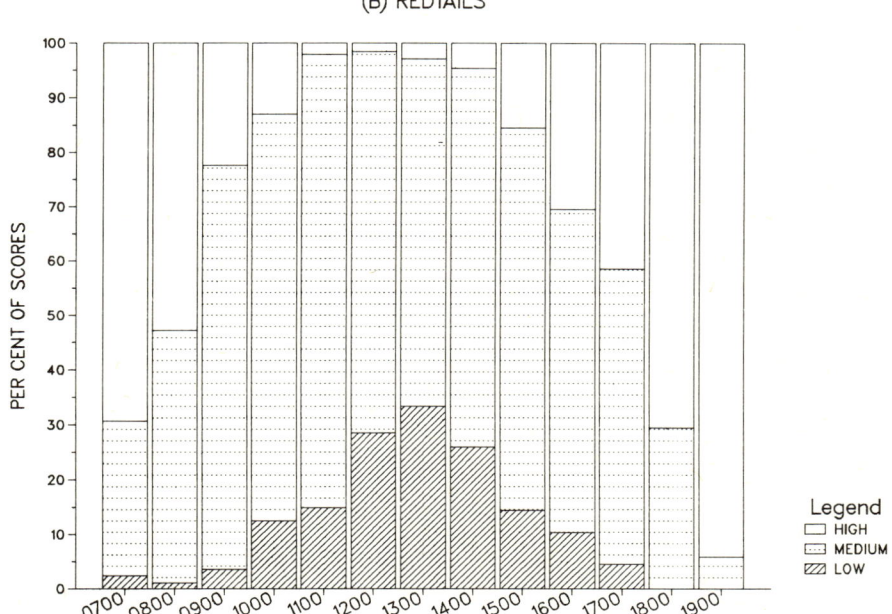

FIGURE 10. Diurnal variation in occupation of different strata.

associated with blues or alone. These results are consistent with the fact that daily path length of blue monkeys is correlated with the amount of time spent in mixed groups per day ($r_s = 0.352$, $p < 0.001$, n = 80 days), whereas no such correlation exists for redtails.

These observations are worth considering in light of the conclusion reached earlier regarding responsibility for mixed-species groups. One could argue that the species that *changes* its behavior when associated is responsible for the association (rather than the species that actively forms the association). On the other hand, the presence of redtails may necessitate more rapid movement by the blues if resources en route are depleted faster in larger groups, or redtails may choose to join blues only when the latter are moving relatively quickly. The latter interpretation is supported by the observation that many disbandments of mixed-species groups occur when the blue monkeys slow down and the redtails pull ahead of them.

Quadrat Use

For blue monkeys, the number of quadrats used per day is positively correlated with the number of hours the group spends in association with redtails ($r_s = 0.198$, $p = 0.036$, n = 84 days). This result is not unexpected given the previously demonstrated correlations of time spent in mixed groups with distance traveled and the number of quadrats entered. For redtails, no such correlation exists.

There is no direct relationship between the amount of time spent in mixed-species groups and the degree of multiple use of quadrats within a given day for either blue monkeys or redtails. Neither the number nor the proportion of recrossed quadrats per day is significantly correlated with the number of hours spent in mixed groups. For redtails, however, the amount of time spent associated with blues is related to the degree of redundancy in quadrat use *between* days in five-day monthly samples. Redtails spend more time with blues in months when the number of different quadrats entered in a five-day period is low ($r_s = 0.67$, $p < 0.05$). Correspondingly, they spend more time with blues when the mean proportion of quadrats used on both of two consecutive days within a five-day sample is high ($r_s = -0.64$, $p = 0.036$). For blues, there is no relationship between the amount of time spent with redtails in a month and the degree of overlap in quadrat use between days. In neither species is the daily or monthly evenness of quadrat use, as indicated by the coefficient of variation, related to the amount of time spent in mixed groups.

Habitat Partitioning

To test whether association with another *Cercopithecus* species influenced the use of different structurally defined types of forest, the 86 full days of blue monkey ranging records and 96 full days of redtail ranging records were each divided into two equal-sized categories: days when the monkeys spent less and days when they spent more than the median number of hours in mixed groups. (The median number of hours spent in mixed groups per day was 7 for blues and 9.5 for redtails.) Within each species (combining all days and months), these samples were compared to see if the number of quadrats used in each of the three classes of canopy A foliage density, canopy B foliage density, and climber abundance differed between days with more or less time spent in association.

For blue monkeys, no significant differences were observed, but redtails entered quadrats with few climbers and with low B canopy foliage density more often in the days with more than 9.5 hours spent with blues than in days when they spent less than 9.5 hours together (Table 20).

TABLE 20
Quadrat Use in Relation to Forest Structure and Time Spent by Redtails with Blue Monkeys

(a) FOLIAGE DENSITY IN THE CANOPY LAYER FROM 7-20 METERS	Low	Med	High	n*
Days with < 9.5 hrs associated	133 (6%)	888 (38%)	1301 (56%)	2322
Days with > 9.5 hrs associated	189 (7%)	1222 (41%)	1553 (52%)	2964

$X^2 = 6.99$, 2 d.f., $p < 0.05$

(b) ABUNDANCE OF CLIMBERS	Few	Med	Many	n*
Days with < 9.5 hrs associated	555 (22%)	730 (29%)	1258 (49%)	2543
Days with > 9.5 hrs associated	740 (27%)	976 (35%)	1044 (38%)	2760

$X^2 = 70.4$, 2 d.f., $p < 0.001$

* Totals for tables (a) and (b) differ because of missing data on forest structure in some quadrats.

Height

Association in mixed-species groups has no effect on the use of different canopy levels. Height records for each hour of the day were tested separately to factor out diurnal variation: there was no significant difference in the heights occupied by either monkey species when alone and when in mixed-species groups.

Some Hypotheses Reconsidered

The data presented in this section are relevant to several hypotheses regarding the adaptive significance of mixed-species groups in the study populations. First is the idea that participation in mixed groups makes certain foraging areas more available to one or both partipating species. The data on use of quadrats with different structural properties are consistent with the hypothesis that redtails are able to increase their use of more open areas (lower foliage density in the middle canopy layer and fewer climbers) when accompanied by blue monkeys for relatively large fractions of the day. Although the mechanism has not been demonstrated, this difference may reflect antipredator advantages derived by mixed-species group members.

Similar observations have been made on *C. cephus* in Gabon in conjunction with its participation in mixed groups with *C. nictitans* and *C. pogonias* (Gautier-Hion et al 1983). In that population, *C. cephus* in monospecific groups prefers low, dense forest; in polyspecific groups, *C. cephus* increases its use of high, more open forest. Gautier-Hion et al. (1981, 1983) propose that low, dense forest provides protection from diurnal predators (crowned hawk eagles and man), and that when associated with other monkeys, *C. cephus* forsakes that protection to some degree because of the watchfulness of companions.

An alternative interpretation of the Kakamega results might reverse the direction of implied causality: perhaps redtails spend more time associated because they use more open forest (for some other reason, e.g.,because food is located there), which the blue monkeys prefer more than they do. That is, the redtails' primary choice is one of habitat, and as a consequence of that choice, they encounter blue monkeys more often in certain habitat types. This interpretation is contradicted by the fact that blue monkey preferences for more open forest do not differ from preferences for the same kind of forest by redtails in monospecific groups; redtails in mixed groups have not simply shifted their use of different forest types so that it more closely resembles that of blue monkeys.

A second hypothesis to which ranging data are pertinent is that members of one species serve as guides to temporally and spatially variable resources. As pointed out earlier, dietary overlap between the two species is a necessary prerequisite for applying this explanation, and blues and redtails in Kakamega do show considerable overlap in diet. Furthermore, we have seen that many of the plant food items eaten by both species are patchily distributed in space and time. Both species move about 1-2 kilometers a day to exploit several such food sources per day at all times of year. Although the daily path length of redtails is longer than that of blues, redtails cover a smaller fraction of their entire home range per day and per five-day period. Therefore it seems likely that blues would be better informed about the location and quality (e.g., in terms of abundance and ripeness of fruit) of resources, and could serve as useful guides of the redtails to high quality feeding sites.

This hypothesis is difficult to disprove (Gautier-Hion et al. 1983) with existing data. One needs detailed information on the feeding efficiency -- relative to ranging -- of the guided species. One might also want to know which species determines the direction of mixed group movements. In Gabon, where the guiding hypothesis may also apply, Gautier-Hion et al (1983) found that the putative guide always followed the species it was supposed to be guiding. However, a species need not be physically the leader to determine group movements (Pook and Pook 1982, Terborgh 1983). In Kakamega, neither blue monkeys nor redtails regularly preceded each other in mixed-species groups at most times, and it was usually not clear if one species or the other was directing group movements.

Related to the guiding hypothesis is the idea that a group of one species can avoid visiting areas already used by groups of other species that share its home range by banding together with them. Thereby all species avoid a duplication (or multiplication) of effort, making it possible to regulate return time to a renewing resource. In this case, the benefits of mixed-species grouping may be similar for all group members. In the present case, however, an asymmetry is introduced by the disparity in size of the blue and redtail home ranges. Blue monkeys, confined to an area that is only a fraction of the redtail home range, will use that area more intensively than redtails, who can, and do, go elsewhere. Consequently when redtails arrive in some part of their range, they are more likely to encounter areas recently used by blues than the blues will encounter areas recently used by redtails. Redtails should gain by joining the "local" group of blue monkeys if they wish to avoid duplication of effort. The facts that blue monkey groups sharing the L home range are bigger than the redtail group, and that these groups move more slowly than redtails, may further contribute to the depleting effect blues have on an area of forest, and hence to the advantages of redtails who avoid such areas.

The time scale on which revisits to certain parts of the home range are avoided can be deduced for the Kakamega *Cercopithecus* from data on overlap of quadrat use within and between days. As indicated on page 64, quadrats are only rarely used more than once on a given day, yet overlap between days is more substantial, especially for blues. Thus it appears that a profitable way of harvesting resources involves visiting several feeding areas each day, but visiting each only once. This pattern results, perhaps, from the monkeys' preference for ripe fruits: both species select ripe fruits when unripe fruits are available as well. Most fruits consumed are produced in fairly large crops that persist on an individual tree for two weeks up to a few months (Cords 1984b). Although ripe fruits are present for a shorter time, individual trees are commonly visited for fruits for periods of 2-6 weeks, during which time relatively few fruits may ripen each day. Thus the hypothesis as it relates to the study population would be that by joining blue monkeys, redtails avoid visiting areas where the blues have been previously on a given day, and thereby the redtails (at least) increase their foraging efficiency, or diet quality, or both. The most effective test of this hypothesis would involve comparing foraging efficiencies and diet quality of redtails in and out of mixed-species groups, or of redtails in trees which had or had not been visited by blues earlier in the day. These data are not available for the study population, but indirect evidence supporting the hypothesis does exist.

Several predictions can be derived from the hypothesis which, if not born out, would give one good reason to reject it. First, when redtails are associated with blues, the groups should not be using areas previously (i.e., earlier on the same day) used by blues. To test this prediction, the amount of time spent by redtails covering "new" and "used" ground (defined in terms of use on the same day) when in mixed groups and when alone was calculated to the nearest half hour for 107 days with at least 11 hours of observation. (This sample is larger than most used in previous analyses because it includes days on which redtail ranging was monitored by my field assistant simultaneously as I followed the blues.) For the 86 days on which I was the only observer in the forest, I kept track of the blue monkey group while watching redtails by monitoring blue vocalizations and other signs of its presence (e.g., fresh dropped seed pods and feces), and by making occasional forays away from the redtails if the blue group was within a few hundred meters. On most days, I had a fairly good idea of where the blue monkeys were ranging, but 15% of the time, it was not possible to determine if an area in use by redtails had been previously occupied by blues. Nevertheless, the overall pattern is clear (Table 21): when in mixed groups, redtails spend much more time covering new ground than used ground. Redtails in mixed-species groups covered new ground 90% of the time (n = 813 hours) that a determination as to previous use could be made.

A second prediction one might derive is that if participation in mixed-species groups reduces duplication of effort, then duplication of (the blue monkeys') effort by redtails should occur less frequently in mixed groups than when the redtails are alone. However,

TABLE 21

Hours Spent by Redtails Covering Ground Used or Not Used Previously on the Same Day by Blue Monkeys

	(a) Mixed		(b) Alone		Total
"New" ground	730	(79%)	220	(70%)	950
"Used" ground	83	(9%)	26	(8%)	109
Ground undetermined	114.5	(12%)	70.5	(22%)	185

Table 21 shows that when the previous use of a quadrat could be ascertained, redtails were equally likely to cover used ground in either situation. This observation should not necessarily weaken our support for the hypothesis under consideration for two reasons: (1) redtails may travel in monospecific groups only when they know that they are not covering used ground. Their knowledge could be based on cues similar to those a human observer can use to determine if blue monkeys were recently present. (2) In the data analysis there is a bias that would tend to reduce the number of hours counted when redtails were *alone* and covering *used* ground, hence reducing the difference between redtails associated and not associated with blue monkeys. Redtails are most likely to be alone at the beginning and end of the day (Figure 3). Early in the day, one is likely to be able to determine whether or not the area in use was occupied previously that day (probably it was not). At the end of the day, however, if redtails are alone, one is less likely to be able to say with confidence that the area in use has not been used previously by blues, though the chances that it has been previously used are higher than earlier in the morning. Therefore the number of hours spent alone covering "undetermined" ground is high (Table 21), and probably contains a large proportion of hours spent covering used ground which are not "counted" when we compare the ratio of used to new ground when the monkeys are in mixed-species groups and alone.

A third prediction of the nonduplication of effort hypothesis is that if blue monkeys backtrack, accompanying redtails should leave the blues in order to avoid areas previously covered. Blue monkey day ranges on 93 days between March 1980 and February 1981 inclusive were examined for occasions on which the blue monkey group retraced its steps by: (i) reversing its direction of movement (literally backtracking), and so returning to an area recently (within an hour) used; (ii) recrossing a part of its path used earlier in the day (more than an hour previously); (iii) ending its daily march in the same area (within 20 meters) from which it started. On each such occasion when redtails were accompanying blues, I noted whether redtails left the blue group when it began to cover ground used

TABLE 22

Redtails' Response to Blue Monkeys' Reuse of Space

	Reds leave	Reds remain
Blues backtrack	26	11
Blues recross path	0	4
Blues' daily path ends meet	2	2

previously that day. As Table 22 shows, when blue monkeys backtrack, accompanying redtails are more likely to leave than to stay (Binomial Test, $p = 0.014$). If the blues simply recross their path or overlap the ends of their daily march, redtails do not leave more often than they stay with the blues. Thus the prediction is upheld only by data on backtracking, and not in cases in which repeated use of an area occurs briefly (recrossing) or after a relatively long time (overlapping of daily march ends).

In sum, the hypothesis that participation in mixed-species groups increases the foraging efficiency or diet quality of redtails because they thereby avoid covering parts of the forest already picked over by blues is supported by several lines of indirect evidence. The possibility that the blue monkeys benefit similarly cannot be ruled out; however, as mentioned previously (page 75), there is reason to believe that these benefits of mixed-species grouping would not be equally distributed. Furthermore, the asymmetry in benefits is congruent with the asymmetry in responsibility for the associations (page 29).

8

DISCUSSION AND CONCLUSIONS

The previous discussion has shown that mixed-species association of *Cercopithecus* monkeys in the Kakamega Forest is not a chance phenomenon, and that several ecological variables play some part in determining its occurrence. Here, the relative importance of different factors and exactly what they explain will be reviewed. It is my hypothesis that in this study population, antipredator benefits, though they may exert an important selective force, are more or less constant and unpredictable in time, and therefore that they do not explain observed temporal variation in association. The dynamics of mixed-species groups are instead related directly to patterns of range use, which are themselves consequences of dietary preference, food availability, interspecific competition and demography. Ranging patterns may also influence the relative responsibilities of the two species in forming and maintaining associations with one another. In addition, antipredatory advantages may not affect the two species equally, and so may contribute to an asymmetrical participation.

Predation

The importance and role of predation by raptors in explaining why mixed-species groups occur have been difficult to specify in this study because none of the 21 attacks observed was successful. The responses of the monkeys to the predator, and to one another's alarm calls, nevertheless suggest that participation in mixed groups could reduce the risk of being preyed upon by facilitating early detection of the predator (and thus early response), confusion of its attack, and dilution of its effect. There is no evidence that either blue or redtail females and juveniles, who elicit the first alarm response with their alarm calls, are intrinsically more alert to predators. Blue adult males, however, do give alarm calls before and call more often than adult male redtails.

To specify how important predation is in explaining mixed-species groups, one should know both the cost associated with predation pressure relative to other factors adversely affecting an animal's fitness, and the degree to which association in mixed groups alleviates that cost. Causes of mortality are poorly known in most monkey populations; however, it is often assumed that predators are an important cause of mortality because of the monkeys' strong alarm responses (Raemakers and Chivers 1980, Struhsaker 1981b). Evidence that participation in mixed-species groups actually does reduce the risk of being preyed upon is equally indirect: in this study, for example, the fact that no successful predation on mixed groups was observed is consistent with the hypothesis that association reduces the risk of being captured, but a difference in predator success on mixed- and single-species groups was not demonstrated. Struhsaker (1981b) has related predator pressure and incidence of mixed grouping on a larger scale, noting that mixed groups appear to involve arboreal monkeys more often than sympatric terrestrial species (the latter presumably being less easily caught, though note that the main prey of the crowned hawk eagle in Kenya is an antelope) and African more often than South American or Asian species; Struhsaker suggests that the latter may suffer lower predator pressure, though there is no firm evidence relevant to this point. (Note that some South American tamarins live in permanent mixed groups; Terborgh (1983) suggests that antipredator benefits in such groups are a major reason for their occurrence.) In short, the power of predation as an explanatory factor for mixed-species associations has not been measured directly in this study, as it has also not been measured in any other study of mixed groups of primates. Based on indirect evidence, however, it is possible, and indeed seems probable, that avoidance of predation is an important benefit derived by mixed group participants in Kakamega, as well as in other populations.

The *role* of predation as an explanatory variable is perhaps somewhat more easily specified. Because eagles are year-round residents that require at least one prey item every 1-3 days (Brown et al. 1982), predation pressure is probably important at all times of year, though it is greatest once every two or so years when young are in the nest. Nevertheless, the monkeys may use a variety of tactics to avoid being victims of a predator's attack, and the tactics they use should depend on other demands they face. For example, when they spend much time foraging for insects, individual vigilance levels must be relatively low (because foraging and vigilance are incompatible activities). The importance of mixed-species grouping as an antipredator tactic might then increase, relative to times when individuals can afford to be more vigilant. Thus the "effective" predator pressure can vary depending on what options the monkeys have for self-defense, and these options may vary temporally with other ecological factors like diet.

In Kakamega, there is little relationship between the prevalence of certain food items in the diet and the tendency to associate with other species. Neither species is more likely to form mixed-species associations when feeding heavily on insects, nor less likely to do so when feeding relatively frequently on quickly and easily harvested fruits (when time available for vigilance should be greater). The one partial exception to this generalization involves the diurnal time scale: both blue monkeys and redtails are more likely to associate in mixed-species groups at times of day when rates of insect foraging, and resting, are high, i.e., at times when individual vigilance levels are probably relatively low. However, this correlation exists in only half of the sample months. In this population, then, varying "effective" predation pressure seems to be an insufficient explanation for most of the temporal variation in the occurrence of mixed-species groups. Additional factors must be considered in order to explain this temporal variation.

Food

One such factor is food. Several hypotheses were considered as to how association in mixed-species groups might improve the foraging ability of blue monkeys, or redtails, or both. For three of them, (1) that the more insectivorous redtails improve their ability to capture mobile insects flushed by blue monkeys, (2) that either species steals prey items from the other, and (3) that either species improves its access to certain foods as a result of the other species' direct action on those foods, there was no supporting evidence.

Two others hypotheses, however, are consistent with observations of feeding and ranging behavior. The first is that participation in mixed-species groups makes certain foraging areas more available to one or both species. When redtails accompanied blue monkeys for more than the median number of hours mixed per day, the redtails used more open parts of the forest more often than they did on days when they associated with blues for less time. This difference may reflect antipredator advantages derived from participation in mixed-species groups. The second hypothesis is that blue monkeys, who cover a larger proportion of their range in less time than do redtails, serve as the redtails' guides to food (especially fruit) sources that have not been recently visited. I suggested that the reason for this behavior is that redtails, like blues, prefer ripe fruits, which often become available over a long time on any one tree. By associating with blues, redtails avoid encountering trees from which most of the ripe fruits have been gathered just previously by blue monkeys. This hypothesis is supported by the fact that redtails in mixed-species groups spend little time visiting areas that were used by blue monkeys earlier on the same day, and by the fact that redtails leave blue monkeys who backtrack on their path. Furthermore, redtails are most likely to associate with blues in months when the proportion of low density but preferred foods in the redtail diet is high; these are exactly the kinds of foods for which a guide would be most useful, because if a source were found that had already been picked over by blues, redtails would have to travel relatively far to find another one.

One might naturally wonder whether the redtails' subordination in competitive situations would not cancel the benefits of finding good food sources. Clearly increased interspecific aggressive competition is one cost the redtails must pay. It seems unlikely that this cost is prohibitive, however, because the great interindividual dispersion of both species ensures that most feeding trees are seldom filled to capacity (i.e., the addition of one more animal invariably precedes the ejection of another). Also, a van of redtails often precedes the bulk of the blue monkey group in a particular feeding tree, and so has some uninterrupted feeding time before the number of blues increases to to the point that the redtails are frequently disturbed and supplanted.

The evidence for the above two foraging advantages in mixed-species groups is clearly indirect: to test these hypotheses directly, the benefits and costs of feeding in mixed groups would have to be quantified in greater detail. Without such detailed information, it is not possible to assess the importance of foraging benefits relative to antipredator benefits in explaining why mixed-species groups occur. It is clear, however, that the foraging advantages suggested above are likely to increase and decrease in importance as a consequence of temporal changes in diet. Dietary changes result, in turn, from fluctuations in availability of different foods and possibly also from fluctuations in the monkeys' preferences. Thus food seems to explain timing of mixed-species associations of Kakamega *Cercopithecus*, at least on a scale of several months, better than predation does.

Competition

Diurnal fluctuations in the occurrence of mixed-species groups may also be due to food, and specifically to the effects of interspecific competition for feeding sites. Redtails, being smaller than blues, are at a competitive disadvantage in feeding trees, from which they are displaced more frequently than they displace blues. During most of the day, redtails in mixed groups seem simply to suffer their disadvantage, either feeding briefly in a tree just before many blues arrive, or waiting their turn. At the ends of the day, redtails are more active than blues, beginning to feed earlier and settling for the night later. At these times they are able to feed essentially uninterrupted and on their own. Struhsaker (1981b) has suggested that blues and redtails in the Kibale Forest also segregate along a temporal niche dimension as described here, though his observation was not related explicitly to the timing of mixed-species groups.

The fact that redtails participate in mixed groups less frequently at the beginning and end of the day could also be related directly to the way they use blues as guides, or to the risk of being caught by a predator, or both. At these times of day, redtails may know better where the local blue monkey group is not, and has not been, either since the previous day or since earlier on the same day. If predators are less likely to hunt successfully at these times, or "effective" predator pressure is less (see page 80), antipredator advantages may also be less important. There is little evidence germane to these explanations, however, and they therefore remain open questions.

Invertebrate Feeding

Throughout this and previous discussions of competition and the nonduplication of effort hypothesis, fruit has been taken as the contested resource. Another dietary item that may be critical is insects, a major source of protein for both monkey species. Insects have not been considered explicitly because there is no information on the dynamics of their abundance, whereas fruit availability is more easily monitored. Blues and redtails overlap considerably in the microhabitats and capture methods used in foraging for insect prey, but the degree to which they compete is not known. Although interference competition for space in fruiting trees is readily observed, competition for insects is more likely to be indirect, because insects are probably more uniformly distributed than fruits (Terborgh 1983). Such exploitative competition would be difficult to see if it occurred.

Data are also lacking on how quickly insects are redistributed and replenished in an area after it has been used by monkeys. Without this information, it is impossible to address explicitly the question of whether redtails improve their insect foraging efficiency by avoiding areas used by blues earlier on a given day. It is conceivable, however, that the monkeys avoid reuse of quadrats within a day at least partly because of a decrease in insect prey abundance in quadrats that have been used recently. The depression of available prey may involve not only a decrease in absolute numbers of prey after harvesting, but also changes in the activities or position of remaining prey which render them more difficult to capture (Charnov et al. 1976).

Even if information on prey abundance and behavior were available, however, it might be difficult to separate the influences of insect and plant food (i.e., fruit) foraging demands on monkey ranging patterns, and hence on participation in mixed-species groups. The areas which monkeys use for insect foraging are probably not independent of the locations of fruit sources (Terborgh 1983). The degree to which insect and fruit foraging demands conflict, and the degree to which either or both are compromised, would be difficult to specify.

Demography

The relevance of demography to mixed groups of blues and redtails is best evaluated by comparing populations with different demographic characteristics. The Kibale Forest primate community provides an interesting basis for comparison because the relative densities of the two species there are opposite to what they are in Kakamega. As Table 23 shows for both species, in the community in which population density is higher, groups are larger, and annual home ranges and daily path lengths are smaller. Most important, the proportion of the total home range used in a one- or five-day period is higher when population density is higher. Therefore, whereas blue monkeys would be expected to be better informed than redtails about the location and quality of resources in their range in Kakamega, blue monkeys in Kibale should be less well informed than sympatric redtails. The degree to which each species may benefit from participation in mixed groups, at least via nonduplication of effort, should differ in the two study sites.

Demographic variables may also affect asymmetries in antipredator advantages of members of mixed-species groups. Any of the mechanisms proposed should benefit both

TABLE 23

Demography and Range Use in Two Study Areas [1]

	Kibale [2]		Kakamega	
	BLUES	REDS	BLUES	REDS
Density (ind/km^2)	42	140	169	72
Group size (n groups)	18.7 ± 8.5 (4)	30-35 (1)	32.6 ± 8.9 (5)	22, 23, 34 (3)
Annual home range (ha) (n ranges)	50.6 ± 14.8 (4)	24 (1)	23 ± 9 (5)	55, 30, 27 [3] (3)
Daily path length (m) [4] (n days)	1298 ± 345 (65)	1447 ± 253 (34)	1136 ± 228 (80)	1543 ± 296 (84)
Mean % annual range [4] used in one day	15	28	34	24
Mean % annual range [4] used in five days	35	68	70	62

[1] Means and standard deviations are given, unless n ≤ 3.
[2] From Rudran (1978), Struhsaker (1978) and unpub., and Butynski (1982) and unpub., Kanyawara area.
[3] The two smaller home range estimates result from position checks of groups not followed as systematically as the L focal group, and are probably underestimates.
[4] Figures calculated based on ranging patterns of focal group only.

participating species; however, members of the species living in smaller groups will gain more from associating with larger groups than vice versa, as long as antipredator advantages of mixed groups increase monotonically with group size in the size range under consideration. Thus, redtails should benefit more (in terms of avoiding predation) from mixed-species associations in Kakamega, while blues should benefit more in Kibale.

One might expect these differences in relative benefits to be reflected in the identity of the species responsible for mixed-species group formation and maintenance. In Kakamega, redtails appear to gain more than blues by joining in mixed groups to avoid areas previously used on a given day, and possibly to avoid predation; redtails are also mainly responsible for forming and maintaining mixed groups. Following the same line of reasoning, one would expect blue monkeys in Kibale to take this responsibility, and although blue monkeys approached redtails on 8 of 12 occasions (Struhsaker 1981b), this tendency is not significant statistically (Binomial Test).

Using an interpopulational comparison of species' roles in the formation and disbandment of mixed-species group as evidence for the role of demography in explaining asymmetrical benefits is not without problems. Other factors may influence the participation of one or both species unequally in the two populations. For example, in Kibale there are four species of diurnal primates that do not occur in Kakamega, but with whom (Kibale) blues and redtails sometimes associate. Interactions with these species could have many effects on the relationship between the two *Cercopithecus*. At present however, such effects cannot be identified because there are too few detailed data on how ecological relationships between particular species pairs affect one another.

A Comparative Perspective

The question naturally arises as to how the results of this study relate to what is known from previous studies of mixed-species groups involving the same or different species in other primate communities. Comparisons of communities are potentially revealing, because one is able, in principle, to relate differences in mixed-species groups to differences in species identity and ecology. As a result, one may be able to generalize about the importance and role of particular ecological factors in explaining why mixed groups of primates occur.

In practice, such comparisons are difficult for two reasons. First, communities often differ in several ways, so that the exact effects of a single ecological variable cannot be isolated. Second, there has been no standardization of methods used to collect information on how mixed-species groups form and disband, how long they last, when they occur, or what the animals are doing when in, and not in, such groups. Struhsaker (1981b) has demonstrated, for example, that estimations of the incidence of mixed-species grouping made by census or by all-day follows can differ greatly. Even if all-day follows are used, the number of days in a monthly sample can affect estimates of association frequency if variation between days is sufficiently great. Day-to-day variation among Kakamega blue monkeys is considerable, for instance (coefficient of variation in time spent in mixed groups per day = 0.66 for blues; for redtails, CV = 0.28).

Here, I would like to take a brief, general comparative view of the importance and role of several ecological factors thought to affect mixed-species associations in three primate communities that have been relatively thoroughly studied: Kibale (Uganda), Makokou (Gabon), and Cocha Cashu (Peru).

Kibale

The Kibale community is interesting relative to Kakamega because it contains the same two *Cercopithecus* species in a rather different ecological setting. The primary difference between the two forests is that Kibale, which is closer to the presumed Pleistocene refugium in eastern Zaire, is richer in plant and animal species. For blues and redtails, this means specifically that there are two or three additional omnivorous primate species, larger than either *Cercopithecus*, and potentially effective competitors for food (Struhsaker 1978).

In Kibale, redtails and blue monkeys spend less time associating with one another than in Kakamega, and the time spent associated varies more from month to month. Blues associate with redtails on 0-29% ($\bar{x}$ = 13%) of scans per month. Blues associate with other monkeys as well, but 68% of all associations are formed with redtails, more than would be expected by virtue of the redtail population density alone (Rudran 1978). Redtails associate with blue monkeys in 0-54% of scans per month, though again they spent additional time with other monkey species. Association with blue monkeys was also variable diurnally, but there was no particular pattern (Struhsaker 1981b).

We have already noted that the different demographic relationships between the two species may affect the degree to which each of them benefits, relative to the other, from participation in mixed groups by avoiding predation or food depressed areas. The absolute difference in the amount of time spent in mixed groups must also be explained. Both species in Kibale spend not only less time in mixed groups with one another but also with all other species combined (Rudran 1978, Struhsaker 1981b). Why should this be so? There are two logical possibilities: either the benefits of associating are less, or the costs are greater, in Kibale than in Kakamega. A comparison of *Cercopithecus* diets in the two study areas (Cords 1986b) suggests that benefits of association may be lower in Kibale because of increased competition in the more complex primate community. In particular, if interspecific competition leads to greater dietary divergence, one species may generally have less to gain from using the other as a guide, because it will be eating fewer of the same foods. Struhsaker (1981b) in fact suggests that associations between Kibale blues and redtails are due mainly to aggregation at shared, low density but popular feeding sites.

At present, it is not possible to assess how the importance of avian predator pressure compares in the two study areas. The greater density of prey in Kibale would mean that the chance of any one individual being taken is lower there if eagle densities are similar to those in Kakamega. There are no data available to allow a comparison of predator densities in Kakamega and Kibale, however, and studies of *Stephanoetus* in Kenya have shown that eagle home ranges decrease with increasing prey density (Brown et al. 1982).

In sum, data on associations of blue and redtail monkeys from Kibale suggest that ecological relations between species, which may be consequences of the number and identity of all species present in the community, have a strong effect on the reasons why species associate in mixed groups. The same species apparently associate to different degrees and for different reasons depending on the exact nature of their relationship to one another.

Makokou

In Makokou, Gabon, *C. pogonias* and *C. nictitans* associate with one another nearly all of the time, and two groups of *C. cephus* were observed associating with the other two species for 6 and 42% of the time respectively (Gautier-Hion et al. 1983). Intermonthly variation in association is not great for *C. nictitans* and *C. pogonias*. On a diurnal time scale, polyspecific association was more frequent in the evening than at other times of day (Gautier and Gautier-Hion 1969), a trend quite opposite to that in Kakamega. These conclusions are based on census records, and may be confounded by diurnal changes in activity, and therefore visibility, of the species considered.

Gautier and Gautier-Hion (1983) identified *C. cephus* as the species that determines how long mixed groups last: like the redtails in Kakamega, *C. cephus* joins and leaves other species more often than it is joined and left; unlike the redtails, *C. cephus* leaves mixed groups by not moving, while the rest of the group moves on. No data are available on relative frequencies of approaching and leaving other species by *C. cephus*. By comparing the behavior and ranging patterns of two *C. cephus* groups, Gautier-Hion et al. (1983) concluded that association with *C. pogonias* and *C. nictitans* benefits *C. cephus* by reducing its vulnerability to predators, and improves foraging efficiency for all three species by increasing the daily foraging area and diet diversity, and avoiding duplication of effort. *C. cephus*, like the redtails (who are closely related taxonomically and ecologically), increases its use of more open but fruit-rich types of forest. *C. pogonias* and *C. nictitans* may benefit from *C. cephus*' probably superior knowledge of its home range.

Gautier-Hion et al. (1983) propose, however, that "predator avoidance constitutes the prime factor in the evolution of polyspecific life." Temporal fluctuations in the formation of mixed-species groups are not related to seasonal variation in fruit supply or to daily feeding activity. Thus mixed groups do not seem to form at times when the proposed foraging benefits would be greatest. How the timing of mixed-species groups is determined remains unclear. The authors do not look for temporal variation in predation risk; they seem to believe such risk is more or less constant or unpredictable.

Although this study seems to show that food, particularly fruit, plays a relatively minor role in determining when and why mixed groups occur, it is unfortunate that seasonal variation in the diet was not explored in greater detail. In Kakamega, the incidence of mixed groups of blues and redtails was also not related to the consumption of any particular dietary constituents, including fruit. In both Kakamega and Kibale, however, when only rare but preferred plant foods were considered, a relationship was found. Thus it is possible that Gautier-Hion et al.'s (1983) implication that food does not explain temporal variation in the occurrence of mixed groups arises from a lack of detailed information.

Cocha Cashu

Whereas the above two studies of *Cercopithecus* indicate some differences from Kakamega in the role of different factors in explaining why mixed-species groups occur, *Cebus* and *Saimiri* associations in Peru (Terborgh 1983) show remarkable parallels with blue monkey and redtail associations in Kakamega. Like the redtails, *Saimiri* initiate and maintain the associations with *Cebus* which last from a few hours to several days. *Saimiri* are smaller than *Cebus* and competitively subordinate, although *Cebus* groups are smaller. *Saimiri* live in larger ranges. There are various indications that *Cebus* are more alert to predators.

According to Terborgh (1983), the "principal motive" behind *Saimiri-Cebus* associations is decreased risk of being captured on the part of the *Saimiri*. Only this explanation is consistent with the year-round attraction *Saimiri* have for *Cebus*. Foraging demands modify this basic attraction, depending on the distribution and abundance of resources being used. The importance of fruiting trees has been emphasized particularly, and Terborgh suggests that the more widely ranging *Saimiri* improve their discovery of resources when in mixed groups by exploiting the more current, detailed knowledge that the *Cebus* have of their own range. When fruits are abundant, associations are frequent and long lasting; during times of fruit scarcity, however, *Saimiri* tend to switch more from one group of *Cebus* to the next.

Here again, then, is a case in which predator avoidance probably provides a benefit for members of mixed-species groups at most times. Temporal variations in the groups' occurrence seem to be due mainly to the effects of changing diet and changing foraging requirements. As Terborgh shows, these result in turn from seasonal phenological patterns and competitive interactions between species.

Conclusions

The main conclusions from this study can be summarized as follows:

1) Blue and redtail monkeys in the Kakamega Forest spend over half their time in mixed groups together. Redtails are primarily responsible for the formation and maintenance of such groups, which suggests that any benefits of participating are not symmetrically distributed.

2) Most behavioral interactions between members of the two species are agonistic. In at least half of the cases observed, aggressive interactions appeared to be contests over food or feeding sites. Relative to redtails, blue monkeys are at a competitive advantage because of their larger body size. The inferior competitive ability of the redtails may lead them to forage on their own early in the morning and late in the evening, times when blue monkeys are still mainly inactive.

3) Mixed-species groups are not the result of chance encounters of groups of both species moving randomly and independently. It is also unlikely that they result from each species independently exploiting food sources shared by both. The fact that redtails are primarily responsible for their formation and persistence suggests that at least this species actively benefits from participation in mixed groups.

4) The importance of predator avoidance as an explanation for the occurrence of mixed-species groups cannot be discounted. Both species behave as if the crowned hawk eagle is well worth avoiding, and their alarm responses suggest several ways in which participation in mixed groups might reduce the risk of being captured. Although predator avoidance may be an important benefit of participation in mixed groups by the Kakamega *Cercopithecus*, it alone does not explain why the occurrence of mixed-species groups varies in time.

5) Redtails, who are more insectivorous than blues, do not appear to benefit by using blue monkeys as "beaters" for insect prey.

6) Neither species benefits by improving its access to particular types of food rendered more available by the other species' direct action on them.

7) Redtails, but not blue monkeys, shift their use of structurally defined forest types

when in mixed groups. When with blues for more than the median length of time per day, redtails use areas with low foliage density in the middle canopy layer and fewer climbers more often. This shift may be related to antipredator advantages of associating with blue monkeys.

8) I hypothesize that redtails use blue monkeys as guides to food sources not recently fed upon. Both species avoid using parts of their ranges twice on a given day. By joining the "local" group of blue monkeys, redtails are also able to avoid areas of the forest exploited by the blues earlier on a given day. Avoiding such areas may be advantageous to redtails if in so doing, they increase their consumption of ripe fruits whose availablity would be depressed by a passing group of blue monkeys. The importance of invertebrate prey in determining this ranging pattern remains to be tested.

9) I suggest that the usual discussion of the relative importance of food and predation in determining mixed-species groups misses the point that while both factors may be important, they may not explain the same aspects of why such groups occur. In Kakamega, the risk of being attacked appeared to be more or less constant in time, though the value of mixed-species association as an antipredator tactic may have varied temporally with activity pattern and type of foraging. An increased tendency to associate at midday could be explained by the value of mixed-species association in reducing the risk of being preyed upon when individual vigilance was probably reduced, but this relationship held in only half of the sample months. Thus antipredator benefits accruing to members of mixed-species groups do not explain most of the temporal variation in association. The dynamics of mixed groups were instead related primarily to patterns of range use, which are themselves consequences of dietary preference, food availability, interspecific competition and demography. A brief look at other primate communities in which mixed groups occur illustrates how these factors influence mixed-species associations.

Appendixes

Corrected page 92 for *Mixed-Species Association of Cercopithecus Monkeys in the Kakamega Forest, Kenya*, by Marina Cords (University of California Publications in Zoology, Volume 117).

Appendix 1. Percent Frequency of Use of Specific Food Items Eaten During 11 Sample Months by Blue Monkeys (n = 8454 records)

SPECIES	Blossom Bud	Blossom	Nectar	Gum	Fruit	Gall	Leaf Bud	Young Leaves	Mature Leaves	Petiole	Seed	Shoot, tendril	Stem	Twig, bark	Unidentified	Total
Teclea nobilis	0.01	0.15			7.27	0.01	0.01	0.30	0.30							8.05
Trema guineense					7.97			0.01	0.01							8.00
Ficus exasperata					6.18		0.13	0.43								6.74
Ficus mallatocarpa					1.77		0.06	4.67		0.01						6.51
Bosqueia phoberos					4.69		0.01	1.08		0.03					0.03	5.84
Ficus thonningi					3.55			1.03								4.62
Albizia gummifera		1.17		1.29			0.34	0.83	0.04	0.09	0.47					4.29
Antiaris toxicaria		0.07			3.96	0.06		0.07	0.03					0.04	0.03	4.10
Bischoffia javanica		0.01			3.61			0.01					0.01		0.01	3.78
Fagara macrophylla	0.20	0.47		0.76	1.73		0.01	0.11	0.03	0.13					0.01	3.41
Celtis africana		0.05			1.65			1.62	0.01	0.09				0.03		3.36
Prunus africana		0.01		0.21	3.10									0.01		3.33
Harungana madagascariensis					3.18					0.01				0.03		3.22
Morus lactea					2.40			0.22								2.63
Celtis durandii	0.01	0.04			2.38		0.01	0.06	0.04	0.01				0.01	0.01	2.51
Funtumia latifolia	0.05				0.12			0.01			1.77				0.01	1.96
Bequaertiodendron oblanceolatum					1.81											1.86
Blighia unijugata		0.20	0.04		1.29	0.01	0.06	1.58	0.03							1.68
Cordia abyssinica					1.54				0.01					0.03		1.57
Croton sylvaticus					1.24			0.01	0.01							1.57
Solanum mauritianum	0.11	0.14			0.13			1.06		0.10			0.11			1.49
Aulacocalyx diervilleoides					1.43				0.03	0.01						1.47
Manilkara buugi					0.19		0.01	0.92	0.03	0.06		0.05			0.03	1.44
Chaetacme aristata	0.21	0.45			0.05			0.03		0.25	0.01	0.01		0.03	0.03	1.32
Markhamia platycalyx		0.07			0.28			0.09	0.36	0.05					0.01	1.04
Olea welwitschii					0.53			0.12	0.01				0.01			0.86
Ficus cyathistipula					0.27			0.38								0.67
Ficus brachylepis		0.03			0.10			0.24	0.19				0.01			0.65
Passiflora edulis					0.50			0.06	0.06							0.57
Margaritaria discoidea											0.54					0.56
Hippocratea spp.																0.54
Premna angolensis		0.25			0.10			0.13	0.03							0.51

Appendix 1. (continued)

SPECIES	Blossom Bud	Blossom	Nectar	Gum	Fruit	Gall	Leaf Bud	Young Leaves	Mature Leaves	Petiole	Seed	Shoot, tendril	Stem	Twig, bark	Unidentified	Total
Cissus oliveri					0.47											0.47
Syzygium guineense					0.44											0.44
Turraea robusta					0.06			0.33								0.39
Aningeria altissima		0.23						0.15								0.38
Casearia battiscombei		0.01						0.13	0.21						0.01	0.37
Ficus ?dawei					0.11		0.01	0.25								0.37
Tiliacora funifera						0.01		0.20				0.12	0.01			0.34
Cassipourea ruwensorensis		0.01			0.22		0.03	0.03	0.01							0.31
Dovyalis macrocalyx		0.03			0.19			0.03	0.01			0.01		0.01		0.28
Lepidotrichilia volkensii		0.03			0.06		0.03	0.05	0.07			0.01			0.03	0.27
Vangueria apiculata	0.01	0.01			0.01			0.11	0.09						0.03	0.27
Bridelia micrantha					0.25		0.01									0.26
Acacia monticola								0.24		0.01						0.25
Croton macrostachyus	0.03				0.19				0.06							0.25
Rothmania urcelliformis	0.06	0.03	0.01					0.11	0.07							0.21
Spathodea nilotica								0.01		0.09						0.20
Mimulopsis solmsii					0.01			0.04	0.14							0.19
Kigelia moosa		0.06						0.03	0.01	0.05						0.15
Phyllanthus ovaliifolius					0.01			0.03	0.01	0.03		0.04	0.03			0.15
Croton megalocarpus					0.03				0.11							0.14
Erthrococca atrovirens								0.04	0.10							0.14
Alangium chineense					0.03			0.10								0.13
Ehretia cymosa		0.04			0.01			0.06		0.01						0.12
Urera cameroonensis					0.09			0.03								0.12
Diospyros abyssinica					0.01			0.09								0.10
Musa paradisiaca					0.05			0.01	0.04							0.10
Piper guineense					0.03			0.04	0.10			0.01				0.09
Albizia grandibracteata		0.06									0.06					0.07
Brillantaisia spp.		0.06											0.01			0.07
Ipomoea acuminata					0.01			0.06								0.07
Trichilia emetica					0.07											0.07
Climber "mujendasi"											0.05					0.07
Acrocarpus fraxinifolius		0.01														0.06
Piper capensis		0.01			0.05											0.06

Appendix 1. (continued)

SPECIES	Blossom Bud	Blossom	Nectar	Gum	Fruit	Gall	Leaf Bud	Young Leaves	Mature Leaves	Petiole	Seed	Shoot, tendril	Stem	Twig, bark	Unidentified	Total
Phyllanthus inflatus					0.06											0.06
Climber P								0.06								0.06
Combretum molle											0.05					0.05
Maerua triphylla					0.01			0.01							0.03	0.05
Acyranthes aspera								0.03	0.01							0.04
Chrysophyllum albidum					0.04											0.04
Deinbollia kilimandscharica					0.05				0.01							0.04
Fagaropsis angolensis								0.01		0.03						0.04
Illigera pentaphylla											0.04					0.04
Persea americana		0.01			0.05											0.04
Acacia abyssinica								0.02								0.03
Dicliptera laxata								0.03								0.03
Ficus "mwanza"					0.03											0.03
Piper umbellatum					0.01				0.01						0.01	0.03
Climber Z		0.01										0.03				0.03
Clerodendrum capitata					0.01				0.01			0.01				0.02
Oncinotis inanandensis										0.01						0.02
Polyscias kikuyensis				0.01				0.01								0.02
Strombosia schaeffleri														0.01		0.02
Clausena anisata					0.01			0.01								0.01
Craibia brownii					0.01											0.01
Dracaena afromontana																0.01
Fagara mildbraedii				0.01												0.01
Khaya spp.				0.01												0.01
Momordica foetida								0.01							0.01	0.01
Pavonia urens									0.01							0.01
Pyrrosia schimperara					0.01											0.01
Rawsonia lucida					0.01											0.01
Sapium ellipticum					0.01											0.01
Solanum giganteum								0.01								0.01
Strychnos spp.																0.01
Vitex fischeri														0.01		0.01
Tree T														0.01		0.01

Appendix 1. (continued)

SPECIES	Blossom Bud	Blossom	Nectar	Gum	Fruit	Gall	Leaf Bud	Young Leaves	Mature Leaves	Petiole	Seed	Shoot, tendril	Stem	Twig, bark	Unidentified	Total
Unidentified climber		0.03			0.04		0.01	0.41	0.32	0.03		0.39	0.17	0.05		1.45
Unidentified tree					0.05			0.03	0.13	0.01		0.01			0.03	0.26
Unidentified fungus															0.25	0.25
Unidentified herb									0.07							0.07
Unidentified orchid		0.01							0.01							0.01
Unidentified epiphyte																0.01
	0.69	3.76	0.05	2.30	65.80	0.03	0.73	17.84	2.75	1.11	2.99	0.70	0.38	0.25	0.62	100.00

Corrected page 96 for *Mixed-Species Association of Cercopithecus Monkeys in the Kakamega Forest, Kenya*, by Marina Cords (University of California Publications in Zoology, Volume 117).

Appendix 2. Percent Frequency of Use of Specific Food Items Eaten During 11 Sample Months by Redtails (n = 6755 records)

SPECIES	Blossom Bud	Blossom	Nectar	Gum	Fruit	Gall	Leaf Bud	Young Leaves	Mature Leaves	Petiole	Seed	Shoot, tendril	Stem	Twig, bark	Unidentified	Total
Bosqueia phoberos					7.15		0.02	0.99	0.02	0.03					0.06	8.27
Ficus exasperata					7.55		0.06	0.07								7.68
Teclea nobilis					6.65			0.01		0.02						6.68
Trema guineense					5.02											5.02
Prunus africana		0.04		0.67	3.98									0.02	0.02	4.80
Ficus mallatocarpa					2.74		0.07	1.89								4.70
Fagara macrophylla		0.04		1.72	2.66			0.04				0.02			0.18	4.66
Harungana madagascariensis		0.02			4.50					0.04				0.03		4.59
Croton sylvaticus					4.40		0.03									4.43
Ficus thonningi					3.60			0.73	0.02						0.02	4.39
Celtis africana					3.23		0.04	0.18	0.01						0.03	3.48
Celtis durandii					3.45			0.01								3.47
Piper guineense					3.01							0.02				3.03
Antiaris toxicaria		0.02			2.93			0.02							0.03	3.00
Morus lactea					2.81					0.02						2.83
Aulacocalyx diervilleoides	0.15				0.25		0.01	2.10	0.06							2.57
Bischoffia javanica					2.38					0.06		0.01				2.45
Bequaeriodendron oblanceolatum					2.06		0.03	0.06								2.15
Albizia gummifera		0.36		1.35			0.18		0.03					0.02		1.96
Cissus oliveri					1.72											1.72
Manilkara butugi					1.26											1.26
Olea welwitschii		0.07			0.65			0.10	0.22	0.04					0.03	1.10
Cordia abyssinica		0.07	0.04		0.89											1.00
Croton macrostachyus					0.99											0.99
Margariaria discoidea					0.99											0.99
Ficus cyathistipula					0.99											0.97
Solanum mauritianum					0.92		0.04	0.03								0.92
Ficus ?dawei					0.74											0.74
Markhamia platycalyx	0.33	0.30								0.04	0.04					0.74
Urera cameroonensis					0.52			0.49								0.53
Ficus "mwanza"					0.02			0.49								0.51
Chaetacme aristata					0.13		0.01	0.31				0.01	0.01		0.03	0.50

Appendix 2. (continued)

	Blossom Bud	Blossom	Nectar	Gum	Fruit	Gall	Leaf Bud	Young Leaves	Mature Leaves	Petiole	Seed	Shoot, tendril	Stem	Twig, bark	Unidentified	Total
Piper capensis		0.31			0.12											0.43
Acrocarpus fraxinifolius											0.40					0.40
Dovyalis macrocalyx		0.01			0.36										0.02	0.39
Spathodea nilotica	0.27	0.02	0.01								0.01				0.06	0.37
Bridelia micrantha	0.02				0.30				0.02							0.34
Syzygium guineense					0.31											0.31
Ficus brachylepis					0.21			0.09								0.30
Trichilia emetica					0.25					0.02						0.27
Cassipourea ruwensorensis		0.02			0.21									0.02		0.25
Ochna holstii					0.25											0.25
Ekebergia capensis				0.02	0.19											0.21
Diospyros abyssinica	0.04	0.07			0.04			0.02							0.02	0.19
Premna angolensis		0.03			0.12											0.15
Psidium guajava					0.15											0.15
Aningeria altissima		0.04			0.02	0.02		0.03	0.02						0.03	0.13
Blighia unijugata					0.06		0.02	0.04							0.01	0.13
Lepidotrichilia volkensii		0.02			0.07					0.02						0.11
Tree MU							0.06			0.04						0.10
Deinbollia kilimandscharica					0.09											0.09
Ehretia cymosa		0.09														0.09
Rothmania urcelliformis		0.04					0.03	0.02								0.09
Impatiens meruensis		0.03						0.02								0.08
Apodytes dimidiata					0.07								0.03			0.07
Funtumia latifolia					0.04									0.02		0.07
Climber PY							0.06				0.01	0.01				0.07
Acacia abyssinica	0.02	0.04														0.06
Casearia battiscombei					0.04				0.02							0.06
Ipomoea acuminata		0.06														0.06
Phyllanthus ovaliifolius					0.02			0.02				0.01				0.05
Afromomum zambesiacum					0.04								0.02			0.04
Brillantaisia spp.								0.02								0.04
Craibia brownii								0.02							0.02	0.04
Strychnos spp.								0.02	0.01							0.04
Turraea robusta					0.02										0.02	0.04

97

Appendix 2. (continued)

SPECIES	Blossom Bud	Blossom	Nectar	Gum	Fruit	Gall	Leaf Bud	Young Leaves	Mature Leaves	Petiole	Seed	Shoot, tendril	Stem	Twig, bark	Unidentified	Total
Vangueria apiculata					0.03										0.01	0.04
Climber "mujendasi"					0.04											0.04
Albizia grandibracteata				0.03												0.03
Combretum molle											0.03					0.03
Croton megalocarpus							0.01				0.02					0.03
Culcasia scandens					0.01											0.03
Ficus storthophylla					0.03											0.03
Mimulopsis solmsii									0.03							0.03
Psychotria bagshawei					0.03											0.03
Tree 44					0.03											0.03
Tree F					0.03											0.03
Acacia monticola									0.02							0.02
Fagaropsis angolensis				0.02												0.02
Kigelia moosa	0.02															0.02
Oxyanthus speciosus					0.02											0.02
Passiflora edulis								0.02								0.02
Pavetta spp.					0.02											0.02
Pavonia urens				0.02												0.02
Polyscias kikuyensis															0.02	0.02
Rhipsalis baccifera					0.02											0.02
Rhamnus prinoides					0.02											0.02
Strombosia scheffleri		0.02														0.02
Tabernaemontana usambarensis		0.02														0.02
Bersama abyssinica											0.01					0.01
Catha edulis												0.01				0.01
Chrysophyllum albidum					0.01											0.01
Clerodendrum capitata					0.01											0.01
Phragmanthera rufescens									0.01							0.01
Vernonia amygdalina					0.01											0.01
Vitex fischeri									0.01							0.01
Climber AA					0.01											0.01
Climber Z					0.01											0.01

Appendix 2. (continued)

SPECIES	Blossom Bud	Blossom	Nectar	Gum	Fruit	Gall	Leaf Bud	Young Leaves	Mature Leaves	Petiole	Seed	Shoot, tendril	Stem	Twig, bark	Unidentified	Total
Unidentified climber	0.01	0.03			0.12		0.01	0.40	0.07	0.03		0.49	0.18			1.34
Unidentified tree					0.13		0.04	0.06		0.03			0.02			0.28
Unidentified fungus															0.09	0.09
	0.86	1.77	0.05	3.83	81.64	0.02	0.69	7.95	0.55	0.44	0.56	0.60	0.26	0.11	0.67	100.00

Literature Cited

Alatalo, R. 1981. Interspecific competition in tits *Parus* spp. and the goldcrest *Regulus regulus*: foraging shifts in multi-specific flocks. *Oikos* 37: 335-44.

Aldrich-Blake, F.P.G. 1968. A fertile hybrid between two *Cercopithecus* species in the Budongo Forest, Uganda. *Folia Primatol.* 9: 15-21.

_____. 1970. The ecology and behaviour of the blue monkey *Cercopithecus mitis stuhlmanni*. Ph. D. dissertation, University of Bristol.

Alexander, R.D. 1974. The evolution of social behavior. *Ann. Rev. Ecol. Syst.* 5: 325-83.

Altmann, J. 1974. Observational study of behavior: sampling methods. *Behaviour* 49: 227-65.

Altmann, S.A., and J. Altmann. 1970. *Baboon Ecology*. Chicago: University of Chicago Press.

Austin, G.T., and E.L. Smith. 1972. Winter foraging ecology of mixed insectivorous bird flocks in an oak woodland in southern Arizona. *Condor* 74: 17-24.

Balph, D.F., and M.H. Balph. 1979. Behavioral flexibility of pine siskins in mixed species foraging flocks. *Condor* 81: 211-12.

Barlow, G.W. 1974. Extraspecific imposition of social grouping among surgeonfishes (Pisces: Acanthuridae). *J. Zool. Lond.* 174: 333-40.

Barnard, C.J., and H. Stephens. 1981. Prey size selection by lapwings in lapwing/gull associations. *Behaviour* 77: 1-22.

_____. 1983. Costs and benefits of single and mixed species flocking in fieldfares (*Turdus pilaris*) and redwings (*T. iliacus*). *Behaviour* 84: 91-123.

Bernstein, I.S. 1967. Intertaxa interactions in a Malayan primate community. *Folia Primatol.* 7: 198-207.

Bertram, B.C. 1978. Living in groups. In Krebs, J.R., and N.B. Davies, eds. *Behavioural Ecology*. Sunderland, MA: Sinauer Associates, pp. 64-96.

Booth, A.H. 1956. The distribution of primates in the Gold Coast. *J. West Afr. Sci. Ass.* 2: 122-33.

Bourliere, F., C. Hunkeler, and M. Bertrand. 1970. Ecology and behavior of Lowe's guenon (*Cercopithecus campbelli lowei*) in the Ivory Coast. In Napier, J.R., and P.H. Napier, eds. *Old World Monkeys*. New York: Academic Press, pp. 297-350.

Breder, C.M. 1959. Studies on social groupings in fishes. *Bull. Am. Mus. Nat. Hist.* 117: 397-481.

Breder, C.M., and F. Halpern. 1946. Innate and acquired behavior affecting aggregation of fishes. *Physiol. Zool.* 19: 154-90.

Brosset, A. 1969. La vie sociale des oiseaux dans une foret equatoriale du Gabon. *Biologia Gabonica* 5: 29-69.

Brown, L. 1970. *African Birds of Prey*. Boston: Houghton Mifflin.

Brown, L.H., E.K. Urban, and K. Newman. 1982. *The Birds of Africa, Vol. I*. New York: Academic Press.

Butynski, T. 1982. Harem-male replacement and infanticide in theblue monkey (*Cercopithecus mitis stuhlmanni*) in the Kibale Forest, Uganda. *Am. J. Primatol.* 3: 1-22.

Caldwell, G.S. 1981. Attraction to tropical mixed-species heron flocks: proximate mechanisms and consequences. *Behav. Ecol. Sociobiol.* 8: 99-103.

Caraco, T. 1979. Time budgeting and group size: a test of theory. *Ecology* 60: 618-27.

Caraco, T., S. Martindale, and H.R. Pulliam. 1980. Avian flocking in the presence of a predator. *Nature* 285: 400-01.

Carcasson, R.H. 1964. A preliminary survey of the zoogeography of African butterflies. *E. Afr. Wildl. J.* 2: 122-157.

Charles-Dominique, P. 1977. *Ecology and Behaviour of Nocturnal Primates*. New York: Columbia University Press.

Charnov, E.L., G.H. Orians, and K. Hyatt. 1976. Ecological implications of resource depression. *Am. Nat.* 110: 247-59.

Cody, M.L. 1971. Finch flocks in the Mojave Desert. *Theor. Pop. Biol.* 2: 142-58.

———. 1974. Optimization in ecology. *Science* 183: 1156-64.

Coelho, A.M., L.S. Coelho, C.A. Bramblett, S.S. Bramblett, and L.B. Quick. 1976. Ecology, population characteristics, and sympatric association in primates: a socio-bioenergetic analysis of howler and spider monkeys in Tikal, Guatemala. *Yrbk. Phys. Anth.* 20: 96-135.

Colwell, R.K., and E.R. Fuentes. 1975. Experimental studies of the niche. *Ann. Rev. Ecol. Syst.* 6: 281-310.

Connell, J.H. 1980. Diversity and the coevolution of competitors, or the ghost of competition past. *Oikos* 35: 131-38.

Cords, M. 1984a. Mating patterns and social structure in redtail monkeys (*Cercopithecus ascanius*). *Z. Tierpsychol.* 64: 313-29.

———. 1984b. Mixed species groups of *Cercopithecus* monkeys in the Kakamega Forest, Kenya. Unpublished Ph.D. dissertation, University of California, Berkeley.

———. 1986a. Interspecific and intraspecific variation in diet of two forest guenons. *J. Anim. Ecol.* 55, 811-28.

———. 1986b. Forest guenons and patas monkeys: male-male competition in one-male groups. In: Smuts, B., D. Cheney, R. Seyfarth, R.Wrangham, T. Struhsaker, eds. *Primate Societies*. Chicago: University of Chicago Press (in press).

Cords, M., B. Mitchell, H.M. Tsingalia, and T.E. Rowell. 1986. Promiscuous mating among blue monkeys of the Kakamega Forest. *Z. Tierpsychol.* 72, 214-26.

Daneel, A. 1979. Prey size and hunting methods of the crowned eagle. *Ostrich* 50: 120-21.

Davis, D.E. 1946. Seasonal analysis of mixed flocks of birds in Brazil. *Ecology* 27: 168-81.

Diamond, J.M. 1978. Niche shifts and the rediscovery of interspecific competition. *Amer. Sci.* 66: 322-31.

———. 1981. Mixed-species foraging groups. *Nature* 292: 408-9.

Diamond, J.M. and J.W. Terborgh. 1967. Observations on bird distribution and feeding assemblages along the Rio Calleria, Department of Loreto, Peru. *Wilson Bull.* 79: 273-82.

Diamond, T. 1979. Kakamega. *Swara* 2: 25-26.

Ehrlich, P.R., and A.H. Ehrlich. 1973. Coevolution: heterotypic schooling of Caribbean reef fishes. *Am. Nat.* 107: 157-60.

Elder, W.H., and N.L. Elder. 1970. Social groupings and primate associations of the bushbuck (*Tragelaphus scriptus*). *Mammalia* 34: 356-62.

Fleagle, J.G., R.A. Mittermeier, and A.L. Skopec. 1981. Differential habitat use by *Cebus apella* and *Saimiri sciureus* in central Surinam. *Primates* 22: 361-67.

Fleming, T.H. 1979. Do tropical frugivores compete for food? *Amer. Zool.* 19: 1157-72.

Fontaine, R. 1980. Observations on the foraging association of double-toothed kites and white-faced capuchin monkeys. *Auk* 97: 94-98.

Freeland, W.J. 1977. Blood-sucking flies and primate polyspecific associations. *Nature* 269: 801-2.

Friedmann, H. 1955. The honey guides. *U.S. Nat. Mus. Bull.* 208: 1-292.

Galat-Luong, A. 1979. Interactions interspecifiques chez les primates diurnes du parc national de Tai, Cote d'Ivoire. Communication au symposium sur la biologie des Cercopitheques, Oxford, 24 Novembre 1978.

Gannon, G.R. 1934. Associations of small insectivorous birds. *Emu* 34: 122-29.

Gartlan, J.S. and T.T. Struhsaker. 1972. Polyspecific associations and niche separation of rain-forest anthropoids in Cameroon, west Africa. *J. Zool. Lond.* 168: 221-66.

Gautier, J.P., and A. Gautier-Hion. 1969. Les associations polyspecifiques chez les Cercopithecidae du Gabon. *Terre et Vie* 2: 164-201.

_____. 1983. Comportement vocal des males adultes et organisation supra-specifique dans les troupes polyspecifiques de Cercopitheques. *Folia Primatol.* 40: 161-74.

Gautier-Hion, A. 1980. Seasonal variations of diet related to species and sex in a community of *Cercopithecus* monkeys. *J. Anim. Ecol.* 49: 237-69.

Gautier-Hion, A., and J.P. Gautier. 1974. Les associations polyspecifiques de Cercopitheques du Plateau de M'passa (Gabon). *Folia Primatol.* 22: 134-77.

Gautier-Hion, A., J.P. Gautier, and R. Quris. 1981. Forest structure and fruit availability as complementary factors influencing habitat use by a troop of monkeys (*Cercopithecus cephus*). *Rev. Ecol. (Terre et Vie)* 35: 511-36.

Gautier-Hion, A., R. Quris, and J.P. Gautier. 1983. Monospecific vs. polyspecific life: a comparative study of foraging and antipredatory tactics in a community of *Cercopithecus* monkeys. *Behav. Ecol. Sociobiol.* 12: 325-35.

Goss-Custard, J.D. 1970. Feeding dispersion in some overwintering birds. In Crook, J.H., ed. *Social Behavior in Birds and Mammals.* New York: Academic Press, pp. 3-36.

Grieg-Smith, P.W. 1978. The formation, structure and function of mixed-species insectivorous bird flocks in west African savanna woodland. *Ibis* 120: 284-312.

Haddow, A.J. 1952. Field and laboratory studies on an African monkey *Cercopithecus ascanius schmidti* Matschie. *Proc. Lond. Zool. Soc.* 122: 297-394.

Haltenorth, T., and H. Diller. 1980. *A Field Guide to the Mammals of Africa including Madagascar.* London: Collins.

Hamilton, A. 1974. The significance of patterns of distribution shown by forest plants and animals in tropical Africa for the reconstruction of upper Pleistocene paleoenvironments: a review. In van Zinderen-Bakker, E.M., ed. *Paleoecol. Afr.*, 63-97.

---. 1981. *A Field Guide to the Trees of Uganda*. Kampala: Makerere University Printery.

Hamilton, W.D. 1971. Geometry for the selfish herd. *J. Theor. Biol.* 31: 295-311.

Hayashi, K. 1975. Interspecific interaction of primate groups in Kibale Forest, Uganda. *Primates* 16: 269-283.

Hinde, R.A., and S. Atkinson. 1970. Assessing the role of social partners in maintaining mutual proximity, as exemplified by mother-infant relations in rhesus monkeys. *Anim. Behav.* 18: 169-76.

Hindwood, K. 1937. The flocking of birds with particular reference to the association of small insectivorous birds. *Emu* 36: 254-61.

Hladik, C.M. 1978. Adaptive strategies of primates in relation to leaf-eating. In Montgomery, G.G., ed. *The Ecology of Arboreal Folivores*. Washington D.C.: Smithsonian Institution Press, pp. 373-95.

Hogstad, O. 1978. Differentiation of foraging niche among tits, *Parus* spp., in Norway during winter. *Ibis* 120: 139-46.

Holmes, R.T., and F.A. Pitelka. 1968. Food overlap among coexisting sandpipers on northern Alaskan tundra. *Syst. Zool.* 17: 305-18.

Itzkowitz, M. 1974. A behavioral reconnaissance of some Jamaican reef fishes. *Zool. J. Linn. Soc.* 55: 87-118.

---. 1977. Social dynamics of mixed-species groups of Jamaican reef fishes. *Behav. Ecol. Sociobiol.* 2: 361-84.

Jolly, A. 1966. *Lemur Behavior*. Chicago: University of Chicago Press.

Karr, J.R. 1976. Within African- and between-habitat avian diveristy in African and Neotropical lowland habitats. *Ecol. Monographs* 46: 457-81.

Keast, A. 1963. Interrelationships of two zebra species in an overlap zone. *J. Mammal.* 46: 53-66.

Kershaw, K.A. 1964. *Quantitative and Dynamic Ecology*. New York: American Elsevier Publishing Company.

Kingdon, J. 1971. *East African Mammals, Vol. I*. New York: Academic Press.

Klein, L.L., and D.J. Klein. 1973. Observations of two types of Neotropical primate intertaxa associations. *Am. J. Phys. Anth.* 38: 649-54.

Krebs, J.R. 1973. Social learning and the significance of mixed species flocks of chickadees (*Parus* spp.). *Can. J. Zool.* 51: 1275-88.

Lamprey, H. 1963. Ecological separation of the large mammal species in the Tarangire Game Reserve, Tanganyika. *E. Afr. Wildl. J.* 1: 63-92.

Lawton, J.H., and D.R. Strong, Jr. 1981. Community patterns and competition in folivorous insects. *Am. Nat.* 118: 317-38.

Livingstone, D.A. 1975. Late quaternatry climatic change in Africa. *Ann. Rev. Ecol. Syst.* 6: 249-81.

Lucas, G.L. 1968. Kenya. In Hedberg, I., and O. Hedberg, eds. Conservation of vegetation in Africa south of the Sahara. *Acta Phytogeographica Suecica* 54: 152-66.

MacArthur, R. 1972. *Geographical Ecology*. New York: Harper and Row.

MacDonald, D.W., and D.G. Henderson. 1977. Aspects of the behaviour and ecology of mixed-species bird flocks in Kashmir. *Ibis* 119: 481-91.

MacLean, S.F., Jr. 1970. Social stimulation modifies the feeding behavior of the American robin. *Condor* 72: 499-500.

McFarland, W.N., and Z. Hillis. 1982. Observations on agonistic behavior between members of juvenile French and white grunts, family Haemulidae. *Bull. Mar. Sci.* 32: 255-68.

McFarland, W.N., and N.M. Kotchian. 1982. Interaction between schools of fish and mysids. *Behav. Ecol. Sociobiol.* 11: 71-76.

Marler, P. 1973. A comparison of vocalizations of red-tailed monkeys and blue monkeys, *Cercopithecus ascanius* and *C. mitis*, in Uganda. *Z. Tierpsychol.* 33: 223-47.

Miller, R.C. 1922. The significance of the gregarious habit. *Ecology* 3: 122-26.

Milton, K. 1980. *The Foraging Strategy of Howler Monkeys*. New York: Columbia University Press.

Morse, D.H. 1970. Ecological aspects of some mixed-species foraging flocks of birds. *Ecol. Monographs* 40: 119-68.

_____. 1977. Feeding behavior and predator avoidance in heterospecific groups. *Bioscience* 27: 332-39.

_____. 1978. Structure and foraging patterns of flocks of tits and associated species in an English woodland during the winter. *Ibis* 120: 298-311.

Moynihan, M. 1962. The organization and probable evolution of some mixed species flocks of neotropical birds. *Smith. Miscell. Collect.* 143: 1-140.

_____. 1970. Some behavior patterns of platyrrhine monkeys: II. *Saguinus geoffroyi* and some other tamarins. *Smith. Contrib. Zool.* 28: 1-77.

Murton, R.K. 1971. Why do some birds feed in flocks? *Ibis* 113: 534-536.

Newton, P.N. 1984. A feeding association between a heteropteran bug and langurs. *J. Bombay Nat. Hist. Soc.* 81: 180-81.

Oruko, J.R. 1979. Unpublished Kakamega Forest Station Annual Report.

Paulsen, D.R. 1969. Commensal feeding in grebes. *Auk* 86: 759.

Pianka, E.R. 1976. Competition and niche theory. In May, R., ed. *Theoretical Ecology*. Philadelphia: W.B. Saunders Company, pp. 114-41.

Pielou, E.C. 1966. Measurement of diversity in different types of biological collections. *J. Theor. Biol.* 13: 131-44.

Pilleri, G., and J. Knuckley. 1969. Behavior patterns of some Delphinidae observed in the western Mediterranean. *Z. Tierpsychol.* 26: 48-72.

Pook, A.G., and G. Pook. 1982. Polyspecific association between *Saguinus fuscicollis*, *Saguinus labiatus*, *Callimico goeldii* and other primates in north-western Bolivia. *Folia Primatol.* 38: 196-216.

Radakov, D.V. 1973. *Schooling in the Ecology of Fish*. Israel transl. prog. New York: John Wiley.

Raemakers, J.J., and D.J. Chivers. 1980. Socioecology of Malayan forest primates. In Chivers, D.J. ed. *Malayan Forest Primates*. New York: Plenum Press, pp. 279-316.

Rand, A.L. 1954. Social feeding behavior of birds. *Fieldiana: Zoology* 36: 5-71.

Rasa, O.A.E. 1983. Dwarf mongoose and hornbill mutualism in the Taru Desert, Kenya. *Behav. Ecol. Sociobiol.* 12: 181-90.

Rasmussen, D.R. 1980. Clumping and consistency in primates' patterns of range use: definitions, sampling, assessment and applications. *Folia Primatol.* 34: 111-39.

Rodman, P.S. 1973. Synecology of Bornean primates I. A test for interspecific interactions in spatial distribution of five species. *Am. J. Phys. Anth.* 38: 655-60.

Rubenstein, D.I., R.J. Barnett, R.S. Ridgely, and P.H. Klopfer. 1977. Adaptive advantages of mixed-species feeding flocks among seed-eating finches in Costa Rica. *Ibis* 119: 10-21.

Rudran, R. 1978. Socioecology of the blue monkeys (*Cercopithecus mitis stuhlmanni*) of the Kibale Forest, Uganda. *Smith. Contrib. Zool.* 249: 1-88.

Salt, G.W. 1983. Roles: their limits and responsibilities in ecological and evolutionary research. *Am. Nat.* 122: 697-705.

Schoener, T.W 1982. The controversy over interspecific competition. *Amer. Sci.* 70: 586-95.

Shaw, E. 1970. Schooling in fishes: critique and review. In Aronson, L.R., E. Tobach, D.S. Lehrman, and J.S. Rosenblatt, eds. *Development and Evolution of Behavior*. San Francisco: W.H. Freeman and Company, pp. 453-80.

Short, L.R., Jr. 1961. Interspecies flocking of birds of montane forest in Oaxaca, Mexico. *Wilson Bull.* 73: 341-47.

Siegfried, W.R. 1971. Feeding associations between *Podiceps ruficollis* and *Anas smithii*. *Ibis* 113: 236-38.

———. 1972. Wilson's phalaropes forming feeding associations with shovelers. *Auk* 89: 667-68.

Sinclair, A.R.E. 1985. Does interspecific competition or predation shape the African ungulate community? *J. Anim. Ecol.* 54: 899-918.

Skorupa, J.P. 1983. Monkeys and matrices: a second look. *Oecologia* 57: 391-96.

Sokal, R.R., and F.J. Rohlf. 1981. *Biometry*. San Francisco: W.H. Freeman and Company.

Spawls, S. 1978. A checklist of the snakes of Kenya. *J. East Afr. Nat. Hist. Soc. and Nat. Mus.* 31: 1-18.

Struhsaker, T.T. 1970. Phylogenetic implications of some vocalizations of *Cercopithecus* monkeys. In Napier, J.R., and P.H. Napier, eds. *Old World Monkeys*. New York: Academic Press, pp. 365-444.

———. 1975. *The Red Colobus Monkey*. Chicago: University of Chicago Press.

———. 1978. Food habits of five monkey species in the Kibale Forest, Uganda. In: Chivers, D.J. and J. Herbert, eds. *Recent Advances in Primatology, Vol. I*. New York: Academic Press, pp. 225-48.

———. 1981a. The socioecology of hybridization between two species of African rain-forest monkeys, redtails (*Cercopithecus ascanius*) and blues (*Cercopithecus mitis*). Abstract of paper presented to International Ethological Congress, Oxford, September 1981.

———. 1981b. Polyspecific associations among tropical rain-forest primates. *Z. Tierpsychol.* 57: 268-304.

———. 1984. Hybrid monkeys of the Kibale Forest. In Macdonald, D., ed. *The Encyclopaedia of Mammals, I*. London: George Allen and Unwin, pp. 396-97.

Struhsaker, T.T., and L. Leland. 1979. Socioecology of five sympatric monkey species in the Kibale Forest, Uganda. *Adv. Stud. Behav.* 9: 159-228.

Swynnerton, C.F.M. 1915. Mixed bird parties. *Ibis* 3: 346-354.

Terborgh, J. 1983. *Five New World Primates*. Princeton: Princeton University Press.

Terborgh, J. and J.M. Diamond. 1970. Niche overlap in feeding assemblages of New Guinea birds. *Wilson Bull.* 82: 24-52.

Tsingalia, H.M. and T.E. Rowell. 1984. The behavior of adult male blue monkeys. *Z. Tierpsychol.* 64: 253-68.

UNESCO/AETFAT (Association pour l'Etude Taxonomique de la Flore d'Afrique Tropicale) Vegetation Map of Africa, draft copy. Compiled by F. White.

Waser, P.M. 1982. Primate polyspecific associations: do they occur by chance? *Anim. Behav.* 30: 1-8.

_____. "Chance" and mixed-species associations. (Unpublished manuscript).

Waser, P.M. and T.J. Case. 1981. Monkeys and matrices: on the coexistence of "omnivorous" forest primates. *Oecologia* 49: 102-8.

Werner, E.E. and D.J. Hall. 1976. Niche shifts in sunfishes: experimental evidence and significance. *Science* 191: 404-6.

Wiens, J.A. 1977. On competition and variable environments. *Amer. Sci.* 65: 590-97.

Willis, E.O. 1966a. Competitive exclusion and birds at fruiting trees in western Columbia. *Auk* 83: 479-480.

_____. 1966b. Interspecific competition and the foraging behavior of Main-Brown woodcreepers. *Ecology* 47: 667-72.

Wilson, E.O. 1975. *Sociobiology.* Cambridge: Harvard University Press.

Winterbottom, J.M. 1949. Mixed bird parties in the tropics, with special reference to northern Rhodesia. *Auk* 66: 258-63.

Wolf, N.G. 1983. Behavioral ecology of herbivorous reef fishes in mixed-species foraging groups. Ph.D. dissertation, Cornell University.

Wolfheim, J.H. 1983. *Primates of the World.* Seattle: University of Washington Press.

Zimmerman, D.A. 1972. The avifauna of the Kakamega Forest, western Kenya, including a bird population study. *Bull. Am. Mus. Nat. Hist.* 149: 257-339.

Forsyth Library

WITHDRAWN

Bloom's Classic Critical Views

PERCY SHELLEY

Bloom's Classic Critical Views

Alfred, Lord Tennyson
Benjamin Franklin
The Brontës
Charles Dickens
Edgar Allan Poe
Geoffrey Chaucer
George Eliot
George Gordon, Lord Byron
Henry David Thoreau
Herman Melville
Jane Austen
John Donne and the Metaphysical Poets
John Milton
Jonathan Swift
Mark Twain
Mary Shelley
Nathaniel Hawthorne
Oscar Wilde
Percy Shelley
Ralph Waldo Emerson
Robert Browning
Samuel Taylor Coleridge
Stephen Crane
Walt Whitman
William Blake
William Shakespeare
William Wordsworth

Bloom's Classic Critical Views

PERCY SHELLEY

Edited and with an Introduction by
Harold Bloom
Sterling Professor of the Humanities
Yale University

Bloom's Classic Critical Views: Percy Shelley

Copyright © 2009 Infobase Publishing

Introduction © 2009 by Harold Bloom

All rights reserved. No part of this publication may be reproduced or utilized in any form or by any means, electronic or mechanical, including photocopying, recording, or by any information storage or retrieval systems, without permission in writing from the publisher. For more information contact:

Bloom's Literary Criticism
An imprint of Infobase Publishing
132 West 31st Street
New York NY 10001

Library of Congress Cataloging-in-Publication Data
Percy Shelley / edited and with an introduction by Harold Bloom ; Melissa Edmundson, volume editor.
    p. cm. — (Bloom's classic critical views)
  Includes bibliographical references and index.
  ISBN 978-1-60413-447-6 (hardcover)
  1. Shelley, Percy Bysshe, 1792–1822—Criticism and interpretation. I. Bloom, Harold. II. Edmundson, Melissa. III. Title. IV. Series.
  PR5438.P435 2009
  821'.7—dc22
                              2009001601

Bloom's Literary Criticism books are available at special discounts when purchased in bulk quantities for businesses, associations, institutions, or sales promotions. Please call our Special Sales Department in New York at (212) 967-8800 or (800) 322-8755.

You can find Bloom's Literary Criticism on the World Wide Web at
http://www.chelseahouse.com

Volume editor: Melissa Edmundson
Series design by Erika K. Arroyo
Cover designed by Takeshi Takahashi
Printed in the United States of America
IBT IBT 10 9 8 7 6 5 4 3 2 1

This book is printed on acid-free paper.

All links and Web addresses were checked and verified to be correct at the time of publication. Because of the dynamic nature of the Web, some addresses and links may have changed since publication and may no longer be valid.

Contents

Series Introduction	ix
Introduction by Harold Bloom	xi
Biography	1
Personal	7
Henry Crabb Robinson (1817)	9
Unsigned (1821)	10
William Hazlitt "On Paradox and Commonplace" (1821–22)	10
Mary Shelley "Preface" (1824)	12
Marguerite, Countess of Blessington (1834)	15
Thomas De Quincey "Notes on Gilfillan's Literary Portraits" (1845–46)	16
Leigh Hunt (1850)	18
Walter Savage Landor "To Shelley" (1853)	21
Samuel Rogers (1855)	22
Thomas Jefferson Hogg (1858)	23
Edward John Trelawny (1858)	26
Jane Shelley "Preface by the Editor" (1859)	30
Thomas Love Peacock "Memoirs of Percy Bysshe Shelley: Part II" (1860)	33
William Cory "Shelley at Eton" (1886)	34
William Graham (1898)	36
General	41
Charles Lamb (1824)	42
Thomas Lovell Beddoes (1824)	43
John Wilson "Preface" (1826)	44
Thomas Babington Macaulay "John Bunyan" (1830)	44
Robert Browning (1833)	45
Leigh Hunt (1844)	46

Edgar Allan Poe "Elizabeth Barrett Browning" (1845)	47
George Meredith "The Poetry of Shelley" (1851)	49
David Macbeth Moir (1851)	49
Walter Bagehot "Percy Bysshe Shelley" (1856)	51
Hippolyte Taine (1871)	53
John Addington Symonds (1878)	55
Dante Gabriel Rossetti "Percy Bysshe Shelley" (1880–81)	56
Harriet Monroe "With a Copy of Shelley" (1889)	57
Edmund Gosse "Shelley in 1892" (1892)	58
Algernon Charles Swinburne "The Centenary of Shelley" (1892)	61
George Saintsbury (1896)	62

Works 67

The Necessity of Atheism 70
Unsigned (1822) 70

Zastrozzi and St. Irvyne 74
Unsigned (1810) 74
John Cordy Jeaffreson (1885) 76
Walter Raleigh (1894) 79

The Revolt of Islam 80
John Gibson Lockhart "Observations on *The Revolt of Islam*" (1819) 80

Prometheus Unbound 82
Unsigned (1821) 82
Unsigned (1821) 83
William P. Trent "Apropos of Shelley" (1899) 84

The Cenci 85
William Godwin (1820) 85
R. Pickett Scott (1878) 86

General Commentary on the Poetry 89
William Hazlitt "Shelley's Posthumous Poems" (1824) 89
Unsigned "Review of New Publications" (1824) 93
Albany Fonblanque "Literary Notices" (1824) 94
Unsigned (1824) 96
Unsigned "Shelley's Posthumous Poems" (1824) 96
Unsigned (1824) 98

Contents

Mary Shelley "Preface" (1839)	100
Edwin P. Whipple "English Poets of the Nineteenth Century" (1845)	104
Robert Browning "Introduction" (1852)	109
Algernon Charles Swinburne "Notes on the Text of Shelley" (1869)	124
Leslie Stephen "Godwin and Shelley" (1879)	127
Stopford A. Brooke "Some Thoughts on Shelley" (1880)	141
Edward Dowden "Last Words on Shelley" (1887)	157
W.B. Yeats "The Philosophy of Shelley's Poetry" (1900)	167
A.C. Bradley "Shelley's View of Poetry" (1904)	171
Chronology	187
Index	189

Series Introduction

Bloom's Classic Critical Views is a new series presenting a selection of the most important older literary criticism on the greatest authors commonly read in high school and college classes today. Unlike the Bloom's Modern Critical Views series, which for more than 20 years has provided the best contemporary criticism on great authors, Bloom's Classic Critical Views attempts to present the authors in the context of their time and to provide criticism that has proved over the years to be the most valuable to readers and writers. Selections range from contemporary reviews in popular magazines, which demonstrate how a work was received in its own era, to profound essays by some of the strongest critics in the British and American tradition, including Henry James, G. K. Chesterton, Matthew Arnold, and many more.

Some of the critical essays and extracts presented here have appeared previously in other titles edited by Harold Bloom, such as the New Moulton's Library of Literary Criticism. Other selections appear here for the first time in any book by this publisher. All were selected under Harold Bloom's guidance.

In addition, each volume in this series contains a series of essays by a contemporary expert, who comments on the most important critical selections, putting them in context and suggesting how they might be used by a student writer to influence his or her own writing. This series is intended above all for students, to help them think more deeply and write more powerfully about great writers and their works.

Introduction by Harold Bloom

I have been in love with Shelley's poetry for seventy years and have taught and written about it for almost the last sixty. Reading through this very useful volume, I keep shaking my head at the quantity of malice and misunderstanding this great lyrical poet has provoked and continues to suffer. The wonderful critic George Saintsbury speaks for me in these pages when he says Shelley "had no parallel and few peers," and goes on to tell us that the two English poets' poets are Edmund Spenser and Shelley. I am also cheered by Swinburne's brilliance: "Shelley outsang all poets on record but some two or three throughout all time." Swinburne might have been thinking of Pindar and Petrarch and perhaps of the lyrical side of Shakespeare.

Shelley transmembers every other genre into the realm of lyric, and, like Pindar, he soars beyond the limits of lyrical art. He defined the Sublime as that which persuaded us to give up easier pleasures for more difficult ones, and he can be a very difficult pleasure indeed. Shelley requires and rewards very close reading as does Hart Crane, the Shelley of twentieth-century America.

As a political and social revolutionary, Shelley goes on shocking many who attempt to read him. He is to the left of wherever you are, whoever you may be. In some respects he was the Leon Trotsky of the romantic period, greatly admired by Karl Marx and by socialist and anarchist rebels until this day. Every historical institution—state, religion, marriage, family—is denounced in Shelley's prophecy, which is Promethean and visionary and radical beyond every expectation.

As a poet, Shelley deeply influenced Robert Browning, Swinburne, Thomas Hardy, William Butler Yeats, and many others down to Hart Crane and Wallace Stevens in the twentieth century. He vastly offended T.S. Eliot and his New Critical disciples—Allen Tate, R.P. Blackmur, Cleanth Brooks—an offense that became questionable when Eliot, in old age, confessed that Shelley's unfinished death poem, *The Triumph of Life*, was better Dantesque verse than Eliot himself was able to write.

The strongest poets in the English language are Chaucer, Shakespeare, and Milton. Just below them in eminence are a group including Spenser, Christopher Marlowe, Ben Jonson, John Donne, Andrew Marvell, Alexander Pope, and the major figures of the romantic tradition: Blake, Wordsworth, Coleridge, Byron, Shelley, Keats, Tennyson, Browning, Yeats. With these you certainly can rank the greater American poets: Walt Whitman, Emily Dickinson, Robert Frost, Stevens, Hart Crane. Of these twenty poets beyond the indisputable Chaucer, Shakespeare, Milton, the sensitive and informed reader is free to choose his or her own favorites. Blake and Wordsworth were once mine. In old age, I choose Shelley and Whitman to kindle me to a perpetual sense of more life, which is what I translate the Hebrew blessing to mean.

BIOGRAPHY

Percy Shelley
(1792–1822)

Percy Bysshe Shelley was born on August 4, 1792, at Field Place, near Horsham, Sussex. The elder son of Timothy Shelley, a Member of Parliament representing Shoreham, and Elizabeth Pilford, and grandson of Bysshe Shelley, a wealthy landowner. Shelley was educated at Sion House Academy in Isleworth (1802–04), Eton College (1804–10), and University College, Oxford (1810–11). In 1810, while still at Eton, he published his first novel, *Zastrozzi*, a gothic romance, followed in that same year by his first collection of verse, *Original Poetry by Victor and Cazire*, written with his sister Elizabeth and later withdrawn. He entered Oxford University in October 1810 and befriended Thomas Jefferson Hogg. The two collaborated on the *Posthumous Fragments of Margaret Nicholson* (1810), mock-revolutionary poems attributed to a mentally unstable washerwoman who had tried to stab George III and "edited" by the fictitious "John Fitzvictor." This was followed by a second gothic novel, *St. Irvyne; or, The Rosicrucian* (1811), and by another collaboration with Hogg, a pamphlet entitled *The Necessity of Atheism* (1811), which led to their expulsion from Oxford.

Shelley's expulsion, coupled with his involvement with Harriet Westbrook, the daughter of an innkeeper, caused his father virtually to disown him. The dismissal also ended Shelley's expectations of becoming his father's heir and of assuming his father's seat in Parliament. In the summer of 1811, he married Harriet Westbrook after eloping with her to Edinburgh. During the next several years, Shelley and his wife traveled throughout Great Britain, involving themselves in a number of political causes, including the emancipation of Ireland. Shelley published many broadsides and pamphlets during this period, such as *An Address to the Irish People* (1812), and also met William Godwin (with whom he began to correspond), Robert Southey, and Thomas Love Peacock. After leaving Ireland, Percy and Harriet moved to Wales and Devon.

In 1813, Shelley published his first major poem, *Queen Mab*, and in that same year his first child, Ianthe, was born. In 1814, Harriet left him, and he went to France with Mary Wollstonecraft Godwin (with whom he had been corresponding and periodically meeting for two years) and her step-sister Claire Clairmont. While the three of them toured continental Europe, Shelley's second child by Harriet, Charles, was born in England. Early in 1815, Shelley's financial position improved after the death of his grandfather, Sir Bysshe Shelley, from whose estate he received an annual income of £1000. Shortly thereafter, his first child by Mary Godwin was born but died two weeks later; a son, William, was born in 1816. In that same year, Shelley published *Alastor; or, The Spirit of Solitude and Other Poems*, spending the summer in Switzerland. There he first met Lord Byron, with whom he quickly formed a close friendship.

Toward the end of 1816, Shelley learned that Harriet, who was pregnant at the time, had drowned herself in the Serpentine, the lake in London's Hyde Park. Shortly afterward, he married Mary Godwin, who in 1817 gave birth to a daughter, Clara. In 1817, Shelley lost custody of his children with Harriet, and he and Mary moved to Marlow, near London. He met Leigh Hunt, and the two become close friends. Shelley published *A Proposal for Putting Reform to the Vote Throughout the Kingdom*, *An Address to the People on the Death of the Princess Charlotte*, and *Laon and Cythna* (later revised as *The Revolt of Islam* and published in 1818). By 1818, Shelley had become troubled by creditors, ill health, and social disapproval. He and his family left England permanently and settled in Naples, Italy. Traveling throughout Italy, he worked on a translation of the *Symposium* and published *Rosalind and Helen* in 1819. In that same year, Shelley wrote *Julian and Maddalo*, *Prometheus Unbound*, the verse tragedy *The Cenci*, "The Mask of Anarchy," the satirical *Peter Bell the Third*, *A Philosophical View of Reform*, and the poems "Ode to the West Wind" and "Sonnet: England in 1819."

After the deaths of both Clara in 1818 and William in 1819, and the birth in November 1819 of another son, Percy Florence, the Shelleys moved to Pisa in 1820. In that year, Shelley wrote "The Sensitive Plant," "Ode to Liberty," and "To a Sky-Lark." The following year, he produced *The Witch of Atlas*, "A Philosophical View of Reform," and more of his better-known shorter poems, including "To a Cloud" and "Ode to Naples." While in Pisa, Shelley met Emilia Viviani, with whom he became infatuated and who inspired him to write his autobiographical poem, *Epipsychidion*, published anonymously in 1821. Also in that year, Shelley produced *A Defence of Poetry* (in response to Thomas Love Peacock's *The Four Ages of Poetry*), *Adonais: An Elegy on the Death of John Keats*, and *Hellas*, a verse drama based in form on Aeschylus's *The Persians*. The following year, Shelley and Mary moved to San Terenzo with Edward and Jane Williams. There, Shelley composed *Charles I* and began work on *The Triumph of Life*, later published posthumously from rough drafts edited by Mary Shelley.

On July 12, 1822, on his way back from meeting Leigh Hunt at Leghorn, Shelley, along with Edward Williams, died when a storm suddenly overturned his boat in the Bay of Spezia. In August, Shelley's body was cremated and, in 1823, his remains were moved to the Protestant Cemetery in Rome. Following Shelley's death, Mary Shelley published many of his poems for the first time in *Posthumous Poems* (1824) and *Poetical Works* (1839) and also collected his prose in *Essays, Letters from Abroad* (1840). Mary Shelley's *Journal* contains much biographical material on her husband, as do her notes to Shelley's poems, which she included in *Poetical Works*.

PERSONAL

Following his death in 1822, there were numerous biographies and memoirs of Shelley produced throughout the rest of the nineteenth century. These include Thomas Medwin's *The Life of Percy Bysshe Shelley* (London: Newby, 1847), Cyrus Redding's *A Brief Sketch of the Life of Percy B. Shelley* (London: James Watson, 1850), Thomas Jefferson Hogg's *The Life of Percy Bysshe Shelley* (London: E. Moxon, 1858), John Todhunter's *A Study of Shelley* (London: C. Kegan Paul, 1880), and John Addington Symonds's *Shelley* (London: Macmillan, 1881). Notable twentieth-century biographies include Newman Ivey White's two-volume *Shelley* (New York: Knopf, 1940), Kenneth Neill Cameron's *The Young Shelley: Genesis of a Radical* (New York: Macmillan, 1950), and Donald H. Reiman's *Percy Bysshe Shelley* (Boston: Twayne, 1990). Recent biographies include James Bieri's *Percy Bysshe Shelley: A Biography* (Newark: University of Delaware Press, 2005).

In addition to biographies, several other studies appeared, including critical biographies such as George Barnett Smith's *Shelley, A Critical Biography* (Edinburgh: Douglas, 1877), as well as early critical studies of Shelley's work: *The Lyrics and Minor Poems of Percy Bysshe Shelley* (New York: James Pott, 1885) by Joseph Skipsey and *Shelley and His Writings* (London: Newby, 1858) by Charles S. Middleton.

Collections of letters appeared by midcentury, in such editions as *Letters of Percy Bysshe Shelley* (London: Moxon, 1852), edited and introduced by Edward Moxon and Robert Browning, and *Select Letters of Percy Bysshe Shelley* (London: K. Paul, Trench, 1884), edited by Richard Garnett. More specialized collections of letters can be found in Thomas James Wise's editions of *Letters from Percy Bysshe Shelley to Elizabeth Hitchener*, published in London in 1890, and *Letters from Percy Bysshe Shelley to J.H. Leigh Hunt*, also published in London in 1894. Additional Shelley letters to Robert Southey were published by Edward Dowden in 1891.

Also in the nineteenth century, authors began to focus on specific areas of Shelley's life and his political and religious philosophies. These take the form of Thomas Jefferson Hogg's *Shelley at Oxford* (London: H. Colburn, 1832–33), Edward John Trelawny's *Recollections of the Last Days of Shelley and Byron* (London: E. Moxon, 1858), George Barnett Smith's *Shelley's Earlier Years* (London: Smith, Elder, 1875), Richard Garnett's *Shelley in Pall Mall* (London: Macmillan, 1860), Charles Sotheran's *Percy Bysshe Shelley as a Philosopher and Reformer* (New York: C.P. Somerby, 1876), Howard S. Pearson's *The Religious Beliefs of Shelley*, privately printed in Birmingham, England, in 1887, and Kineton Parkes's *Shelley's Faith, Its Development and Relativity*, also privately printed by Richard Clay and Sons in 1888.

The organization of the Shelley Society in March 1886 led to the publication of various author-related books, including H. Buxton Forman's *The Shelley Library* (1886), Robert Browning and W. Tyas Harden's *An Essay on Percy Bysshe Shelley* (1888), and the Shelley Society's *Papers* (1888–91), all published in London for the society by Reeves and Turner. There are also numerous studies of contemporary criticism of Shelley's works, including Newman I. White's *The Unextinguished Hearth: Shelley and His Contemporary Critics* (New York: Octagon Books, 1966), Theodore Redpath's *The Young Romantics and Critical Opinion, 1807–1824* (London: Harrap, 1973), Stephen C. Behrendt's *Shelley and His Audiences* (Lincoln: University of Nebraska Press, 1989), Kim Wheatley's *Shelley and His Readers: Beyond Paranoid Politics* (Columbia: University of Missouri Press, 1999), and Donald H. Reiman's indispensible multivolume collection, *The Romantics Reviewed* (New York: Garland Publishing, 1972). For an annotated listing of Shelley criticism between the years 1822–60, see Karsten Klejs Engelberg's *The Making of the Shelley Myth* (London: Mansell, 1988).

The following passages display the wide range of opinions held on Shelley during his life, as well as memoirs by those who knew him personally or knew people who were close to Shelley. William Hazlitt, who met Shelley a number of times, mainly at the residence of their mutual friend, Leigh Hunt, gives a brief physical description of Shelley and then questions Shelley's motivations and his penchant to "shock" his audience: "It would seem that he wished not so much to convince or inform as to shock the public by the tenor of his productions." Mary Shelley, in her 1824 "Preface" to Shelley's works, defended her husband's reputation and seeks to retrieve his good name from "the ungrateful world [which] did not feel his loss." Considering Shelley's current status as one of the greatest of the romantic poets, her comments are prophetic: "Hereafter men will lament that his transcendent powers of intellect were extinguished before they had bestowed on them their choicest treasures. To his friends his loss is irremediable: the wise, the brave, the gentle, is gone for ever!" In the passage by the Countess of Blessington, Byron apparently lamented Shelley's death, and Thomas De Quincey praises Shelley's "moral nature." Samuel Rogers states that "both in appearance and in manners Shelley was the perfect gentleman." Leigh Hunt remembers his close friend by relating an instance when Shelley tried to help the less fortunate, while Thomas Jefferson Hogg, Edward John Trelawny, and Thomas Love Peacock recall humorous stories about their friend. Trelawny also remarks that "Shelley's mental activity was infectious; he kept your brain in constant action" and goes on to say, "The truth was, Shelley loved everything better than himself."

HENRY CRABB ROBINSON (1817)

Henry Crabbe Robinson (1775–1867) was acquainted with many of the leading poets of his day, including William Wordsworth, Samuel Taylor Coleridge, and Charles Lamb. He resided in Germany from 1800–05 and became acquainted with Goethe and Schiller. He worked as a lawyer and as a foreign correspondent for *The Times*. In 1824, Robinson became one of the founders of the Athenaeum Club. He also contributed to the creation of University College, London. Robinson kept numerous accounts of his daily life but never published them. They were collected and edited by Dr. Thomas Sadler and published as *Diary, Reminiscences, and Correspondences of Henry Crabbe Robinson* in 1869.

I went to Godwin's. Mr. Shelley was there. I had never seen him before. His youth, and a resemblance to Southey, particularly in his voice, raised a pleasing impression, which was not altogether destroyed by his conversation, though it is vehement, and arrogant, and intolerant. He was very abusive towards Southey, whom he spoke of as having sold himself to the Court. And this he maintained with the usual party slang. His pension and his Laureateship, his early zeal and his recent virulence, are the proofs of gross corruption. On every topic but that of violent party feeling, the friends of Southey are under no difficulty in defending him. Shelley spoke of Wordsworth with less bitterness, but with an insinuation of his insincerity, &c.

—HENRY CRABB ROBINSON, *Diary*,
November 6, 1817, Boston:
Houghton, Mifflin, 1898, p. 369

Unsigned (1821)

We have spoken of Shelley's genius, and it is doubtless of a high order; but when we look at the purposes to which it is directed, and contemplate the infernal character of all its efforts, our souls revolt with tenfold horror at the energy it exhibits, and we feel as if one of the darkest of the fiends had been clothed with a human body, to enable him to gratify his enmity against the human race, and as if the supernatural atrocity of his hate were only heightened by his power to do injury. So strongly has this impression dwelt upon our minds, that we absolutely asked a friend who had seen this individual, to describe him to us—as if a cloven foot, or horn, or flames from the mouth, must have marked the external appearance of so bitter an enemy to mankind. We were almost disappointed to learn that the author was only a tall, boyish looking man, with eyes of unearthly brightness, and a countenance of the wildest cast: that he strode about with a hurried and impatient gait, and that a perturbed spirit seemed to preside over all his movements. It is not then in his outward semblance but in his inner man, that the explicit demon is seen; and it is a frightful supposition, that his own life may have been a fearful commentary upon his principles—principles, which in the balance of law and justice, happily deprived him of the superintendance of his infants, while they plunged an unfortunate wife and mother into ruin, prostitution, guilt, and suicide.

We are aware, that ordinary criticism has little or nothing to do with the personal conduct of authors; but when the most horrible doctrines are promulgated with appalling force, it is the duty of every man to expose, in every way, the abominations to which they irresistibly drive their odious professors. We declare against receiving our social impulses from a destroyer of every social virtue; our moral creed, from an incestuous wretch; or our religion, from an atheist, who denied God, and reviled the purest institutes of human philosophy and divine ordination, did such a demon exist.

—Unsigned, review of *Queen Mab*,
Literary Gazette, May 19, 1821, pp. 305–308

William Hazlitt "On Paradox and Commonplace" (1821–22)

Son of a Unitarian minister, William Hazlitt (1778–1830) was heavily influenced by his father's radical political views. He attended Hackney College in London from 1793–95. As a political reporter for the *Morning Chronicle*, Hazlitt expanded his talents to include essayist, reviewer, lecturer, and critic. Some of his most recognizable works include *The Round Table* and *Characters of Shakespeare's Plays*, both published in 1817, *Political Essays* (1819), *Table Talk* (1821), *The Spirit of the Age* (1825), and *The Life of Napoleon* (1828–30). Hazlitt, a close friend of Leigh Hunt, had ongoing disagreements with Shelley over his radical philosophies. The two met often at Hunt's residence and reportedly disagreed over Napoléon, of whom Hazlitt was a strong supporter. On several occasions, Hazlitt attacked Shelley in writing.

The author of the *Prometheus Unbound* . . . has a fire in his eye, a fever in his blood, a maggot in his brain, a hectic flutter in his speech, which mark out the philosophic fanatic. He is sanguine-complexioned, and shrill-voiced. As is often observable in the case of religious enthusiasts, there is a slenderness of constitutional stamina, which renders the flesh no match for the spirit. His bending, flexible form appears to take no strong hold of things, does not grapple with the world about him, but slides from it like a river—

> And in its liquid texture mortal wound Receives no more than can the fluid air.

The shock of accident, the weight of authority make no impression on his opinions, which retire like a feather, or rise from the encounter unhurt, through their own buoyancy. He is clogged by no dull system of realities, no earth-bound feelings, no rooted prejudices, by nothing that belongs to the mighty trunk and hard husk of nature and habit, but is drawn up by irresistible levity to the regions of mere speculation and fancy, to the sphere of air and fire, where his delighted spirit floats in 'seas of pearl and clouds of amber.' There is no caput mortuum of worn-out, thread-bare experience to serve as ballast to his mind; it is all volatile intellectual salt of tartar, that refuses to combine its evanescent, inflammable essence with any thing solid or any thing lasting. Bubbles are to him the only realities:—touch them, and they vanish. Curiosity is the only proper category of his mind, and though a man in knowledge, he is a child in feeling. Hence he puts every thing into a metaphysical crucible to judge of it himself and exhibit it to others as a subject of interesting experiment, without first making it over to the ordeal of his common sense or trying it on his heart. This faculty of speculating at random on all questions may in its overgrown and uninformed state do much mischief without intending it, like an overgrown child with the power of a man. Mr. Shelley has been accused of vanity—I think he is chargeable with extreme levity; but this levity is so great, that I do not believe he is sensible of its consequences. He strives to overturn all established creeds and systems: but this is in him an effect of constitution. He runs before the most extravagant opinions, but this is because he is held back by none of the merely mechanical checks of sympathy and habit. He tampers with all sorts of obnoxious subjects, but it is less because he is gratified with the rankness of the taint, than captivated with the intellectual phosphoric light they emit. It would seem that he wished not so much to convince or inform as to shock the public by the tenor of his productions, but I suspect he is more intent upon startling himself with his electrical experiments in morals and philosophy; and though they may scorch other people, they are to him harmless amusements, the coruscations of an Aurora Borealis, that 'play round the head, but do not reach the heart.' Still I could wish that he would put a stop to the incessant, alarming whirl of his Voltaic battery. With his zeal, his talent, and his fancy, he would do more good and less harm, if he were to give up his wilder theories, and if he took less pleasure in feeling his heart flutter in unison with the panic-struck apprehensions of his readers.

—WILLIAM HAZLITT, "On Paradox and Commonplace," *Table-Talk*, 1821–22

MARY SHELLEY "PREFACE" (1824)

Mary Wollstonecraft Godwin Shelley (1797–1851) was born in London, the daughter of William Godwin and Mary Wollstonecraft. After her mother's death less than two weeks after Mary's birth, William Godwin raised her and provided her with a solid, diverse education. From an early age, she was influenced by the reformist theories of her parents. Other radical writers of the day flocked to the Godwin home and, in November 1812, Mary met Percy Shelley during one of his many visits to her father's house. Despite Shelley's existing marriage to Harriet Westbrook, the next year he and Mary eloped to continental Europe. They returned to England in 1814, and Mary gave birth to a daughter in 1815. The infant lived twelve days. In 1816, William Shelley was born, and Percy and Mary were married later that year. They traveled to Geneva, where they met Byron, and later moved on to Italy. During this time, Mary Shelley completed *Frankenstein* (1818) and *Matilda* (1819, not published until 1959). In 1819, tragedy again struck, as their son William died of malaria. After the child's death, Mary fell into a deep depression and became increasingly distant from Percy. After Shelley's death in 1822, Mary took it upon herself to support herself and their son, Percy Florence, by writing. Her financial situation was straitened by Sir Timothy Shelley's reluctance to provide for Mary and her son. After she published Shelley's *Posthumous Poems* in 1824, Sir Timothy ended his financial support and would not allow Mary to publish any further works of Percy Shelley. She then supported herself by publishing short pieces in various periodicals and also continued her work on novels, including *Valperga* (1823), *The Last Man* (1826), *The Fortunes of Perkin Warbeck* (1830), *Lodore* (1835), and *Falkner* (1837). In 1839, she finally obtained permission to publish the four-volume *Poetical Works of Percy Bysshe Shelley*, which included sections of notes that served as an alternative to the Shelley biography she was not allowed to write, and the two-volume *Essays, Letters from Abroad, Translations and Fragments by Percy Bysshe Shelley*. Her final years were spent with Percy Florence and his wife, Jane Gibson St. John. With Sir Timothy's death in 1844, Mary began her biography of Percy, but it remained uncompleted because of her failing health and eventual death in 1851.

The comparative solitude in which Mr. Shelley lived was the occasion that he was personally known to few; and his fearless enthusiasm in the cause which he considered the most sacred upon earth, the improvement of the moral and physical state of mankind, was the chief reason why he, like other illustrious reformers, was pursued by hatred and calumny. No man was ever more devoted than he, to the endeavour of making those around him happy; no man ever possessed friends more unfeignedly attached to him. The ungrateful world did not feel his loss, and the gap it made seemed to close as quickly over his memory as the murderous sea above his living frame. Hereafter men will lament that his transcendent powers of intellect were extinguished before they had bestowed on them their choicest treasures. To his friends his loss is irremediable: the wise, the brave, the gentle, is gone for ever! He is to them as a bright vision, whose radiant track, left behind in the memory, is worth all the realities that society can afford. Before the critics contradict me, let them appeal to any one who had ever known him. To see him was to love him; and his presence, like Ithuriel's spear, was alone sufficient to disclose the falsehood of the tale which his enemies whispered in the ear of the ignorant world.

His life was spent in the contemplation of nature, in arduous study, or in acts of kindness and affection. He was an elegant scholar and a profound metaphysician; without possessing much scientific knowledge, he was unrivalled in the justness and extent of his observations on natural objects; he knew every plant by its name, and was familiar with the history and habits of every production of the earth; he could interpret without a fault each appearance in the sky; and the varied phenomena of heaven and earth filled him with deep emotion. He made his study and reading-room of the shadowed copse, the stream, the lake, and the waterfall. Ill health and continual pain preyed upon his powers; and the solitude in which we lived, particularly on our first arrival in Italy, although congenial to his feelings, must frequently have weighed upon his spirits; those beautiful and affecting "Lines written in Dejection near Naples" were composed at such an interval; but, when in health, his spirits were buoyant and youthful to an extraordinary degree.

Such was his love for nature, that every page of his poetry is associated, in the minds of his friends, with the loveliest scenes of the countries which he inhabited. In early life he visited the most beautiful parts of this country and Ireland. Afterwards the Alps of Switzerland became his inspirers. *Prometheus Unbound* was written among the deserted and flower-grown ruins of Rome; and, when he made his home under the Pisan hills, their roofless recesses harboured him as he composed the *Witch of Atlas, Adonais,* and *Hellas.* In the wild but beautiful Bay of Spezzia, the winds and waves which he loved became his playmates. His days were chiefly spent on the water; the management of his boat, its alterations and improvements, were his principal occupation. At night, when the unclouded moon shone on the calm sea, he often went alone in his little shallop to the rocky caves that bordered it, and, sitting beneath their shelter, wrote the *Triumph of Life,* the last of his productions. The beauty but strangeness of this lonely place, the refined pleasure which he felt in the companionship of a few selected friends, our entire sequestration from the rest of the world, all contributed to render this period of his life one of continued enjoyment. I am convinced that the two months we passed there were the happiest he had ever known: his health even rapidly improved, and he was never better than when I last saw him, full of spirits and joy, embark for Leghorn, that he might there welcome Leigh Hunt to Italy. I was to have accompanied him; but illness confined me to my room, and thus put the seal on my misfortune. His vessel bore out of sight with a favourable wind, and I remained awaiting his return by the breakers of that sea which was about to engulph him.

He spent a week at Pisa, employed in kind offices towards his friend, and enjoying with keen delight the renewal of their intercourse. He then embarked with Mr. Williams, the chosen and beloved sharer of his pleasures and of his fate, to return to us. We waited for them in vain; the sea by its restless moaning seemed to desire to inform us of what we would not learn:— but a veil may well be drawn over such misery. The real anguish of those moments transcended all the fictions that the most glowing imagination ever pourtrayed; our seclusion, the savage nature of the inhabitants of the surrounding villages, and our immediate vicinity to the troubled sea, combined to imbue with strange horror our days of uncertainty. The truth was at last known,—a truth that made our loved and lovely Italy appear a tomb, its sky a pall. Every heart echoed the deep lament, and my only consolation was in the praise and earnest love that each voice bestowed and each countenance demonstrated for him we had lost,—not, I fondly hope, for ever: his unearthly and elevated nature is a pledge of the continuation of his being, although in an altered form. Rome received his ashes; they are deposited beneath its weed-grown wall, and "the world's sole monument" is enriched by his remains.

—MARY SHELLEY, "Preface" to
Posthumous Poems, 1824

MARGUERITE, COUNTESS OF BLESSINGTON (1834)

Margaret Gardiner (1789–1849) was born near Clonmel, County Tipperary, Ireland. After two early failed relationships with military men that left her estranged from her family, she met Charles John Gardiner, the first earl of Blessington, whom she married in 1818. The couple settled in St. James's Square, and their residence became a popular gathering spot for the leading literati of the day. Margaret, who had changed her name to Marguerite, began her publishing career with the anonymous *Magic Lantern, or, Sketches of Scenes in the Metropolis*, followed by *Sketches and Fragments*, and *Journal of a Tour Through the Netherlands to Paris in 1821*, all published in 1822. Her acquaintance with Lord Byron in Genoa in 1823 led to one of her most famous works, *Journal of Conversations with Lord Byron*, which was first published in the *New Monthly Magazine* from July 1832 to December 1833 and then published as a book in 1834. She also published *Grace Cassidy, or, The Repealers* (1834), *The Victims of Society* (1837), and *The Governess* (1839), among other novels, as well as numerous stories in contemporary periodicals. She is perhaps best known today for her contributions to the annual gift books, which were popular during her time. Blessington edited and wrote items for *The Book of Beauty* and *The Keepsake*.

On looking out from the balcony this morning with Byron, I observed his countenance change, and an expression of deep sadness steal over it. After a few minutes' silence he pointed out to me a boat anchored to the right, as the one in which his friend Shelley went down, and he said the sight of it made him ill.—"You should have known Shelley," said Byron, "to feel how much I must regret him. He was the most gentle, most amiable, and *least* worldly-minded person I ever met; full of delicacy, disinterested beyond all other men, and possessing a degree of genius, joined to a simplicity, as rare as it is admirable. He had formed to himself a *beau ideal* of all that is fine, high-minded, and noble, and he acted up to this ideal even to the very letter. He had a most brilliant imagination, but a total want of worldly-wisdom. I have seen nothing like him, and never shall again, I am certain. I never can forget the night that his poor wife rushed into my room at Pisa, with a face pale as marble, and terror impressed on her brow, demanding, with all the tragic impetuosity of grief and alarm, where was her husband! Vain were all our efforts to calm her; a desperate sort of courage seemed to give her energy to confront the horrible truth that awaited her; it was the courage of despair. I have seen nothing in tragedy on the stage so powerful, or so affecting, as her appearance, and it often presents itself to my memory. I knew nothing then of the catastrophe, but the vividness of her terror communicated itself to me, and I feared the worst, which fears were, alas! too soon fearfully realized."

—Marguerite, Countess of Blessington,
Conversations of Lord Byron, 1834, 2nd edition,
1850, London, Henry Colburn, pp. 75–76

Thomas De Quincey "Notes on Gilfillan's Literary Portraits" (1845–46)

Born in Manchester to a textile merchant, Thomas De Quincey (1785–1859) left school in 1802. A year later, he entered Oxford University. De Quincey was friends with Coleridge and Wordsworth, and after leaving the university without a degree in 1808, he moved to Grasmere, where he became the inhabitant of Dove Cottage, the previous residence of William and Dorothy Wordsworth. From 1818 to 1819, De Quincey edited the *Westmorland Gazette*. In 1821, he would write the work that would win him lasting literary fame, *Confessions of an English Opium-Eater*, based largely on his own experiences using the drug. Following a move to Edinburgh in 1820, De Quincey continued writing and publishing, including several contributions to *Blackwood's*, the *London Magazine* (the first publisher of *Confessions*), and *Tait's Magazine*. Despite *Blackwood's* Tory leanings, De Quincey viewed Shelley's work favorably.

Shelley, it must be remembered, carried his irreligion to a point beyond all others. Of the darkest beings we are told that they "believe and tremble"; but

Shelley believed and *hated,* and his defiances were meant to show that he did *not* tremble. Yet, has he not the excuse of something like *monomania* upon this subject? I firmly believe it. But a superstition, old as the world, clings to the notion that words of deep meaning, uttered even by lunatics or by idiots, execute themselves, and that also, when uttered in presumption, they bring round their own retributive chastisements.

On the other hand, however shocked at Shelley's obstinate revolt from all religious sympathies with his fellow-men, no man is entitled to deny the admirable qualities of his moral nature, which were as striking as his genius. Many people remarked something seraphic in the expression of his features; and something seraphic there was in his nature. No man was better qualified to have loved Christianity; and to no man, resting under the shadow of that one darkness, would Christianity have said more glady—*talis cum sis, utinam noster esses!* Shelley would, from his earliest manhood, have sacrificed all that he possessed to any comprehensive purpose of good for the race of man. He dismissed all injuries and insults from his memory. He was the sincerest and the most truthful of human creatures. He was also the purest. If he denounced marriage as a vicious institution, *that* was but another phasis of the partial lunacy which affected him; for to no man were purity and fidelity more essential elements in his idea of real love. I agree, therefore, heartily with Mr. Gilfillan, in protesting against the thoughtless assertion of some writer in the *Edinburgh Review* that Shelley at all selected the story of his *Cenci* on account of its horrors, or that he has found pleasure in dwelling on those horrors. Far from it! Indeed, he has retreated so entirely from the most shocking feature of the story—viz. the incestuous violence of Cenci the father—as actually to leave it doubtful whether the murder were in punishment of the last outrage committed or in repulsion of a menace continually repeated. The true motive of the selection of such a story was—not its darkness, but (as Mr. Gilfillan, with so much penetration, perceives) the light which fights with the darkness: Shelley found the whole attraction of this dreadful tale in the angelic nature of Beatrice, as revealed in local traditions and in the portrait of her by Guido. Everybody who has read with understanding the *Wallenstein* of Schiller is aware of the repose and the divine relief arising upon a background of so much darkness, such a tumult of ruffians, bloody intriguers, and assassins, from the situation of the two lovers, Max Piccolomini and the Princess Thelka, both yearning so profoundly after peace, both so noble, both so young, and both destined to be so unhappy. The same fine relief, the same light shining in darkness, arises here from the touching beauty of Beatrice, from her noble aspirations after deliverance, from the remorse which reaches her in the midst of real innocence, from her meekness, and from the depth of her inexpressible affliction. Even the murder, even the parricide, though proceeding from herself, do but deepen that background of darkness which throws into fuller revelation the glory of that suffering face immortalised by Guido.

Something of a similar effect arises to myself when reviewing the general abstract of Shelley's life—so brief, so full of agitation, so full of strife. When one thinks of the early misery which he suffered, and of the insolent infidelity which, being yet so young, he wooed with a lover's passion, then the darkness of midnight begins to form a deep, impenetrable background, upon which the phantasmagoria of all that is to come may arrange itself in troubled phosphoric streams, and in sweeping processions of woe. Yet, again, when one recurs to his gracious nature, his fearlessness, his truth, his purity from all fleshliness of appetite, his freedom from vanity, his diffusive love and tenderness, suddenly out of the darkness reveals itself a morning of May, forests and thickets of roses advance to the foreground, and from the midst of them looks out "the eternal child," cleansed from his sorrow, radiant with joy, having power given him to forget the misery which he suffered, power given him to forget the misery that he caused, and leaning with his heart upon that dove-like faith against which his erring intellect had rebelled.

—THOMAS DE QUINCEY, "Notes on Gilfillan's
Literary Portraits," 1845–46, *Collected Writings*,
ed. David Masson, vol. 11, pp. 375–377

LEIGH HUNT (1850)

A longtime friend of Shelley's, (James Henry) Leigh Hunt (1784–1859) gained literary fame in 1808 when he founded the *Examiner*. As editor, he established the journal as one of the leading liberal voices in England. Hunt's radical views often riled his Tory adversaries, and after an article in the *Examiner* labeling the Prince Regent as a "libertine," Hunt was sent to prison. However, in keeping with his flamboyant nature, he quickly turned his cell into one of the most fashionable salons in England. Hunt was also responsible for supporting and encouraging the writing careers of Shelley, Byron, Keats, Hazlitt, and Lamb, a group that would come to be known as the Cockney School. Because of Hunt's support in the pages of the *Examiner*, Shelley quickly gained fame as a rising young poet, and Hunt remained a stalwart supporter of Shelley and frequently defended him against Tory attacks. The two men collaborated on a new magazine called *The Liberal*; however after Shelley's death and because of the lack of support by Byron, the magazine issued only four numbers. Hunt mentioned Shelley in his *Lord Byron and Some of His Contemporaries* (1828). In

1850, Hunt published his *Autobiography*, which includes reminiscences of his friendship with Shelley, as well as his thoughts on witnessing Shelley's cremation.

―――

I first saw Shelley during the early period of the *Examiner*, before its indictment on account of the Regent; but it was only for a few short visits, which did not produce intimacy. He was then a youth, not come to his full growth; very gentlemanly, earnestly gazing at every object that interested him, and quoting the Greek dramatists. Not long afterwards he married his first wife; and he subsequently wrote to me while I was in prison . . . I renewed the correspondence a year or two afterwards, during which period one of the earliest as well as most beautiful of his lyric poems, the "Hymn to Intellectual Beauty," had appeared in the *Examiner*. Meantime, he and his wife had parted; and now he re-appeared before me at Hampstead, in consequence of the calamity which I am about to mention. . . .

Shelley often came there to see me, sometimes to stop for several days. He delighted in the natural broken ground, and in the fresh air of the place, especially when the wind set in from the north-west, which used to give him an intoxication of animal spirits. Here also he swam his paper boats on the ponds, and delighted to play with my children, particularly with my eldest boy, the seriousness of whose imagination, and his susceptibility of a "grim" impression (a favourite epithet of Shelley's), highly interested him. He would play at "frightful creatures" with him, from which the other would snatch "a fearful joy", only begging him occasionally "not to do the horn", which was a way that Shelley had of screwing up his hair in front, to imitate a weapon of that sort. This was the boy (now the man of forty-eight, and himself a fine writer) to whom Lamb took such a liking on similar accounts, and addressed some charming verses as his "favourite child." I have already mentioned him during my imprisonment.

As an instance of Shelley's playfulness when he was in good spirits, he was once going to town with me in the Hampstead stage, when our only companion was an old lady, who sat silent and still after the English fashion. Shelley was fond of quoting a passage from *Richard the Second*, in the commencement of which the king, in the indulgence of his misery, exclaims—

> For Heaven's sake! let us sit upon the ground, And tell sad stories
> of the death of kings.

Shelley, who had been moved into the ebullition by something objectionable which he thought he saw in the face of our companion, startled her into a look of the most ludicrous astonishment, by suddenly calling this passage to mind, and, in his enthusiastic tone of voice, addressing me by name with the first two lines. "Hunt!" he exclaimed,—For Heaven's sake! let us sit upon the ground, And tell sad stories of the death of kings. The old lady looked on the coach floor, as if expecting to see us take our seats accordingly.

But here follows a graver and more characteristic anecdote. Shelley was not only anxious for the good of mankind in general. We have seen what he proposed on the subject of Reform in Parliament, and he was always very desirous of the national welfare. It was a moot point when he entered your room, whether he would begin with some half-pleasant, half-pensive joke, or quote something Greek, or ask some question about public affairs. He once came upon me at Hampstead, when I had not seen him for some time; and after grasping my hands with both his, in his usual fervent manner, he sat down, and looked at me very earnestly, with a deep, though not melancholy, interest in his face. We were sitting with our knees to the fire, to which we had been getting nearer and nearer, in the comfort of finding ourselves together. The pleasure of seeing him was my only feeling at the moment; and the air of domesticity about us was so complete, that I thought he was going to speak of some family matter, either his or my own, when he asked me, at the close of an intensity of pause, what was "the amount of the national debt."

I used to rally him on the apparent inconsequentiality of his manner upon those occasions, and he was always ready to carry on the jest, because he said that my laughter did not hinder my being in earnest.

But here follows a crowning anecdote, with which I shall close my recollections of him at this period. We shall meet him again in Italy, and there, alas! I shall have to relate events graver still.

I was returning home one night to Hampstead after the opera. As I approached the door, I heard strange and alarming shrieks, mixed with the voice of a man. The next day it was reported by the gossips that Mr. Shelley, no Christian (for it was he who was there), had brought some "very strange female" into the house, no better, of course, than she ought to be. The real Christian had puzzled them. Shelley, in coming to our house that night, had found a woman lying near the top of the hill, in fits. It was a fierce winter night, with snow upon the ground; and winter loses nothing of its fierceness at Hampstead. My friend, always the promptest as well as most pitying on these occasions, knocked at the first houses he could reach, in order to have the woman taken in. The invariable answer was, that they could not do it. He asked for an outhouse to put her in, while he went for a doctor. Impossible! In vain he assured them she was no impostor. They would not dispute the point with him; but doors were closed, and windows were shut down. Had he lit upon worthy Mr. Park, the philologist, that gentleman would assuredly have come, in spite of his Calvinism. But he lived too far off. Had he lit upon my friend Armitage Brown, who lived on another side of the Heath; or on his friend and neighbour Dilke; they would either of them have jumped up from amidst their books or their bed-clothes, and have gone out with him. But the paucity of Christians is astonishing, considering the number of them. Time flies; the poor woman is in convulsions; her son, a young man, lamenting over her. At last my friend sees a carriage driving up to a house at a little distance. The knock is given; the warm door opens; servants and lights pour forth. Now, thought he, is the time. He puts on his best address, which anybody might recognize for that of the highest gentleman as well as of an interesting individual, and plants himself in the way of an elderly person, who is stepping out of the carriage with his family. He tells his story. They only press on the faster. "Will you go and see her?" "No, sir; there's no necessity for that sort of thing, depend on it. Impostors swarm everywhere: the thing cannot be done; sir, your conduct is extraordinary." "Sir," cried Shelley, assuming a very different manner, and forcing the flourishing householder to stop out of astonishment, "I am sorry to say that *your* conduct is *not* extraordinary; and if my own seems to amaze you, I will tell you something which may amaze you a little more, and I hope will frighten you. It is such men as you who madden the spirits and the patience of the poor and wretched; and if ever a convulsion comes in this country (which is very probable), recollect what I tell you: —you will have your house, that you refuse to put the miserable woman into, burnt over your head." "God bless me, sir! Dear me, sir!" exclaimed the poor, frightened man, and fluttered into his mansion. The woman was then brought to our house, which was at some distance, and down a bleak path (it was in the Vale of Health); and Shelley and her son were obliged to hold her till the doctor could arrive. It appeared that she had been attending this son in London, on a criminal charge made against him, the agitation of which had thrown her into the fits on her return. The doctor said that she would have perished, had she laid there a short time longer. The next day my friend sent mother and son comfortably home to Hendon, where they were known, and whence they returned him thanks full of gratitude.

—Leigh Hunt, *Autobiography*, 1850, chapter 15

Walter Savage Landor "To Shelley" (1853)

Born to the wealthy doctor Walter Landor and the heiress Elizabeth Savage Landor, Walter Savage Landor (1775–1864) attended Rugby School and

Trinity College, Oxford. Early in his career, Landor worked at the *Morning Chronicle* and, in 1808, fought against Napoléon with the Spanish Army. After marrying Julia Thuillier in 1811, the Landors moved to France and then to Italy, where they lived until 1829. His works include the tragedy *Count Julian* (1812), *Imaginary Conversations* (1824–1853, a work that established Landor's literary reputation), *Last Fruit off an Old Tree* (1853), and *Heroic Idyls* (1863). In a life that spanned both the romantic and Victorian eras, Landor counted Robert Southey and Robert Browning as his friends. He died in Florence in 1864.

Shelley! whose song so sweet was sweetest here, We knew each other little; now I walk Along the same green path, along the shore Of Lerici, along the sandy plain Trending from Lucca to the Pisan pines, Under whose shadow scatter'd camels lie, The old and young, and rarer deer uplift Their knotty branches o'er high-feather'd fern. Regions of happiness! I greet ye well; Your solitudes, and not your cities, stay'd My steps among you; for with you alone Converst I, and with those ye bore of old. He who beholds the skies of Italy Sees ancient Rome reflected, sees beyond, Into more glorious Hellas, nurse of Gods And godlike men: dwarfs people other lands. Frown not, maternal England! thy weak child Kneels at thy feet and owns in shame a lie.

—Walter Savage Landor, "To Shelley," 1853

SAMUEL ROGERS (1855)

Born in Middlesex, Samuel Rogers (1763–1855) was educated in private schools and was encouraged by his father to enter the banking profession. He read widely and was influenced by the major eighteenth-century writers of the day: Johnson, Goldsmith, and Gray. Rogers began his own writing career contributing essays to the *Gentlemen's Magazine*. He later turned to poetry and published *An Ode to Superstition* in 1786. In 1792, he published *The Pleasures of Memory*, which became a popular work. Rogers was known throughout the literary world for his generosity and counted among his friends Charles Fox, Richard Brinsley Sheridan, William Wordsworth, Thomas Moore, and Lord Byron (who would later turn against Rogers). He continued to publish poetry, including *Human Life* (1819) and *Italy* (1822, 1828, republished with illustrations in 1830). Rogers became acquainted with Shelley while in Italy in 1822. In 1850, he declined the offer to become the nation's poet laureate.

One day, during dinner, at Pisa, when Shelley and Trelawney were with us, Byron chose to run down Shakespeare (for whom he, like Sheridan, either had, or pretended to have, little admiration). I said nothing. But Shelley immediately took up the defence of the great poet, and conducted it in his usual meek yet resolute manner, unmoved by the rude things with which Byron interrupted him,—"Oh, that's very well *for an atheist*, &c." Before meeting Shelley in Italy, I had seen him only once. It was at my own house in St. James's Place, where he called upon me,—introducing himself,—to request the loan of some money which he wished to present to Leigh Hunt; and he offered me a bond for it. Having numerous claims upon me at that time, I was obliged to refuse the loan. Both in appearance and in manners Shelley was the perfect gentleman.

—Samuel Rogers, *Table Talk*, circa 1855

Thomas Jefferson Hogg (1858)

Born in 1792, Hogg studied and practiced law but is perhaps best known today as Shelley's biographer. The two met as students at Oxford and collaborated on various letters critiquing Church doctrine. Hogg continued this religious critique in his novel, *Leonora*. For their collaboration on *The Necessity of Atheism*, both were expelled from Oxford in 1811. Their friendship was strained after Hogg attempted to court Harriet Westbrook, after her marriage to Shelley, but was renewed by Shelley in 1812. Through Shelley, Hogg met Leigh Hunt and Thomas Love Peacock. After Shelley began his affair with Mary Godwin, Hogg repeated his old behavior, this time declaring his love for Mary. Hogg and Shelley continued their correspondence throughout Shelley's life. Following Shelley's death, Hogg turned his romantic attentions to Jane Williams. Most scholars today consider it a partnership of necessity (Williams, who was the wife of John Edward Johnson and later the common law wife of Edward Williams could not legally marry Hogg) because Jane needed financial support for her children and was also pregnant with Hogg's child. In his remaining years, Hogg worked as a barrister in Northumberland and Berwick. In 1857, Shelley's son and his daughter-in-law requested that Hogg write a biography of Shelley, using both the letters in his possession, as well as many documents, journals, and letters that they held. The two-volume *Life of Percy Shelley* appeared in 1858; however, the Shelleys soon regretted their decision, when Hogg's work was found to contain deliberate misstatements, omitted information (including entire letters written by Shelley, as well as the identities of most of his correspondents), and altered versions of Shelley's letters. Sir Percy stopped the publication of additional volumes of the biography and requested that all documents pertaining to his father be returned to him. Because of adverse reaction to his *Life*, Hogg lost contact with several friends. He died in 1862.

I was surprised at the contrast between the general indifference of Shelley for the mechanical arts, and his intense admiration of a particular application of one of them the first time I noticed the latter peculiarity. During our residence at Oxford, I repaired to his rooms one morning at the accustomed hour, and I found a tailor with him. He had expected to receive a new coat on the preceding evening; it was not sent home, and he was mortified, I know not why, for he was commonly altogether indifferent about dress, and scarcely appeared to distinguish one coat from another. He was now standing erect in the middle of the room in his new blue coat, with all its glittering buttons, and to atone for the delay the tailor was loudly extolling the beauty of the cloth and the felicity of the fit; his eloquence had not been thrown away upon his customer, for never was man more easily persuaded than the master of persuasion. The man of thimbles applied to me to vouch his eulogies; I briefly assented to them. He withdrew, after some bows, and Shelley, snatching his hat, cried with shrill impatience:

'Let us go!'

'Do you mean to walk in the fields in your new coat?' I asked.

'Yes certainly', he answered, and we sallied forth.

We sauntered for a moderate space through lanes and byeways, until we reached a spot near a farm-house, where the frequent trampling of much cattle had rendered the road almost impassable, and deep with black mud; but by crossing the corner of a stack yard, from one gate to another, we could tread upon clean straw, and could wholly avoid the impure and impracticable slough.

We had nearly effected the brief and commodious transit, I was stretching forth my hand to open the gate that led us back into the lane, when a lean, brindled, and most ill-favoured mastiff, that had stolen upon us softly over the straw unheard, and without barking, seized Shelley suddenly by the skirts. I instantly kicked the animal in the ribs with so much force, that I felt for some days after the influence of his gaunt bones on my toe. The blow caused him to flinch towards the left, and Shelley, turning round quickly, planted a kick in his throat, which sent him away sprawling, and made him retire hastily among the stacks, and we then entered the lane. The fury of the mastiff, and the rapid turn, had torn the skirts of the new blue coat across the back, just about that part of the human loins which our tailors for some wise, but inscrutable purpose, are wont to adorn with two buttons. They were entirely severed from the body, except a narrow strip of cloth on the left side, and this Shelley presently rent asunder.

I never saw him so angry either before or since; he vowed that he would bring his pistols and shoot the dog, and that he would proceed at law against the owner. The fidelity of the dog towards his master is very beautiful in theory, and there is much to admire and to revere in this ancient and venerable alliance; but, in practice, the most unexceptionable dog is a nuisance to all mankind, except his master, at all times, and very often to him also, and a fierce surly dog is the enemy of the whole human race. The farm-yards, in many parts of England, are happily free from a pest that is formidable to everybody but thieves by profession; in other districts savage dogs abound, and in none so much, according to my experience, as in the vicinity of Oxford. The neighbourhood of a still more famous city, of Rome, is likewise infested by dogs, more lowering, more ferocious, and incomparable more powerful.

Shelley was proceeding home with rapid strides, bearing the skirts of his new coat on his left arm, to procure his pistols, that he might wreak his vengeance upon the offending dog. I disliked the race, but I did not desire to take an ignoble revenge upon the miserable individual.

'Let us try to fancy, Shelley', I said to him, as he was posting away in indignant silence, 'that we have been at Oxford, and have come back again, and that you have just laid the beast low—and what then?'

He was silent for some time, but I soon perceived, from the relaxation of his pace, that his anger had relaxed also.

At last he stopped short, and taking the skirts from his arm, spread them upon the hedge, stood gazing at them with a mournful aspect, sighed deeply, and after a few minutes continued his march.

'Would it not be better to take the skirts with us?' I inquired.

'No', he answered despondingly, 'let them remain as a spectacle for men and gods!'

We returned to Oxford, and made our way by back streets to our College. As we entered the gates, the officious scout remarked with astonishment Shelley's strange spenser, and asked for the skirts, that he might instantly carry the wreck to the tailor. Shelley answered, with his peculiarly pensive air, 'They are upon the hedge.'

The scout looked up at the clock, at Shelley, and through the gate into the street as it were at the same moment and with one eager glance, and would have run blindly in quest of them, but I drew the skirts from my pocket, and unfolded them, and he followed us to Shelley's rooms.

We were sitting there in the evening, at tea, when the tailor who had praised the coat so warmly in the morning, brought it back as fresh as ever and apparently uninjured. It had been fine-drawn; he showed how skilfully the wound had been healed, and he commended, at some length, the artist who had effected the cure. Shelley was astonished and delighted: had the tailor consumed the new blue coat in one of his crucibles, and suddenly raised it, by magical incantation, a fresh and purple Phoenix from the ashes, his admiration could hardly have been more vivid. It might be, in this instance, that his joy at the unexpected restoration of a coat, for which, although he was utterly indifferent to dress, he had, through some unaccountable caprice, conceived a fondness, gave force to his sympathy with art; but I have remarked in innumerable cases, where no personal motive could exist, that he was animated by all the ardour of a maker in witnessing the display of the creative energies.

—THOMAS JEFFERSON HOGG, *The Life of Percy Bysshe Shelley*, 1858, pp. 217–221

EDWARD JOHN TRELAWNY (1858)

A close friend of both Shelley and Byron who, by some accounts, planned to sail with Shelley the morning of his death, Edward John Trelawny (1792–1881) is the author of one of the earliest accounts of Shelley. His *Recollections of the Last Days of Shelley and Byron* was published in 1858 and revised in 1878 as *Records of Shelley, Byron, and the Author*, with the editorial help of William Michael Rossetti. Many critics argue over the validity of Trelawny's claims in the memoir, which included many factual changes in the revised edition. Byron even once remarked that Trelawny was unable to tell the truth. The discrepancies in both accounts also make it more difficult for scholars to have a clear understanding of the events immediately preceding Shelley's death. After the poet's demise, Trelawny tried to establish a closer relationship with Mary Shelley, intending marriage, but was refused by her. This most likely accounts for the harsh treatment Mary received in the 1878 edition. He also fell in love with Mary's step-sister, Clair Clairmont, but was likewise rejected by her. Trelawny claims to have been the last person to see Percy Shelley and Edward Williams in the boat before it capsized in the storm off Leghorn; he led the search for their bodies. He also was responsible for rescuing Shelley's heart from his funeral pyre, as well as moving Shelley's remains to the Protestant Cemetery. Trelawny died on August 13, 1881, and is buried in an adjoining plot to Shelley's in the Protestant Cemetery in Rome.

Shelley's mental activity was infectious; he kept your brain in constant action. Its effect on his comrade was very striking. Williams gave up all his accustomed sports for books, and the bettering of his mind; he had excellent natural ability; and the poet delighted to see the seeds he had sown, germinating. Shelley said he was the sparrow educating the young of the cuckoo. After a protracted labour Ned was delivered of a five-act play. Shelley was sanguine that his pupil would succeed as a dramatic writer. One morning I was in Mrs. Williams's drawing-room, by appointment, to hear Ned read an act of his drama. I sat with an aspect as caustic as a critic who was to decide his fate. Whilst thus intent Shelley stood before us with a most woeful expression.

Mrs. Williams started up, exclaiming: 'What's the matter, Percy?'

'Mary has threatened me.'

'Threatened you with what?'

He looked mysterious and too agitated to reply.

Mrs. Williams repeated: 'With what? to box your ears?'

'Oh, much worse than that; Mary says she will have a party; there are English singers here, the Sinclairs, and she will ask them, and every one she or you know—oh, the horror!' We all burst into a laugh except his friend Ned.

'It will kill me.'

'Music kill you!' said Mrs. Williams. 'Why, you have told me, you flatterer, that you loved music.'

'So I do. It's the company terrifies me. For pity go to Mary and intercede for me; I will submit to any other species of torture than that of being bored to death by idle ladies and gentlemen.'

After various devices it was resolved that Ned Williams should wait upon the lady—he being gifted with a silvery tongue, and sympathizing with the poet in his dislike of fine ladies—and see what he could do to avert the threatened invasion of the poet's solitude. Meanwhile, Shelley remained in a state of restless ecstasy; he could not even read or sit. Ned returned with a grave face; the poet stood as a criminal stands at the bar, whilst the solemn arbitrator of his fate decides it. 'The lady', commenced Ned, 'has set her heart on having a party, and will not be baulked'; but, seeing the poet's despair, he added: 'It is to be limited to those here assembled, and some of Count Gamba's family; and instead of a musical feast—as we have no souls—we are to have a dinner'. The poet hopped off, rejoicing, making a noise I should have thought whistling, but that he was ignorant of that accomplishment. . . .

To know an author personally is too often but to destroy the illusion created by his works; if you withdraw the veil of your idol's sanctuary, and see him in his night-cap, you discover a querulous old crone, a sour pedant, a supercilious coxcomb, a servile tuft-hunter, a saucy snob, or, at best, an ordinary mortal. Instead of the high-minded seeker after truth and abstract knowledge, with a nature too refined to bear the vulgarities of life, as we had imagined, we find him full of egotism and vanity, and eternally fretting and fuming about trifles. As a general rule, therefore, it is wise to avoid writers whose works amuse or delight you, for when you see them they will delight you no more. Shelley was a grand exception to this rule. To form a just idea of his poetry you should have witnessed his daily life; his words and actions best illustrated his writings. If his glorious conception of gods and men constituted an atheist, I am afraid all that listened were little better. Sometimes he would run through a great work on science, condense the author's laboured exposition, and by substituting simple words for the jargon of the schools, make the most abstruse subject transparent. The cynic Byron acknowledged him to be the best and ablest man he had ever known. The truth was, Shelley loved everything better than himself. Self-preservation is, they say, the first law of nature, with him it was the last; and the only pain he ever gave his friends arose from the utter indifference with which he treated everything concerning himself. I was bathing one day in a deep pool in the Arno, and astonished the poet by performing a series of aquatic gymnastics, which I had learnt from the natives of the South Seas. On my coming out, whilst dressing, Shelley said mournfully:

'Why can't I swim, it seems so very easy?'

I answered: 'Because you think you can't. If you determine, you will; take a header off this bank, and when you rise turn on your back, you will float like a duck; but you must reverse the arch in your spine, for it's now bent the wrong way'.

He doffed his jacket and trousers, kicked off his shoes an socks, and plunged in, and there he lay stretched out on the bottom like a conger eel, not making the least effort or struggle to save himself. He would have been drowned if I had not instantly fished him out. When he recovered his breath, he said:

'I always find the bottom of the well, and they say Truth lies there. In another minute I should have found it, and you would have found an empty shell. It is an easy way of getting rid of the body.'

'What would Mrs. Shelley have said to me if I had gone back with your empty cage?'

'Don't tell Mary—not a word!' he rejoined, and then continued: 'It's a great temptation; in another minute I might have been in another planet'.

'But as you always find the bottom,' I observed, "you might have sunk "deeper than did ever plummet sound"'.

'I am quite easy on that subject,' said the bard. 'Death is the veil, which those who live call life: they sleep, and it is lifted. Intelligence should be imperishable; the art of printing has made it so in this planet'.

'Do you believe in the immortality of the spirit?'

He continued: 'Certainly not; how can I? We know nothing; we have no evidence; we cannot express our inmost thoughts. They are incomprehensible even to ourselves'.

'Why', I asked, 'do you call yourself an atheist? it annihilates you in this world'.

'It is a word of abuse to stop discussion, a painted devil to frighten the foolish, a threat to intimidate the wise and good. I used it to express my abhorrence of superstition; I took up the word, as a knight took up a gauntlet, in defiance of injustice. The delusions of Christianity are fatal to genius and originality: they limit thought'.

Shelley's thirst for knowledge was unquenchable. He set to work on a book, or a pyramid of books; his eyes glistening with an energy as fierce as that of the most sordid gold-digger who works at a rock of quartz, crushing his way through all impediments, no grain of the pure ore escaping his eager scrutiny. I called on him one morning at ten, he was in his study with a German folio open, resting on the broad marble mantelpiece, over an old-fashioned fireplace, and with a dictionary in his hand. He always read standing if possible. He had promised overnight to go with me, but now begged me to let him off. I then rode to Leghorn, eleven or twelve miles distant, and passed the day there; on returning at six in the evening to dine with Mrs. Shelley and the Williams's, as I had engaged to do, I went into the poet's room and found him exactly in the position in which I had left him in the morning, but looking pale and exhausted.

'Well,' I said, 'have you found it?'

Shutting the book and going to the window, he replied: 'No, I have lost it': with a deep sigh: 'I have lost a day'.

'Cheer up, my lad, and come to dinner.'

Putting his long fingers through his masses of wild tangled hair, he answered faintly: 'You go, I have dined—late eating don't do for me'.

'What is this?' I asked as I was going out of the room, pointing to one of his bookshelves with a plate containing bread and cold meat on it.

'That,' colouring, 'why that must be my dinner. It's very foolish; I thought I had eaten it.'

Saying I was determined that he should for once have a regular meal, I lugged him into the dining-room, but he brought a book with him and read more than he ate. He seldom ate at stated periods, but only when hungry—and then like the birds, if he saw something edible lying about—but the cupboards of literary ladies are like Mother Hubbard's, bare. His drink was water, or tea if he could get it, bread was literally his staff of life; other things he thought superfluous. An Italian who knew his way of life, not believing it possible that any human being would live as Shelley did, unless compelled by poverty, was astonished when he was told the amount of his income, and thought he was defrauded or grossly ignorant of the value of money. He, therefore, made a proposition which much amused the poet, that he, the friendly Italian, would undertake for ten thousand crowns a year to keep Shelley like a grand seigneur, to provide his table with luxuries, his house with attendants, a carriage and opera box for my lady, besides adorning his person after the most approved Parisian style. Mrs. Shelley's toilette was not included in the wily Italian's estimates. The fact was, Shelley stinted himself to bare necessaries, and then often lavished the money, saved by unprecedented self-denial, on selfish fellows who denied themselves nothing; such as the great philosopher had in his eye, when he said: 'It is the nature of extreme self-lovers, as they will set a house on fire, an it were only to roast their own eggs.' Byron on our voyage to Greece, talking of England, after commenting on his own wrongs, said: 'And Shelley too, the best and most benevolent of men; they hooted him out of his country like a mad dog, for questioning a dogma. Man is the same rancorous beast now that he was from the beginning, and if the Christ they profess to worship reappeared they would again crucify him'.

—EDWARD JOHN TRELAWNY, *Recollections of the Last Days of Shelley and Byron*, 1858, reprinted 1906, London, Henry Frowde, pp. 36–42

JANE SHELLEY
"PREFACE BY THE EDITOR" (1859)

Before her marriage to Percy Florence Shelley in 1848 (which several people believe was arranged by Mary Shelley), Jane Gibson was married to Charles Robert St. John, the son of the third Viscount Bolingbroke. Sir

Percy Florence and Lady Jane had no children, and the couple remained dedicated to Mary Shelley until her death in 1851. The responsibility of faithfully retelling her in-laws' history fell to Lady Jane Shelley because of the false accounts of Shelley's life given by Shelley's cousin Thomas Medwin in his *The Life of Percy Bysshe Shelley* (London: Newby, 1847) and by Shelley's longtime friend Thomas Jefferson Hogg in *The Life of Percy Bysshe Shelley* (London: E. Moxon, 1858). Lady Shelley was also responsible for the reinterment of William Godwin and Mary Wollstonecraft, moving their remains from St. Pancras to St. Peter's Church in Bournemouth and burying Mary Shelley alongside her parents in 1851. Sir Percy Florence and Lady Jane lived at Boscombe Manor in Bournemouth and are buried with the family at St. Peter's (dying in 1889 and 1892, respectively). The couple also collaborated on *Shelley and Mary*, published for private circulation by Chiswick Press in 1882.

Had it been left entirely to the uninfluenced wishes of Sir Percy Shelley and myself, we should have preferred that their publication of the materials for a life of Shelley which we possess should have been postponed to a later period of our lives; but, as we had recently noticed, both in French and English magazines, many papers on Shelley, all taking for their text Captain Medwin's *Life of the Poet* (a book full of errors), and as other biographies had been issued, written by those who had no means of ascertaining the truth, we were anxious that the numerous misstatements which had gone forth should be corrected.

For this purpose, we placed the documents in our possession at the disposal of a gentleman whose literary habits and early knowledge of the poet seemed to point him out as the most fitting person for bringing them to the notice of the public. It was clearly understood, however, that out wishes and feelings should be consulted in all the details.

We saw the book for the first time when it was given to the world. It was impossible to imagine beforehand that from such materials a book could have been produced which has astonished and shocked those who have the greatest right to form an opinion on the character of Shelley; and it was with the most painful feelings if dismay that we perused what we could only look upon as a fantastic caricature, going forth to the public with my apparent sanction,—for it was dedicated to myself.

Our feelings of duty to the memory of Shelley left us no other alternative than to withdraw the materials which we had originally entrusted to his early friend, and which we could not but consider had been strangely misused; and to take upon ourselves the task of laying them before the public, connected only by as slight a thread of narrative as would suffice to make them intelligible to the reader.

I have condensed as much as possible the details of the early period of Shelley's life, for I am aware that a great many of them have already appeared in print. The repetition of some, however, was considered advisable, since it is very probable that this volume will be read by many who have not seen, nor are likely to see, any other work giving an account of the writings and actions of Shelley.

I little expected that this task would devolve on me; and I am fully sensible how unequal I am to its proper fulfillment. To give a truthful statement of long-distorted facts, and to clear away the mist in which the misrepresentations of foes and professed friends have obscured the memory of Shelley, have been my only objects. My labours have been greatly assisted by the help of an intimate and valued friend of ours. Shelley, and by Mr. Edmund Ollier, whose father (the publisher of Shelley's works) at once freely offered me the use of some most interesting letters written to himself. I regret to say that this gentleman died while the present work was passing through the printer's hands.

It is needless to say that the authenticity of all the documents contained in this volume is beyond question; but the public would do well to receive with the utmost caution all letters purporting to be by Shelley, which have not some indisputable warrant.*

The art of forging letters purporting to be relics of men of literary celebrity, and therefore apparently possessing a commercial value, has been brought to a rare perfection by those who have made Mr. Shelley's handwriting the object of their imitation. Within the last fourteen years, on no less than three occasions, have forged letters been presented to our family for purchase. In December, 1851, Sir Percy Shelley and the late Mr. Moxon bought several letters, all of which proved to be forgeries, though, on the most careful inspection, we could scarcely detect any difference between these and the originals; for some were exact copies of documents in our possession. The water-mark on the paper was generally, though not always, the mark appropriate to the date; and the amount of ingenuity exercised was the most extraordinary. Mr. Moxon published what he had bought in a small volume, but recalled the work shortly afterwards, on discovering that some of the letters had been manufactured from articles and reviews, written long after Shelley's death.

The letter to Lord Ellenborough has never before been published; but I regard it as too extraordinary a production for a youth of eighteen to feel myself justified in suppressing it.

The fragmentary Essay on Christianity, published at the end of this volume, was found among Shelley's papers in the imperfect state in which it is now produced.

* Those printed in the work to which allusion has already been made have never, for the most part, been seen by any other person than the author of that work; and the erasures which he has already made in them, together with the arrangement of their paragraphs, render them of doubtful value, however authentic may be the originals which that gentleman asserts he possesses.

—Jane Shelley,
"Preface by the Editor," *Shelley Memorials*, 1859, pp. iii–vi

Thomas Love Peacock "Memoirs of Percy Bysshe Shelley: Part II" (1860)

The son of a London glass merchant, Peacock (1785–1866) was largely self-educated, having left school at age thirteen. In 1812, Peacock met Shelley through their mutual acquaintance, the bookseller Thomas Hookham. Despite tensions surrounding the death of Harriet Shelley, of whom Peacock was also a friend, the two remained close throughout Shelley's life, and Peacock often advised Shelley on his manuscripts. The two also had one of the most famous literary debates of the nineteenth century, as Peacock's 1820 *The Four Ages of Poetry* led to Shelley posthumously published rebuttal in *A Defense of Poetry*. Peacock shared responsibilities as Shelley's literary executor with Byron until the latter's death in 1824, and Peacock was also charged with establishing financial support for Mary and Percy Florence from Shelley's generally disapproving father, Sir Timothy Shelley. As a result of his close friendship with Shelley, several of Peacock's works portray thinly disguised Shelleyan characters, particularly *Headlong Hall* (1815), *Melincourt* (1817), and *Nightmare Abbey* (1818).

There was not much comedy in Shelley's life; but his antipathy to 'acquaintance' led to incidents of some drollery. Amongst the persons who called on him at Bishopgate, was one whom he tried hard to get rid of, but who forced himself on him in every possible manner. He saw him at a distance one day, as he was walking down Egham Hill, and instantly jumped through a hedge, ran across a field, and laid himself down in a dry ditch. Some men and women, who were haymaking in the field, ran up to see what was the matter, when he said to them, 'Go away, go away: don't you see it's a bailiff?' On which they left him, and he escaped discovery.

After he had settled himself at Marlow, he was in want of a music-master to attend a lady staying in his house, and I inquired for one at Maidenhead. Having found one, I requested that he would call on Mr. Shelley. One morning Shelley rushed into my house in great trepidation, saying: 'Barricade the doors; give orders that you are not at home. He is in the town.' He passed the whole day with me, and we sat in expectation that the knocker or the bell would announce the unwelcome visitor; but the evening fell on the unfulfilled fear. He then ventured home. It turned out that the name of the music-master very nearly resembled in sound the name of the obnoxious gentleman; and when Shelley's man opened the library door and said, 'Mr., sir,' Shelley, who caught the name as that of his *Monsieur Tonson*, exclaimed, 'I would just as soon see the devil!', sprang up from his chair, jumped out of the window, ran across the lawn, climbed over the garden-fence, and came round to me by a back-path: when we entrenched ourselves for a day's siege. We often laughed afterwards at the thought of what must have been his man's astonishment at seeing his master, on the announcement of the musician, disappear so instantaneously through the window, with the exclamation, 'I would just as soon see the devil!' and in what way he could explain to the musician that his master was so suddenly 'not at home'.

Shelley, when he did laugh, laughed heartily, the more so as what he considered the perversions of comedy excited not his laughter but his indignation, although such disgusting outrages on taste and feeling as the burlesques by which the stage is now disgraced had not then been perpetrated. The ludicrous, when it neither offended good feeling, nor perverted moral judgement, necessarily presented itself to him with greater force.

—THOMAS LOVE PEACOCK,
"Memoirs of Percy Bysshe Shelley: Part II,"
Fraser's Magazine, January 1860, pp. 92–109

WILLIAM CORY "SHELLEY AT ETON" (1886)

William Johnson Cory (1823–1892) was born William Johnson in Devon and educated at Eton and King's College, Cambridge, earning his B.A. in 1845. In that same year, he took the position of assistant master at Eton, a post he would retain for twenty-six years. While there, he wrote pamphlets on education, including *Eton Reform, Eton Reform II,* and the posthumously published *Hints for Eton Masters.* He also published collections of poetry.

Amid a scandal involving possible intimate relationships with students at Eton, Johnson resigned and changed his last name to Cory. His letters and journals were collected and published in 1897.

I was one day in South Meadow, a field adjoining the well-known Brocas, and used in winter as a football and hurdle-race ground by Eton boys. I was with Mr. Edward Coleridge, nephew of the poet, brother, ten years younger, of Sir John Coleridge the judge, who was at one time editor of the *Quarterly Review*. We were standing near one of the pollard willows which line many a ditch in the Thames Valley. It was a wretched tree, with only half a trunk; it was black inside. Mr. Coleridge said to me: "This is the tree that Shelley blew up with gunpowder: that was his last bit of naughtiness at school." He went on to say that his brother John was of Shelley's standing at Eton, and used to say that he never joined in teasing Shelley, but he did not know any one else that did not tease him: there used to be a "Shelley-bait" every day about noon: the boys hunted Shelley up the street; he was known for not wearing strings in his shoes.

I got nothing else out of Mr. Edward Coleridge on this subject. I have stayed in Sir John Coleridge's house and heard him talk of literary men that he had known, but not of Shelley.

If I remember right, it is in the *Revolt of Islam* that the poet writes of the cruelties of schools. I believe boys suffer more from mortification than from rough usage, and that a life may be poisoned by insulting notice taken of deficiencies in dress. I consider the shoe-strings in this case not to have been trifles.

Shelley's use of gunpowder reminds me of the tradition which seems to be well known, of his amusing his companions with a frictional electric machine in his own room, and charging the door-handle, and failing in his dutiful attempt to warn his tutor, Mr. Bethell, against opening the door when he came to stop the noise caused by the electric shocks. This Mr. Bethell was, to boys, famous for inefficiency as a classical teacher; but he was a true gentleman, a cadet of a good Yorkshire family; he was known *to men* as a modest but stedfast vindicator of the "statutable rights of the scholars" of Eton College against the iniquitous usurpations of the Provost and Fellows. He was a just and also a courteous man.

In a recent paper on Eton Buildings in the *Saturday Review* it was erroneously said that the picturesque house standing in a corner of the playing-fields was the house in which Shelley boarded. This house, which the governors lately wished, but no longer wish, to destroy, was twenty years ago graced by the presence of a boy-poet who had a singular influence, and died the most poetical of deaths at the age of nineteen; but the house in which Shelley gave Bethell the shock was a lower house standing at the corner of the road, and it was taken down about twenty-five years ago; it was next door to a shop well known fifty years back—a shop kept by some elderly women called Spire or Spires. At the end of the college precinct, or that part of the village of Eton in which the school-boys lived, there was at the same time a shop kept by people named Towers. I dare say Shelley may, like me, have heard Gray's line quoted thus: "Ye ancient Spires, ye distant Towers," a "derangement" of epithets made to suit the visitor coming from Salt Hill, not from Windsor.

—WILLIAM CORY, "Shelley at Eton," *The Shelley Society's Note-Book*, part I, 1888, pp. 14–15

WILLIAM GRAHAM (1898)

In the "Introduction" to *Last Links with Byron, Shelley, and Keats*, published in London in 1898, William Graham states, "I have been asked to write an introduction to the following articles, appearing now for the first time in book form—an introduction rendered doubtless to some extent necessary by the numerous controversies they aroused, and the great attention they claimed at the time of their appearance in monthly review form" (ix). According to the rest of the "Introduction," the majority of these "controversies" arose from Shelley's characterization in the book: "but certain weak-minded Shelley enthusiasts took another tack and attacked me bitterly for what they were pleased to consider the vilification of their idol's character" (x). Graham does not indeed help himself among the so-called "Shelley enthusiasts" with his claims that "with all his genius and obscurity, [Shelley] was, to my mind, not particularly complex" (xi), as well as by his description of "the dogmas and dreams of Shelley's overwrought brain" (xi). Graham frequently goes on the defensive elsewhere and dismisses his critics, saying:

> There is, however, a clique which had made what Mr Rudyard Kipling would term "a little tin god" of Shelley; and the members of this absurd coterie, in affecting to raise their idol above ordinary human nature, really do his fame nothing but great disservice in depicting him as what that very caustic and sarcastic lady Miss Clairmont termed "an inspired idiot." Some of these good people seemed to have contracted the idea that Shelley is their exclusive property, and have attacked me furiously because in my pages he has been set before the world as a reasonable being, upon the authority of one who had the best opportunities of judging. (xii)

He later addresses the rumors surrounding Shelley's sexual relationship with Clairmont:

> Another strange complaint made by Shelleyolaters against me is that I have given it to be understood that Shelley and his sister-in-law ... were—what shall I say? How shall I put it? My natural timidity comes in here again—well, not absolutely on such platonic terms as might be desired. I beg your pardon, my dear Shelleyolaters; I made no such rash statement or insinuation. I have been merely the humble reporter of what Miss Clairmont said to me; and if you choose to put nasty ungenerous constructions upon this, that is no fault of mine. (xv)

Such statements give a hint as to the reception of Graham's work among Shelley's followers near the turn of the century. The first two chapters of the book, which include his interviews with the reclusive Claire Clairmont in Italy in 1878, were first published in the November 1893 and January 1894 issues of *Nineteenth Century*.

Of the Shelley *entourage*, Miss (Jane) Clermont appeared to like her namesake, Jane Williams, the best. A charming woman, she said; both Shelleys were devoted to her. This lady's grief at the terrible news of the disaster which involved the death of both Shelley and her husband, Miss Clermont described to me as pitiful to witness.

"All you ladies," I remarked, "seem to have formed a kind of adoring circle around Shelley."

"Yes," she said; "Shelley had an irresistible attraction for all women; his nature was so pure and noble; the tone of his poetry whenever a woman is mentioned is of an almost unearthly purity. Instead of holding with Byron that woman is inferior to man, he looked up to woman as something higher and nobler. Many of his poems express this feeling most forcibly.

The desire of the Moth for the Star, The desire of the night for the morrow, The devotion to something afar."

"I can imagine Shelley," I said, "almost like a pretty girl himself. I am sure that poetical epistle to Maria Gisborne is most ladylike."

She replied indignantly, "Not at all; there was no lack of manliness about Shelley. He was utterly without any sense of fear; always in the open air, yachting, or taking strong physical exertion. He was the finest walker of any man of the Byron-Shelley *clique*, and could tire out almost any of the others."

—WILLIAM GRAHAM, *Last Links with Byron, Shelley, and Keats*, 1898, pp. 56–57

GENERAL

The following passages discuss the changing nature of Shelley's literary reputation during the course of the nineteenth century. Most critics rebuke Shelley for being too complex and elusive in his poetry, while a few recognize a true poetic gift in his writings. Several authors complain that they simply do not understand Shelley, his poetry, or his ideas. Others, such as John Wilson and David Macbeth Moir, put forward the idea of a "mad Shelley" who was completely out of touch with anything earthly and rational. Wilson claims that "[Shelley was] scarcely in his right mind," while Moir asserts that Shelley negatively influenced contemporary poets and adds that "his mind was diseased." Still others lament the loss of poetic potential at Shelley's early death. Critics such as Thomas Babington Macaulay and Leigh Hunt speculate on what Shelley could have offered had he lived. As Hunt says, "for assuredly, had he lived, he would have been the greatest dramatic writer since the days of Elizabeth." There are scholarly disagreements throughout these passages. For example, Edgar Allan Poe calls Shelley "profoundly original" and believes "he was at all times sincere," while Moir is adamant in his notion that Shelley "bowed down to the idols of affectation and false taste."

Several authors compare Shelley to his contemporaries. Walter Bagehot, Hippolyte Taine, and John Addington Symonds compare him favorably to Wordsworth. For Bagehot, it was Shelley's otherworldly qualities in his poetry that set him apart from Wordsworth. After briefly quoting the latter poet and discussing Wordsworth's love of the simple and plain, Bagehot goes on to say:

> Shelley had nothing of this. The essential feelings he hoped to change; the eternal facts he struggled to remove. Nothing in human life to him was inevitable or fixed; he fancied he could alter it all.... Wordsworth describes this earth as we know it, with all its peculiarities; where there

are moors and hills, where the lichen grows, where the slate-rock juts out. Shelley describes the universe. He rushes away among the stars; this earth is an assortment of imagery, he uses it to deck some unknown planet. He scorns 'the smallest light that twinkles in the heavens.' His theme is the vast, the infinite, the immeasurable. He is not of our home, nor homely; he describes not our world, but that which is common to all worlds—the Platonic idea of a world.

Likewise, Symonds declares, "In none of Shelley's greatest contemporaries was the lyrical faculty so paramount; ... In range of power he was the loftiest and the most spontaneous singer of our language. Not only did he write the best lyrics, but the best tragedy, the best translations, and the best familiar poems of his century." George Saintsbury goes back even further into English literary history with his conclusion that "there are two English poets, and two only, in whom the purely poetical attraction, exclusive of and sufficient without all others, is supreme, and these two are Spenser and Shelley."

In this section, authors also recognize Shelley's improving literary reputation, which grew increasingly throughout the nineteenth and early twentieth centuries. Edmund Gosse remarks that Shelley can best be appreciated by the young but refuses to "rank" him among other poets. He notes that the public is finally appreciating Shelley as he deserves and tells his 1892 audience: "we are gathered here as a sign that the period of prejudice is over, that England is in sympathy at last with her beautiful wayward child, understands his great language, and is reconciled to his harmonious ministry." In similar appreciation, Algernon Charles Swinburne, in his poem celebrating Shelley's shift from being criticized to embraced by his readers, describes the poet as "One whom hate once hailed as now love hails by name."

CHARLES LAMB (1824)

Charles Lamb (1775–1834) was born in London, the son of John and Elizabeth Lamb. He left school in 1789 and supported himself as a clerk in the East India Company until 1824. Lamb's life was met with hardships, perhaps most famously when his mentally ill sister Mary stabbed their mother to death in 1796, leading to Charles's role as her guardian. Plagued by continuing poverty, alcoholism, and the responsibilities of caring for his sister, Lamb, nonetheless, embarked on a writing career, which began when four of his poems were published in *Poems on Various Subjects* (1796) by his friend Coleridge. His plays, including *John Woodvil* (1802) and *Mr. H* (1806), met with little public success. Charles and Mary collaborated on

Tales from Shakespeare (1807), *Mrs. Leicester's School* (1809), and *Poetry for Children* (1809), all of which were interpreted for children. Lamb became increasingly respected as a literary critic and also published essays in Leigh Hunt's *Reflector*. This experience led to the popular series of essays (1820–25) for the *London Magazine*, which Lamb wrote using the pseudonym "Elia." Lamb, like William Hazlitt, disagreed with many of Shelley's philosophies, although the two were frequently brought into contact by their mutual friend, Leigh Hunt.

I can no more understand Shelley than you can. His poetry is 'thin sewn with profit or delight.' Yet I must point to your notice a sonnet conceivd and expressed with a witty delicacy. It is that addressed to one who hated him, but who could not persuade him to hate *him* again. His coyness to the other's passion (for hate demands a return as much as Love, and starves without it) is most arch and pleasant. Pray, like it very much.

For his theories and nostrums they are oracular enough, but I either comprehend 'em not, or there is miching malice and mischief in 'em. But for the most part ringing with their own emptiness. Hazlitt said well of 'em—Many are wiser and better for reading Shakspeare, but nobody was ever wiser or better for reading Shelley.

—Charles Lamb, letter to
Bernard Barton, August 17, 1824

Thomas Lovell Beddoes (1824)

Beddoes (1803–49) is best known for *The Bride's Tragedy* (1822), and *Death's Jest-Book; or, The Fool's Tragedy* (1850), which was published a year after his death. Many of his other plays, including *The Second Brother, Torrismond, Love's Arrow Poisoned*, and *The Last Man*, survive only in fragmentary form. He attended Pembroke College, Oxford, but was eventually expelled in 1829 because of excessive drinking and unruly behavior. In 1824, he served as guarantor to the publishing firm of Procter, Kelsall, and Waller for the edition of Shelley's *Posthumous Poems*.

The disappearance of Shelley from the world, seems, like the tropical setting of that luminary (*aside* I hate that word) to which his poetical genius can alone be compared with reference to the companions of his day, to have been followed by instant darkness and owl-season; whether the vociferous Darley is to be the comet, or tender fullfaced L. E. L. the milk-and-watery moon of our darkness, are questions for the astrologers: if I were the literary weather-guesser for 1825 I would safely prognosticate fog, rain, blight in due succession for it's dullard months.

—THOMAS LOVELL BEDDOES, letter to
Thomas Forbes Kelsall, August 25, 1824

JOHN WILSON "PREFACE" (1826)

Born in Scotland, John Wilson (1785–1854) was an acquaintance of several of the Lake Poets and is best known for his collaboration, under the pseudonym "Christopher North," with John Gibson Lockhart in *Blackwood's Edinburgh Magazine*. From 1822–35, Wilson authored many of the periodical's popular *Noctes Ambrosianae*. He wrote poetry, including *The Isle of Palms* (1812) and *The City and the Plague* (1816), as well as prose works, such as *Lights and Shadows of Scottish Life* (1822), *Trials of Margaret Lyndsay* (1823), and *The Foresters* (1825). Wilson also had a long career as professor of moral philosophy at Edinburgh from 1820 to 1851.

Percy Bysshe Shelley was a man of far superior powers to Keats. He had many of the faculties of a great poet. He was however, we verily believe it now, scarcely in his right mind. His errors in private life had been great, but not *prodigious*, as the *Quarterly Review* represented them; and they brought evils along with them which Shelley bore with fortitude and patience. He had many noble qualities; and thus gifted, thus erring, and thus an outcast, we spoke of him with kindness and with praise. He felt, and gratefully acknowledged both; and was proud to know, that some of the articles in our work on his poetry, were written by a poet whose genius he admired and imitated.

—JOHN WILSON, "Preface" to
Blackwood's Edinburgh Magazine,
January–June 1826, p. xxix

THOMAS BABINGTON MACAULAY "JOHN BUNYAN" (1830)

Thomas Babington Macaulay (1800–59) attended Trinity College, Cambridge, and earned his B.A. in 1822. While indifferently pursuing a law degree, Macaulay began writing for *Knight's Quarterly Magazine* in 1824 and the *Edinburgh Review* in 1825. He was elected to Parliament in 1830 and soon after began the speeches that would make him famous. In

1832, he was elected official spokesman on Indian matters in the House of Commons, and two years later he was appointed to the Indian council. He remained in India until 1838 and, during that time, had influence on several laws pertaining to British imperial rule, ranging from Indian higher education to the penal code. In 1839, he began writing his *History of England*. That same year he was elected to Parliament from Edinburgh and during his eight-year tenure argued for changes in copyright laws, voted against the Chartist petition, and entered the debate on the Maynooth grant. In 1843, Longman published Macaulay's periodical contributions, which sold well. This success continued in 1848, when the first two volumes of *History of England* were published, selling three thousand copies in the first two weeks and quickly going through second and third editions. The third and fourth volumes of the *History* appeared in 1855. Macaulay died in 1859 and is buried in Poets' Corner in Westminster Abbey.

Some of the metaphysical and ethical theories of Shelley were certainly most absurd and pernicious. But we doubt whether any modern poet has possessed in an equal degree some of the highest qualities of the great ancient masters. The words bard and inspiration, which seem so cold and affected when applied to other modern writers, have a perfect propriety when applied to him. He was not an author, but a bard. His poetry seems not to have been an art, but an inspiration. Had he lived to the full age of man, he might not improbably have given to the world some great work of the very highest rank in design and execution.

—Thomas Babington Macaulay,
"John Bunyan," 1830, *Critical, Historical, and Miscellaneous Essays*, 1860, vol. 2, p. 257

Robert Browning (1833)

Sun-treader—life and light be thine for ever; Thou art gone from us—years go by—and spring Gladdens, and the young earth is beautiful, Yet thy songs come not—other bards arise, But none like thee—they stand—thy majesties, Like mighty works which tell some Spirit there Hath sat regardless of neglect and scorn, Till, its long task completed, it hath risen And left us, never to return: and all Rush in to peer and praise when all in vain. The air seems bright with thy past presence yet, But thou art still for me, as thou hast been When I have stood with thee, as on a throne With all thy dim creations gathered round Like mountains,—and I felt of mould like them, And creatures of my own were mixed with them, Like things half-lived, catching and giving life. But thou art still for me, who have adored, Tho' single, panting but to hear thy name, Which I believed a spell to me alone, Scarce deeming thou wert as a star to men.

—ROBERT BROWNING, *Pauline*, 1833, lines 151–171

LEIGH HUNT (1844)

The finest poetry of Shelley is so mixed up with moral and political speculation, that I found it impossible to give more than the following extracts, in accordance with the purely poetical design of the present volume. Of the poetry of reflection and tragic pathos, he has abundance; but even such fanciful productions as the "Sensitive Plant" and the *Witch of Atlas* are full of metaphysics, and would require a commentary of explanation. The short pieces and passages, however, before us, are so beautiful, that they may well stand as the representatives of the whole powers of his mind in the region of pure poetry. In sweetness (and not even there in passages) the "Ode to the Skylark" is inferior only to Coleridge,—in rapturous passion to no man. It is like the bird it sings,—enthusiastic, enchanting, profuse, continuous, and alone,—small, but filling the heavens. One of the triumphs of poetry is to associate its remembrance with the beauties of nature. There are probably no lovers of Homer and Shakspeare, who, when looking at the moon, do not often call to mind the descriptions in the eighth book of the *Iliad* and the fifth act of the *Merchant of Venice*. The nightingale (in England) may be said to have belonged exclusively to Milton, till a dying young poet of our own day partook of the honour by the production of his exquisite Ode: and notwithstanding Shakspeare's lark singing "at heaven's gate," the longer effusion of Shelley will be identified with thoughts of the bird hereafter, in the minds of all who are susceptible of its beauty. What a pity he did not live to produce a hundred such! or to mingle briefer lyrics, as beautiful as Shakspeare's, with tragedies which Shakspeare himself might have welcomed! for assuredly, had he lived, he would have been the greatest dramatic writer since the days of Elizabeth, if indeed he has not abundantly proved himself such in his tragedy of the *Cenci*. Unfortunately, in his indignation against every conceivable form of oppression, he took a subject for that play too much resembling one which Shakspeare had taken in his youth, and still more unsuitable to the stage; otherwise, besides grandeur and terror, there are things in it lovely as heart can worship; and the author showed himself able to draw both men and women, whose names would have become "familiar in our mouths as household words." The utmost might of gentleness, and of the sweet habitudes of domestic affection, was never more balmily impressed through the tears of the reader, than in the unique and divine close of that dreadful tragedy. Its loveliness, being that of the highest reason, is superior to the madness of all the crime that has preceded it, and leaves nature in a state of reconcilement with her ordinary course. The daughter, who is going forth with her mother to execution, utters these final words:— Give yourself *no unnecessary pain,* My dear Lord Cardinal. Here, mother, tie My girdle for me, and bind up this hair In any simple knot. Ay, that does well; *And yours, I see, is coming down. How often Have we done this for one another! now We shall not do it any more.* My Lord, We are quite ready. Well,—*'t is very well.* The force of simplicity and moral sweetness cannot go further than this. But in general, if Coleridge is the sweetest of our poets, Shelley is at once the most ethereal and most gorgeous; the one who has clothed his thoughts in draperies of the most evanescent and most magnificent words and imagery. Not Milton himself is more learned in Grecisms, or nicer in etymological propriety; and nobody, throughout, has a style so Orphic and primaeval. His poetry is as full of mountains, seas, and skies, of light, and darkness, and the seasons, and all the elements of our being, as if Nature herself had written it, with the creation and its hopes newly cast around her; not, it must be confessed, without too indiscriminate a mixture of great and small, and a want of sufficient shade,—a certain chaotic brilliancy, "dark with excess of light." Shelley (in the verses to a Lady with a Guitar) might well call himself Ariel. All the more enjoying part of his poetry is Ariel,—the "delicate" yet powerful "spirit," jealous of restraint, yet able to serve; living in the elements and the flowers; treading the "ooze of the salt deep," and running "on the sharp wind of the north;" feeling for creatures unlike himself; "flaming amazement" on them too, and singing exquisitest songs. Alas! and he suffered for years, as Ariel did in the cloven pine: but now he is out of it, and serving the purposes of Beneficence with a calmness befitting his knowledge and his love.

—LEIGH HUNT, *Imagination and Fancy*, 1844

EDGAR ALLAN POE
"ELIZABETH BARRETT BROWNING" (1845)

Born in Boston, Edgar Allan Poe (1809–49) attended the University of Virginia but, because of increasing debt, had to leave before earning a degree. He was also dismissed from West Point and lived briefly in New York City before settling with his aunt in Baltimore. Poe had published a few poetry collections without notice before he began concentrating on short stories, a literary form at which he excelled. His first stories appeared in the *Saturday Courier* of Philadelphia and the *Saturday Visitor* of Baltimore. In 1835, Poe became editor of *The Southern Literary Messenger*, published in Richmond. He worked as both an author of original stories and poems as well as a literary critic. Over the next decade, Poe edited *Burton's Gentlemen's Magazine*, *Graham's Magazine* and the *Broadway Journal*. He married his cousin, Virginia Clemm, in 1836, and the two remained together until her death in 1847. In September 1849, Poe died of a brain lesion in Baltimore. He is best known today for his gothic stories, including "The Tell-Tale Heart," "The Black Cat," "The Cask of Amontillado," and "The Fall of the House of Usher." His poems include "To Helen," "Lenore," and "The Raven." His novel, *The Narrative of Arthur Gordon Pym* (1838), is a gothic classic. Poe is also noted as one of the first detective story writers, for his trilogy featuring the French detective Inspector Dupin: "The Murders in the Rue Morgue," "The Purloined Letter," and "The Mystery of Marie Roget."

If ever mortal "wreaked his thoughts upon expression" it was Shelley. If ever poet sang (as a bird sings)—impulsively—earnestly—with utter abandonment—to himself solely—and for the mere joy of his own song—that poet was the author of the "Sensitive Plant." Of Art—beyond that which is the inalienable instinct of Genius—he either had little or disdained all. He *really* disdained that Rule which is the emanation from Law, because his own soul was law in itself. His rhapsodies are but the rough notes—the stenographic memoranda of poems—memoranda which, because they were all-sufficient for his own intelligence, he cared not to be at the trouble of transcribing in full for mankind. In his whole life he wrought not thoroughly out a single conception. For this reason it is that he is the most fatiguing of poets. Yet he wearies in having done too little, rather than too much; what seems in him the diffuseness of one idea, is the conglomerate concision of many;—and this concision it is which renders him obscure. With such a man, to imitate was out of the question; it would have answered no purpose—for he spoke to his own spirit alone, which would have comprehended no alien tongue;—he was, therefore, profoundly original. His quaintness arose from intuitive perception of that truth to which Lord Verulam alone has given distinct voice:—"There is no exquisite beauty which has not some strangeness in its proportion." But whether obscure, original, or quaint, he was at all times sincere. He had no *affectations*.

—Edgar Allan Poe, "Elizabeth Barrett Browning," 1845, *Essays and Reviews*, ed. G. R. Thompson, 1984, pp. 139–140

GEORGE MEREDITH
"THE POETRY OF SHELLEY" (1851)

Victorian novelist and journalist George Meredith (1828–1909) was born in Portsmouth, England. Some of his well-known novels include *The Ordeal of Richard Feveral* (1859), *Evan Harrington* (1860), *Rhoda Fleming* (1865), *The Egoist* (1879), and *Diana of the Crossways* (1885). He also established his name as a poet with the publication of *Modern Love* (1862), a sonnet cycle many consider to be based on his own failed first marriage. Meredith was also a noted journalist, working for such magazines as the *Westminster Review, Morning Post, Pall Mall Gazette, Fortnightly Review*, and the *Graphic*. He also founded the *Monthly Observer*.

See'st thou a Skylark whose glistening winglets ascending
Quiver like pulses beneath the melodious dawn?
Deep in the heart-yearning distance of heaven it flutters—
Wisdom and beauty and love are the treasures it brings down at eve.

—GEORGE MEREDITH, "The Poetry of Shelley,"
1851, *The Poetical Works of George Meredith*,
New York: Scribner's, 1912, p. 15

DAVID MACBETH MOIR (1851)

David Moir (1798–1851) was born in Musselburgh, Midlothian, and earned his medical degree from the University of Edinburgh in 1816. Though he was well known in literary circles of his day, Moir also continued his study of medicine, which included the published works, *Outlines of the Ancient History of Medicine* (1831) and *Practical Observations on Malignant Cholera* (1832), the latter being influenced by his firsthand observations of the cholera epidemic that affected his hometown of Musselburgh that same year. He published in *Scots Magazine, Edinburgh Magazine, Edinburgh Literary Gazette, Fraser's Magazine*, and *Blackwood's Magazine*. While writing for *Blackwood's*, Moir established himself under the pseudonym "Delta" and became part of the inner circle of *Blackwood*'s writers, including William Maginn, John Wilson, and founder William Blackwood. Moir's other published works include *The Bombardment of Algiers, and Other Poems* (1816) and the lighthearted *Autobiography of Mansie Waugh* (1828).

Such were Shelley's powers, when legitimately directed, but unfortunately it is rarely that he thus writes; and a much higher place has been claimed for the great mass of his verse than it seems to me to be at all entitled to. Gorgeous, graceful, and subtle qualities it indeed invariably possesses, and no one can be more ready to admit them than I am; but he had only a section of the essential properties necessary to constitute a master in the art. The finest poetry is that (whatever critical coteries may assert to the contrary, and it is exactly the same with painting and sculpture) which is most patient to the general understanding, and hence to the approval or disapproval of the common sense of mankind. We have only to try the productions of Shakspeare, of Milton, of Dryden, of Pope, of Gray and Collins, of Scott, Burns, Campbell, and Byron, indeed, of any truly great writer whatever in any language, by this standard, to be convinced that such must be the case. Verse that will not stand being read aloud before a jury of common-sense men, is—and you may rely upon the test—wanting in some great essential quality. It is here that the bulk of the poetry of Shelley—and not of him only, but of most of those who have succeeded him in his track as poets—is, when weighed in the balance, found wanting. And why? Because these writers have left the highways of truth and nature, and, seeking the bye-lanes, have there, mistaking the uncommon for the valuable, bowed down to the idols of affectation and false taste.

I make this remark here, because I think that Shelley had much to do in the indoctrinating of those principles which have mainly guided our poetical aspirants of late years—sadly to their own disadvantage and the public disappointment. Shelley was undoubtedly a man of genius—of very high genius—but of a peculiar and unhealthy kind. It is needless to disguise the fact, and it accounts for all—his mind was diseased: he never knew, even from boyhood, what it was to breathe the atmosphere of healthy life, to have the *mens sana in corpore sano.* His sensibilities were over acute; his morality was thoroughly morbid; his metaphysical speculations illogical, incongruous, incomprehensible—alike baseless and objectless. The suns and systems of his universe were mere nebulae; his continents were a chaos of dead matter; his oceans "a world of waters, and without a shore." For the law of gravitation—that law which was to preserve the planets in their courses—he substituted some undemonstrable dreamlike reflection of a dream, which he termed intellectual beauty. Life, according to him, was a phantasmagorial pictured vision, mere colours on the sunset clouds; and earth a globe hung on nothing—self-governing, yet, strange to say, without laws. It is gratuitous absurdity to call his mystical speculations a search after truth; they are no such thing; and are as little worth the attention of reasoning and responsible man as the heterogeneous reveries of nightmare. They are a mere flaring up in the face of all that Revelation has mercifully disclosed, and all that sober Reason has confirmed. Shelley's faith was a pure psychological negation, and cannot be confuted, simply because it asserts nothing; and, under the childish idea that all the crime, guilt, and misery of the world resulted from—what?—not the depravity of individuals, but from the very means, civil and ecclesiastical, by which these, in all ages and nations, have been at least attempted to be controlled, he seemed to take an insane delight in selecting, for poetical illustration, subjects utterly loathsome and repulsive; and which religion and morality, the virtuous and the pure, the whole natural heart and spirit of upright man, either rises up in rebellion against, or shrinks back from instinctively, and with horror.

—David Macbeth Moir,
*Sketches of the Poetical Literature of the
Past Half-Century*, 1851, pp. 227–229

Walter Bagehot
"Percy Bysshe Shelley" (1856)

Bagehot (1826–77) was born in Somerset, England. He attended University College, London, and briefly worked in the legal field before pursuing a career in banking. Bagehot found the work tedious and turned increasingly to writing, although he continued his work with the Bristol and London branches of Stuckey's Bank and later in his career published the economic study *Lombard Street* (1873). He cofounded, along with his friend Richard Holt Hutton, the *National Review* and wrote several items for the journal, including essays on the leading writers of the eighteenth and nineteenth centuries. Bagehot later wrote for *The Economist* and *Fortnightly Review*, producing a series of essays in the latter journal that would become *The English Constitution* (1867), his best-known book.

The excellence of Shelley does not, however, extend equally over the whole domain of lyrical poetry. That species of art may be divided—not perhaps with the accuracy of science, but with enough for the rough purposes of popular criticism—into the human and the abstract. The sphere of the former is of course the actual life, passions, and actions of real men,— such are the war-songs of rude nations especially; in that early age there is no subject for art but natural life and primitive passion. At a later time, when from the deposit of the *debris* of a hundred philosophies, a large number of half-personified abstractions are part of the familiar thoughts and language of all mankind, there are new objects to excite the feelings,—we might even say there are new feelings to be excited; the rough substance of original passion is sublimated and attenuated till we hardly recognise its identity. Ordinarily and in most minds the emotion loses in this process its intensity or much of it; but this is not universal. In some peculiar minds it is possible to find an almost dizzy intensity of excitement called forth by some fancied abstraction, remote altogether from the eyes and senses of men. The love-lyric in its simplest form is probably the most intense expression of primitive passion; yet not in those lyrics where such intensity is the greatest,—in those of Burns, for example,—is the passion so dizzy, bewildering, and bewildered, as in the *Epipsychidion* of Shelley, the passion of which never came into the real world at all, was only a fiction founded on fact, and was wholly—and even Shelley felt it—inconsistent with the inevitable conditions of ordinary existence. In this point of view, and especially also taking account of his peculiar religious opinions, it is remarkable that Shelley should have taken extreme delight in the Bible as a composition. He is the least biblical of poets. The whole, inevitable, essential conditions of real life—the whole of its plain, natural joys and sorrows—are described in the Jewish literature as they are described nowhere else. Very often they are assumed rather than delineated; and the brief assumption is more effective than the most elaborate description. There is none of the delicate sentiment and enhancing sympathy which a modern writer would think necessary: the inexorable facts are dwelt on with a stern humanity, which recognises human feeling though intent on something above it. Of all modern poets, Wordsworth shares the most in this peculiarity; perhaps he is the only recent one who has it at all. He knew the hills beneath whose shade 'the generations are prepared:' Much did he see of men,

> Their passions and their feelings; chiefly those
> Essential and eternal in the heart,
> That mid the simpler forms of rural life
> Exist more simple in their elements,
> And speak a plainer language.

Shelley has nothing of this. The essential feelings he hoped to change; the eternal facts he struggled to remove. Nothing in human life to him was inevitable or fixed; he fancied he could alter it all. His sphere is the 'unconditioned;' he floats away into an imaginary Elysium or an expected Utopia; beautiful and excellent, of course, but having nothing in common with the absolute laws of the present world. Even in the description of mere nature the difference may be noted. Wordsworth describes this earth as we know it, with all its peculiarities; where there are moors and hills, where the lichen grows, where the slate-rock juts out. Shelley describes the universe. He rushes away among the stars; this earth is an assortment of imagery, he uses it to deck some unknown planet. He scorns 'the smallest light that twinkles in the heavens.' His theme is the vast, the infinite, the immeasurable. He is not of our home, nor homely; he describes not our world, but that which is common to all worlds—the Platonic idea of a world. Where it can, his genius soars from the concrete and real into the unknown, the indefinite, and the void.

—WALTER BAGEHOT, "Percy Bysshe Shelley,"
1856, *Collected Works,* ed. Norman
St. John-Stevas, vol. 1, pp. 466–468

HIPPOLYTE TAINE (1871)

Hippolyte Adolphe Taine (1828–93) was born in the town of Vouziers in the Ardennes Mountains. He went to Paris for his education and attended the Collège Bourbon and the École Normale Supérieure from 1841 to 1851, where he studied English literature and culture. He published several works on England, including the four-volume *Historie de la littérature anglaise* (1863–64) and *Notes sur l'Angleterre* (1871). During his few short visits to England, Taine became acquainted with such political and literary figures as Lord Houghton, Richard Monckton Milnes, Benjamin Jowett, and Mark Pattison.

When a certain phasis of the human intelligence comes to light, it does so from all sides; there is no part where it does not appear, no instincts which it does not renew. It enters simultaneously the two opposite camps, and seems to undo with one hand what it has made with the other. If it is, as it was formerly, the oratorical style, we find it at the same time in the service of cynical misanthropy, and in that of decorous humanity, in Swift and Addison. If it is, as now, the philosophical spirit, it produces at once conservative harangues and socialistic Utopias, Wordsworth and Shelley. The latter, one of the greatest poets of the age, son of a rich baronet, beautiful as an angel, of extraordinary precocity, sweet, generous, tender, overflowing with all the gifts of heart, mind, birth, and fortune, marred his life, as it were, wantonly, by introducing into his conduct the enthusiastic imagination which he should have kept for his verses. From his birth he had "the vision" of sublime beauty and happiness, and the contemplation of the ideal world set him in arms against the actual. Having refused at Eton to be the fag of the big boys, he was treated by the boys and their masters with a revolting cruelty; suffered himself to be made a martyr, refused to obey, and, falling back into forbidden studies, began to form the most immoderate and most poetical dreams. He judged society by the oppression which he underwent, and man by the generosity which he felt in himself; thought that man was good, and society bad, and that it was only necessary to suppress established institutions to make earth "a paradise." He became a republican, a communist, preached fraternity, love, even abstinence from flesh, and as a means the abolition of kings, priests, and God. Fancy the indignation which such ideas roused in a society so obstinately attached to established order—so intolerant, in which, above the conservative and religious instincts, Cant spoke like a master. He was expelled from the university; his father refused to see him; the Lord Chancellor, by a decree, took from him, as being unworthy, the custody of his two children; finally, he was obliged to quit England. I forgot to say that at eighteen he married a girl of mean birth; that they had been separated, that she committed suicide, that he had undermined his health by his excitement and sufferings, and that to the end of his life he was nervous or sick. Is not this the life of a genuine poet? Eyes fixed on the splendid apparitions with which he peopled space, he went through the world not seeing the highroad, stumbling over the stones of the roadside. That knowledge of life which most poets have in common with novelists, he had not. Seldom has a mind been seen in which thought soared in loftier regions, and more far from actual things. When he tried to create characters and events—in *Queen Mab,* in *Alastor,* in *The Revolt of Islam,* in *Prometheus*—he only produced unsubstantial phantoms. Once only, in the *Cenci,* did he inspire a living figure worthy of Webster or old Ford; but in some sort in spite of himself, and because in it the sentiments were so unheard of and so strained that they suited superhuman conceptions. Elsewhere his world is throughout beyond our own. The laws of life are suspended or transformed. We move in this world between heaven and earth, in abstraction, dreamland, symbolism: the beings float in it like those fantastic figures which we see in the clouds, and which alternately undulate and change form capriciously, in their robes of snow and gold.

For souls thus constituted, the great consolation is nature. They are too fairly sensitive to find a distraction in the spectacle and picture of human passions. Shelley instinctively avoided it; this sight reopened his own wounds. He was happier in the woods, at the seaside, in contemplation of grand landscapes. The rocks, clouds, and meadows, which to ordinary eyes seem dull and insensible, are to a wide sympathy, living and divine existences, which are an agreeable change from men. No virgin smile is so charming as that of the dawn, nor any joy more triumphant than that of the ocean when its waves sleep and tremble, as far as the eye can see, under the prodigal splendor of heaven. At this sight the heart rises unwittingly to the sentiments of ancient legends, and the poet perceives in the inexhaustible bloom of things the peaceful soul of the great mother by whom everything grows and is supported. Shelley spent most of his life in the open air, especially in his boat; first on the Thames, then on the Lake of Geneva, then on the Arno, and in the Italian waters. He loved desert and solitary places, where man enjoys the pleasure of believing infinite what he sees, infinite as his soul. And such was this wide ocean, and this shore more barren than its waves. This love was a deep Germanic instinct, which, allied to pagan emotions, produced his poetry, pantheistic and yet pensive, almost Greek and yet English, in which fancy plays like a foolish, dreamy child, with the splendid skein of forms and colors. A cloud, a plant, a sunrise,—these are his characters: they were those of the primitive poets, when they took the lightning for a bird of fire, and the clouds for the flocks of heaven. But what a secret ardor beyond these splendid images, and how we feel the heat of the furnace beyond the colored phantoms, which it sets afloat over the horizon! Has any one since Shakespeare and Spenser lighted on such tender and such grand ecstasies? Has any one painted so magnificently the cloud which watches by night in the sky, enveloping in its net the swarm of golden bees, the stars:

> The sanguine sunrise, with his meteor eyes, And his burning plumes outspread. Leaps on the back of my sailing rack, When the morning star shines dead . . That orbed maiden, with white fire laden, Whom mortals call the moon, Glides glimmering o'er my fleece-like floor, By the midnight breezes strewn.

—HIPPOLYTE TAINE, *History of English Literature,* tr. H. Van Laun, 1871, book 4, ch. 1

JOHN ADDINGTON SYMONDS (1878)

Symonds was born in 1840 in Bristol, England. He attended the Harrow School and later Balliol College, Oxford. He published biographies of Shelley, Ben Jonson, and Sir Phillip Sydney. Symonds also translated Michaelangelo's sonnets and wrote a seven-volume history of *The Renaissance in Italy* (1875–1886). In *A Problem in Greek Ethics* and *A Problem in Modern Ethics*, Symonds collaborated with Havelock Ellis and Carl Ulrichs to address male homosexuality. He later worked with Ellis again on the controversial *Sexual Inversion*. These studies were influenced by Symonds's lifelong struggle to come to terms with his own homosexual impulses. Plagued by ill health his entire life, Symonds died in Rome in 1893.

As a poet, Shelley contributed a new quality to English literature—a quality of ideality, freedom, and spiritual audacity, which severe critics of other nations think we lack. Byron's daring is in a different region: his elemental worldliness and pungent satire do not liberate our energies, or cheer us with new hopes and splendid vistas. Wordsworth, the very antithesis to Shelley in his reverent accord with institutions, suits our meditative mood, sustains us with a sound philosophy, and braces us by healthy contact with the Nature he so dearly loved. But in Wordsworth there is none of Shelley's magnetism. What remains of permanent value in Coleridge's poetry—such work as *Christabel*, the *Ancient Mariner*, or *Kubla Khan*—is a product of pure artistic fancy, tempered by the author's mysticism. Keats, true and sacred poet as he was, loved Nature with a somewhat sensuous devotion. She was for him a mistress rather than a Diotima; nor did he share the prophetic fire which burns in Shelley's verse, quite apart from the direct enunciation of his favourite tenets. In none of Shelley's greatest contemporaries was the lyrical faculty so paramount; and whether we consider his minor songs, his odes, or his more complicated choral dramas, we acknowledge that he was the loftiest and the most spontaneous singer of our language. In range of power he was also conspicuous above the rest. Not only did he write the best lyrics, but the best tragedy, the best translations, and the best familiar poems of his century. As a satirist and humourist, I cannot place him so high as some of his admirers do; and the purely polemical portions of his poems, those in which he puts forth his antagonism to tyrants and religions and custom in all its myriad forms, seem to me to degenerate at intervals into poor rhetoric.

—John Addington Symonds,
Shelley, 1878, pp. 183–184

Dante Gabriel Rossetti "Percy Bysshe Shelley" (1880–81)

Rossetti (1828–82) was born in London and was a sibling of Christina and William Michael Rossetti. He, along with the members of William Holman Hunt's studio, founded the Pre-Raphaelite Brotherhood in 1848. His most famous paintings include *Dante's Dream* (1856) and *Beata Beatrix* (1863). Later in life, Rossetti focused increasingly on his poetry, collected in *Ballads and Sonnets* (1881). His poem to Shelley is included in his sonnets to "Five English Poets": Chatterton, Blake, Coleridge, Keats, and Shelley.

(Inscription for the Couch, Still Preserved, on which He Passed the Last Night of His Life.)
'Twixt those twin worlds,—the world of Sleep, which gave
No dream to warn,—the tidal world of Death,
Which the earth's sea, as the earth, replenisheth,—
Shelley, Song's orient sun, to breast the wave,
Rose from this couch that morn. Ah! did he brave
Only the sea?—or did man's deed of hell
Engulph his bark 'mid mists impenetrable?
No eye discerned, nor any power might save.

When that mist cleared, O Shelley? what dread veil
Was rent for thee, to whom far-darkling Truth
Reigned sovereign guide through thy brief ageless youth?
Was the Truth *thy* Truth, Shelley!—Hush? All-Hail,
Past doubt, thou gav'st it; and in Truth's bright sphere
Art first of praisers, being most praised here.

—DANTE GABRIEL ROSSETTI,
"Percy Bysshe Shelley," from
"Five English Poets," 1880–81

HARRIET MONROE
"WITH A COPY OF SHELLEY" (1889)

Monroe (1860–1936) was born in Chicago, Illinois. She founded *Poetry: A Magazine of Verse* in 1912 and served as its editor until 1936. Earlier in her career she worked as a reporter and critic for the *Chicago Tribune* and the *New York Tribune*. In 1893, she was commissioned to write "The Columbian Ode," the official poem of the Chicago world's fair. She published several volumes of poetry and verse drama, including *Valeria and Other Poems* (1891), *After All* (1900), *The Passing Show: Five Modern Plays in Verse* (1903), *The Dance of Seasons* (1911), *You and I* (1914), and *The Difference and Other Poems* (1924). Along with Alice Corbin Henderson, she edited *The New Poetry: An Anthology* (1917, revised and expanded in 1923). Her autobiography, *A Poet's Life: Seventy Years in a Changing World,* was published in 1938. Monroe's first published work, "With a Copy of Shelley," appeared in the *Century* magazine in 1889.

Behold I send thee to the heights of song,
My brother! Let thine eyes awake as clear
As morning dew, within whose glowing sphere
Is mirrored half a world; and listen long,
Till in thine ears, famished to keenness, throng
The bugles of the soul, till far and near
Silence grows populous, and wind and mere
Are phantom-choked with voices. Then be strong—
Then halt not till thou seest the beacons flare
Souls mad for truth have lit from peak to peak.
Haste on to breathe the intoxicating air—
Wine to the brave and poison to the weak—
Far in the blue where angels' feet have trod,
Where earth is one with heaven and man with God.

—Harriet Monroe, "With a Copy of Shelley,"
Century Magazine, Dec. 1889, p. 313

Edmund Gosse "Shelley in 1892" (1892)

Born in London, Sir Edmund William Gosse (1849–1928) was known for his talent as a literary historian, biographer, and translator. He became librarian at the House of Lords in 1904, and his 1907 autobiography, *Father and Son: A Study of Two Temperaments*, remains a classic in the genre. Gosse translated Ibsen and wrote influential biographies of John Donne, William Congreve, Thomas Gray, and Algernon Charles Swinburne, among others.

In Shelley we see a certain type of revolutionist, born out of due time, and directed to the bloodless field of literature. The same week that saw the downfall of La Fayette saw the birth of Shelley, and we might believe the one to be an incarnation of the hopes of the other. Each was an aristocrat, born with a passionate ambition to play a great part in the service of humanity; in neither was there found that admixture of the earthly which is needful for sustained success in practical life. Had Shelley taken part in active affairs, his will and his enthusiasm must have broken, like waves, against the coarser type of revolutionist, against the Dantons and the Robespierres. Like La Fayette, Shelley was intoxicated with virtue and glory; he was chivalrous, inflammable, and sentimental. Happily for us, and for the world, he was not thrown into a position where these beautiful qualities could be displayed only to be shattered like a dome of many-coloured glass. He was the not unfamiliar figure of revolutionary times, the *grand seigneur* enamoured of democracy. But he was much more than this; as Mr. Swinburne said long ago, Shelley "was born a son and soldier of light, an archangel winged and weaponed for angel's work." Let us attempt to discover what sort of prophecy it was that he blew through his golden trumpet.

It is in the period of youth that Shelley appeals to us most directly, and exercises his most unquestioned authority over the imagination. In early life, at the moment more especially when the individuality begins to assert itself, a young man or a young woman of feeling discovers in this poet certain qualities which appear to be not merely good, but the best, not only genuine, but exclusively interesting. At that age we ask for light, and do not care how it is distributed; for melody, and do not ask the purpose of the song; for colour, and find no hues too brilliant to delight the unwearied eye. Shelley satisfies these cravings of youth. His whole conception of life is bounded only by its illusions. The brilliancy of the morning dream, the extremities of radiance and gloom, the most pellucid truth, the most triumphant virtue, the most sinister guilt and melodramatic infamy, alone contrive to rivet the attention. All half-lights, all arrangements in grey or russet, are cast aside with impatience, as unworthy of the emancipated spirit. Winged youth, in the bright act of sowing its intellectual wild oats, demands a poet, and Horsham, just one hundred years ago, produced Shelley to satisfy that natural craving.

It is not for grey philosophers, or hermits wearing out the evening of life, to pass a definitive verdict on the poetry of Shelley. It is easy for critics of this temper to point out weak places in the radiant panoply, to say that this is incoherent, and that hysterical, and the other an ethereal fallacy. Sympathy is needful, a recognition of the point of view, before we can begin to judge Shelley aright. We must throw ourselves back to what we were at twenty, and recollect how dazzling, how fresh, how full of colour, and melody, and odour, this poetry seemed to us—how like a May-day morning in a rich Italian garden, with a fountain, and with nightingales in the blossoming boughs of the orange-trees, with the vision of a frosty Apennine beyond the belt of laurels, and clear auroral sky everywhere above our heads. We took him for what he seemed, "a pard-like spirit, beautiful and swift," and we thought to criticise him as little as we thought to judge the murmur of the forest or the reflections of the moonlight on the lake. He was exquisite, emancipated, young like ourselves, and yet as wise as a divinity. We followed him unquestioning, walking in step with his panthers, as the Bacchantes followed Dionysus out of India, intoxicated with enthusiasm.

If our sentiment is no longer so rhapsodical, shall we blame the poet? Hardly, I think. He has not grown older, it is we who are passing further and further from that happy eastern morning where the light is fresh, and the shadows plain and clearly defined. Over all our lives, over the lives of those of us who may be seeking to be least trammelled by the commonplace, there creeps ever onward the stealthy tinge of conventionality, the admixture of the earthly. We cannot honestly wish it to be otherwise. It is the natural development, which turns kittens into cats, and blithe-hearted lads into earnest members of Parliament. If we try to resist this inevitable tendency, we merely become eccentric, a mockery to others, and a trouble to ourselves. Let us accept our respectability with becoming airs of gravity; it is another thing to deny that youth was sweet. When I see an elderly professor proving that the genius of Shelley has been overrated, I cannot restrain a melancholy smile. What would he, what would I, give for that exquisite ardour, by the light of which all other poetry than Shelley's seemed dim? You recollect our poet's curious phrase, that to go to him for common sense was like going to a gin-palace for mutton chops. The speech was a rash one, and has done him harm. But it is true enough that those who are conscious of the grossness of life, and are over-materialised, must go to him for the elixir and ether which emancipate the senses.

If I am right in thinking that you will all be with me in considering this beautiful passion of youth, this recapturing of the illusions, as the most notable of the gifts of Shelley's poetry to us, you will also, I think, agree with me in placing only second to it the witchery which enables this writer, more than any other, to seize the most tumultuous and agitating of the emotions, and present them to us coloured by the analogy of natural beauty. Whether it be the petulance of a solitary human being, to whom the little downy owl is a friend, or the sorrows and desires of Prometheus, on whom the primal elements attend as slaves, Shelley is able to mould his verse to the expression of feeling, and to harmonise natural phenomena to the magnitude or the delicacy of his theme. No other poet has so wide a grasp as he in this respect, no one sweeps so broadly the full diapason of man in nature. Laying hold of the general life of the universe with a boldness that is unparalleled, he is equal to the most sensitive of the naturalists in his exact observation of tender and humble forms.

And to the ardour of fiery youth and the imaginative sympathy of pantheism, he adds what we might hardly expect from so rapt and tempestuous a singer, the artist's self-restraint. Shelley is none of those of whom we are sometimes told in these days, whose mission is too serious to be transmitted with the arts of language, who are too much occupied with the substance to care about the form. All that is best in his exquisite collection of verse cries out against this wretched heresy. With all his modernity, his revolutionary instinct, his disdain of the unessential, his poetry is of the highest and most classic technical perfection. No one, among the moderns, has gone further than he in the just attention to poetic form, and there is so severe a precision in his most vibrating choruses that we are taken by them into the company, not of the Ossians and the Walt Whitmans, not of those who feel, yet cannot control their feelings, but of those impeccable masters of style,

> who dwelt by the azure sea
> Of serene and golden Italy,
> Or Greece the mother of the free.

And now, most inadequately and tamely, yet, I trust, with some sense of the greatness of my theme, I have endeavoured to recall to your minds certain of the cardinal qualities which animated the divine poet whom we celebrate to-day. I have no taste for those arrangements of our great writers which assign to them rank like schoolboys in a class, and I cannot venture to suggest that Shelley stands above or below this or that brother immortal. But of this I am quite sure, that when the slender roll is called of those singers, who make the poetry of England second only to that of Greece (if even of Greece), however few are named, Shelley must be among them. To-day, under the auspices of the greatest poet our language has produced since Shelley died, encouraged by universal public opinion and by dignitaries of all the professions, yes, even by prelates of our national church, we are gathered here as a sign that the period of prejudice is over, that England is in sympathy at last with her beautiful wayward child, understands his great language, and is reconciled to his harmonious ministry. A century has gone by, and once more we acknowledge the truth of his own words:

> The splendours of the firmament of time
>   May be eclipsed, but are extinguished not; Like stars to their appointed height they climb.

—EDMUND GOSSE, "Shelley in 1892,"
Questions at Issue, 1893, pp. 208–215

ALGERNON CHARLES SWINBURNE
"THE CENTENARY OF SHELLEY" (1892)

Born in London, Swinburne (1837–1909) was both an accomplished lyric poet and a prolific literary critic. He attended Oxford, where he met and befriended the Rossetti brothers, but left without a degree. Although criticized for its indecent themes when first published, *Poems and Ballads* (1866) is considered by many to be his best work and prefigured the "art for art's sake" movement of the late nineteenth century. He also authored several dramas during his career, including *The Queen-Mother* (1860), *Rosamond* (1860), *Atalanta in Calydon* (1865), and *Erechtheus: A Tragedy* (1876). His study, *Percy Bysshe Shelley*, was published in 1903.

Now a hundred years agone among us came
Down from some diviner sphere of purer flame,
Clothed in flesh to suffer; maimed of wings to soar,
One whom hate once hailed as now love hails by name,
Chosen of love as chosen of hatred. Now no more
Ear of man may hear or heart of man deplore
Aught of dissonance or doubt that mars the strain
Raised at last of love where love sat mute of yore.
Fame is less than love, and loss is more than gain,
When the sweetest souls and strongest, fallen in fight,
Slain and stricken as it seemed in base men's sight,
Rise and lighten on the graves of foeman slain,
Clothed about with love of all men as with light.
Suns that set not, stars that know not day from night.

—ALGERNON CHARLES SWINBURNE,
"The Centenary of Shelley," 1892

GEORGE SAINTSBURY (1896)

George Edward Bateman Saintsbury (1845–1933) was born in Southampton, England. He attended King's College School in London and Merton College, Oxford. After working as a schoolmaster in Manchester and Guernsey, he moved to Moray, Scotland, in 1874 and, two years later, to London. Saintsbury contributed to *The Academy* and the *Fortnightly Review*, mostly writing on French literature. He also published French literature essays for the *Encyclopedia Britannica* (1875–89). He would later contribute to several more journals, including *Macmillan's Magazine*, the *Pall Mall Gazette*, the *St. James's Gazette*, and the *Saturday Review*, where he worked as assistant editor from 1883–94. His works on English literature are numerous. They include *Short History of English Literature* (1898), the three-volume *A History of Criticism and Literary Taste in Europe from the Earliest Texts to the Present Day* (1900–04), the three-volume *History of English Prosody from the Twelfth*

Century to the Present Day (1906–10), The Later Nineteenth Century (1907), and The English Novel (1913), as well as more than twenty chapters in the Cambridge History of English Literature (1907–16).

Shelley has been foolishly praised, and it is very likely that the praise given here may seem to some foolish. It is as hard for praise to keep the law of the head as for blame to keep the law of the heart. He has been mischievously and tastelessly excused for errors both in and out of his writings which need only a kindly silence. In irritation at the "chatter" over him some have even tried to make out that his prose—very fine prose indeed, and preserved to us in some welcome letters and miscellaneous treatises, but capable of being dispensed with—is more worthy of attention than his verse, which has no parallel and few peers. But that one thing will remain true in the general estimate of competent posterity I have no doubt. There are two English poets, and two only, in whom the purely poetical attraction, exclusive of and sufficient without all others, is supreme, and these two are Spenser and Shelley.

—GEORGE SAINTSBURY, *A History of Nineteenth Century Literature,* 1896, p. 86

WORKS

❖

Collections of Shelley's work, along with the works of other romantic poets, appeared as early as the 1820s. A year following his death, *Elegant Extracts in Prose and Verse* included selections from Shelley, and in 1829, *The Poetical Works of Coleridge, Shelley, and Keats* was published in Paris, followed by the Philadelphia publication by Crissy and Markley in 1847. David Masson's *Wordsworth, Shelley, and Keats* appeared in 1874, published by Macmillan, and George Henry Calvert's *Coleridge, Shelley, and Goethe* was published in Boston by Lee and Shepard in 1880. In the last decade of the nineteenth century, there was a renewed interest in Shelley's work, evident in such publications as George Edward Woodberry's *The Complete Poetical Works of Percy Bysshe Shelley* (Cambridge [Mass.]: Riverside, 1892), Frederick Henry Sykes's *Select Poems of Goldsmith, Wordsworth, Scott, Keats, Shelley, Byron* (Toronto: W.J. Gage, 1896), A. Ellis's *Chosen English Selections from Wordsworth, Byron, Shelley, Lamb, Scott, Prepared with Short Biographies and Notes for the Use of Schools* (London and New York: Macmillan, 1896), and Joseph Forster's *Great Teachers: Burns, Shelley, Coleridge, Tennyson, Ruskin, Carlyle, Emerson, Browning* (London: G. Redway, 1898).

Although Shelley received much public praise and blame for his often-termed "radical" writing, his fame as one of the greatest of the romantic poets would not have been as certain if it were not for Mary Shelley's initiative and dedication to publishing her husband's works. The collections and her notes to the poems provided the only public account of their relationship, as Sir Timothy Shelley forbade the publication of a biography by Mary. This would leave Thomas Medwin (1788–1869) to publish the first biography, *The Life of Percy Bysshe Shelley*, in two volumes in 1847.

In 1824, Mary Shelley published *Posthumous Poetry*, the first edition of Shelley's poetry from his unfinished drafts and notebooks. In this edition, Mary sanitized Shelley's image, language, and ideas and chose to omit any scandalous material within the poems. As a result, a full picture of the poet was kept from the reading public. Mary's plans for a prose edition were thwarted when Timothy Shelley gave his son's manuscripts to Thomas Love Peacock. Sir Timothy also forbade the publication of any other Shelley writings during his lifetime. This created a literary free-for-all in terms of Shelley's work. There were more than twenty pirated editions of his poetry in the twenty years following his death. In 1839, Mary finally gained permission to publish *Poetical Works* in four volumes. This publication also marked the first time her name was listed on the title page as editor, and included the following dedication: "To Percy Florence Shelley, The Poetical Works of His Illustrious Father Are Dedicated, by His Affectionate Mother, Mary Wollstonecraft Shelley."

During their marriage, Mary would often copy extra versions of Shelley's letters before sending them, letters that would eventually make it into his collected works. However, Mary was also careful to avoid any potential scandals surrounding her deceased husband's reputation by frequently omitting parts of his letters that might shock her Victorian audiences. For instance, in a letter to Peacock written from Italy, Shelley mentions taking mercury for his ongoing health problems, but this reference to the drug was omitted in his collected letters because mercury was used to treat venereal disease. She also attempted to lessen the force of Shelley's depression while in Naples by omitting some of the poems written in 1819–20, as well as placing the poems throughout the collection, instead of as one concise group. Her prefatory comments about this time in Shelley's life are also intentionally vague. She mentions that his health was not good but does not admit the extent of his depression. She also censored Shelley's political poems; his attack on Sidmouth and Castlereagh became "Similes for Two Political Characters of 1819," and she omitted entire lines in other political poems. She deleted stanzas from "The Sensitive Plant," as well as parts of "I fear thy kisses gentle maiden," written for Sophia Stacey. She also chose to leave out stanzas in *The Witch of Atlas*, a poem whose main character Mary believed to be influenced by her own emotional and physical coldness to Shelley.

The writers represented in this section differ greatly in their appraisal of Shelley's work. The commentary on his early gothic romance, *Zastrozzi* (1810), is mainly negative, reviewers put off by the novel's weak storyline, the characters' moral corruption, and the excessive violence. *Prometheus Unbound* and *The Revolt of Islam* also met with mixed reviews. William P. Trent felt that *Prometheus Unbound* was overrated, but John Gibson Lockhart, never one to miss a chance to condemn one of Leigh Hunt's radical protégées, shows his talent and finesse as a literary critic in his ability to separate the man from his poem. While scolding Shelley for his radical views and the company he keeps, Lockhart nevertheless shows his appreciation of Shelley's poetic achievement and writes a favorable review of *The Revolt of Islam*.

At the publication of Shelley's *Posthumous Poems* in 1824, critics were again divided about the poet's work. William Hazlitt, as with his previous discussion of Shelley (included in the first section of this volume), had a conflicted view of his contemporary that he was never able to fully reconcile in his writing. In his review of *Posthumous Poems*, Hazlitt admits Shelley's talent but criticizes his propensity to complicate needlessly his subjects. He believed that Shelley was at his best when he attempted the least:

> Yet Mr Shelley, with all his faults, was a man of genius; and we lament that uncontrollable violence of temperament which gave it a forced and false direction. He has single thoughts of great depth and force, single images of rare beauty, detached passages of extreme tenderness; and, in his smaller pieces, where he has attempted little, he has done most. If some casual and interesting idea touched his feelings or struck his fancy, he expressed it in pleasing and unaffected verse: but give him a larger subject, and time to reflect, and he was sure to get entangled in a system.

However, Hazlitt's criticism of Shelley is tempered by an underlying appreciation of his character. He remembers Shelley as "honest" and "sincere": "He thought and acted logically, and was what he professed to be, a sincere lover of truth, of nature, and of human kind." Hazlitt goes on to say, "We wish to speak of the errors of a man of genius with tenderness. His nature was kind, and his sentiments noble; but in him the rage of free inquiry and private judgment amounted to a species of madness."

It is precisely this kindness in Shelley's character that Mary Shelley stresses to her readers in the preface to her husband's *Works* (1839). Despite her insistence to "abstain from any remark on the occurrences of his private life," which was brought about by Timothy Shelley's refusal to let his daughter-in-law issue a biography of the poet, Mary Shelley frequently focused on the good qualities of Shelley's personality. She also circumvented her inability to release details of his life by providing lengthy notes in each section of the *Works* that discussed the circumstances under which the specific poems were written:

> In the notes appended to the poems, I have endeavoured to narrate the origin and history of each. The loss of nearly all letters and papers which refer to his early life, renders the execution more imperfect than it would otherwise have been. I have, however, the liveliest recollection of all that was done and said during the period of my knowing him. Every impression is as clear as if stamped yesterday,

and I have no apprehension of any mistake in my statements as far as they go. In other respects, I am, indeed, incompetent; but I feel the importance of the task, and regard it as my most sacred duty. I endeavour to fulfil it in a manner he would himself approve; and hope in this publication to lay the first stone of a monument due to Shelley's genius, his sufferings, and his virtues.

Mary Shelley's two editions of Percy's poetry did, perhaps more than anything else, serve to revive, establish, and preserve his reputation as a poet. In the remainder of the passages in this section, the success of this project is evident. Edwin P. Whipple discusses Shelley's unpopular character and asserts that contemporary critics failed to recognize his true poetic worth: "Poetasters and rhyme-stringers without number were published, puffed, patronized, paid, and forgotten, during the period when the *Revolt of Islam* and *Prometheus Unbound* were only known by garbled extracts which gleamed amid the dull malice of unscrupulous reviews." Robert Browning's introduction to a collection of Shelley's letters was so full of praise and informed understanding of his forebear's poetic worth that it was later reprinted in an edition of five hundred copies by the Shelley Society in 1888. Other poets also took up the call to praise Shelley's literary merit and carry that appreciation into the twentieth century. Algernon Charles Swinburne takes issue with Matthew Arnold's criticism of Shelley's ineffectiveness as a poet and disagrees with Arnold's comparison of Shelley to Keats. In comparison to Byron, Swinburne also believed that Shelley was the more gifted "singer": "Shelley outsang all poets on record but some two or three throughout all time; his depths and heights of inner and outer music are as divine as nature's, and not sooner exhaustible. He was alone the perfect singing-god; his thoughts, words, deeds, all sang together." At the turn of the century, W.B. Yeats also praised Shelley's poetic ideas. In his discussion of the poetic symbolism of Blake, Keats, and Shelley, Yeats praises the latter's intuitive use of stars, saying that "The most important, the most precise of all Shelley's symbols, the one he uses with the fullest knowledge of its meaning, is the Morning and Evening Star. It rises and sets for ever over the towers and rivers, and is the throne of his genius."

THE NECESSITY OF ATHEISM
Unsigned (1822)

The name of Percy Bysshe Shelly is not prefixed to these tracts, but they are well known to be the production of his pen; and we have selected them in our first notice of his works, as with them he commenced his literary career. In this view they are extraordinary, not as efforts of genius, but as indications of that bold and daring insubordination of mind, which led the writer, at a very early age, to trample both on human and divine authority. The Necessity of Atheism contains a distinct negation of a Deity; and the Declaration of Rights is an attempt to subvert the very foundations of civil government. Were not the subject far too grave for pleasantry, we might amuse ourselves with the idea of a stripling, an under-graduate, commencing hostilities against heaven and earth, and with the utmost self-satisfaction exulting that he had vanquished both.

Some of our readers are aware, that for the first of these performances, (after every persuasion from his superiors to induce him to retract it had been urged in vain,) Mr. Shelly was expelled from college; and that for posting up the second on the walls of a provincial town, his servant was imprisoned; and, from these facts, they may perhaps imagine that they are remarkably effective engines of atheism and democracy. But, in truth, they are below contempt,—they rather insult than support the bad cause to which they are devoted.

To maintain the Necessity of Atheism, is, perhaps, the wildest and most extravagant effort of a perverted understanding; and to consider this as achieved by a mere boy in thirteen widely-printed pages of a duodecimo than any recorded in the Scriptures. Had we not of late been accustomed to witness the arrogance and presumption of impiety; had not the acuteness of our sensibility been somewhat deadened by familiar acquaintance with the blasphemies of the school in which this young man is now become a professor, we could not trust our feelings even with a remote reference to his atrocious, yet most imbecile, production. It is difficult, on such a subject, to preserve the decorum of moral tolerance, and to avoid a severity of indignation incompatible with the office of Christian censors.

Mr. Shelly oddly enough denominates belief a passion; then he denies that it is ever active; yet he tells us that it is capable of excitement, and that the degrees of excitement are three. But lest we should be suspected of misrepresentation, Mr. Shelly shall speak for himself.

"The senses are the sources of all knowledge to the mind, consequently their evidence claims the strongest assent. The decision of the mind, founded upon our own experience derived from these sources, claims the next degree; the experience of others, which addresses itself to the former one, occupies the lowest degree. Consequently, no testimony can be admitted which is contrary to reason; reason is founded on the evidence of our senses."

"Every proof may be referred to one of these three divisions; we are naturally led to consider what arguments we receive from each of them, to convince us of the existence of a Deity."

These sentences embrace a page of the pamphlet, and immediately succeed a general introduction occupying eight more; and of course the whole investigation is despatched in less than four. Its result is summed up in the following words:

"From this it is evident, that having no proofs from any of the three sources of conviction, the mind cannot believe the existence of a God. It is also evident, that as belief is a passion of the mind, no degree of criminality can be attached to disbelief. *They only are reprehensible who willingly neglect to remove the false medium through which their mind views the subject.* It is almost unnecessary to observe, that the general knowledge of the deficiency of such proof cannot be prejudicial to society. Truth has always been found to promote the best interests of mankind. Every reflecting mind must allow, that there is no proof of the existence of a Deity."

Such is the jargon of the new philosophy. "The satanic school" maintains, that belief cannot be virtuous; yet, that it may be reprehensible, and therefore vicious; and that the greatest crime of which a rational creature can be guilty, is to admit the being of a God. Such is the logic of Mr. Shelly. To discuss the question at issue between atheists and theists with such a writer, would be extreme folly; nor should we have drawn from oblivion this extravagant freak of his boyhood, had he not by subsequent writings, and a matured period of his life, avowed the same sentiments, and obtruded them upon the world with an effrontery unexampled in the annals of impiety. But on this strange intellectual and moral phenomenon we shall take occasion to offer a few remarks. In what light are we to consider the intellectual qualities and attainments of an individual, who denies the existence of a Deity, on the supposition that he has discovered a great and momentous truth? But he has explored the universe, and not only cannot find a God, but can demonstrate the impossibility of his existence. How surprisingly great must be his understanding! how stupendous and overpowering his knowledge! For as this is a fact that requires demonstration, no inferior degree of evidence can be admitted as conclusive. What wondrous Being then presents himself before us in all the confidence of absolute persuasion, founded on irrefragable evidence, declaring that there is no God? And how has he grown to this immense intelligence? Yesterday he was an infant in capacity, and humble; and now he is invested with the attributes of the very Divinity whose existence he denies.

* * *

To us there is something fearful and even terrific in the state of mind which can delight in the renunciation of a Deity—which can derive satisfaction from the feeling that the infinite Sprit is gone, that the only solid foundation of virtue is wanting; which can enjoy pleasure in renouncing that system of doctrine of which a God is the great subject, and that train of affections and conduct of which HE is the supreme object.

* * *

But "Truth," says Mr. Shelly, "has always been found to promote the best interests of mankind." We admit the proposition, and therefore maintain that that which is subversive of their best interests, cannot be truth. We may confidently ask, in what possible way can Atheism secure the well-being of society?

If we grant that the belief of a Deity operates as a very slight restraint on vice, in individual cases where the character has become utterly depraved, yet its general influence must be mighty, interwoven as it is with the whole civil and social economy of man. It must act powerfully as an incentive to whatever is good, and as a check to whatever is evil; and, it can only fail in particular instances of atrocious obduracy. But, what offences against himself or his fellow-creatures, may not an Atheist perpetrate with conscious impunity, without regret, and without a blush? What protection can his principles afford to confiding innocence and beauty? What shall deter him from dooming an amiable and lovely wife to penury, to desolation, and an untimely grave? What shall make seduction and adultery criminal in his eyes, or induce him when she is in his power, to spare the victim of unhallowed and guilty passions? What can he know of honour, of justice, and integrity? What friend will he not pursue to utter destruction? What lawless gratification will he not indulge, when its indulgence does not compromise his personal safety? Who, we may ask, are those that set the decencies of life at defiance, that laugh at virtue, and riot in epicurean debauchery? Are they not the base apostates from God, who boast of their impiety, and write themselves "ATHEISTS" to their own disgrace, and the scandal of the country that gave them birth? These are the questions which we put to what was once a conscience in the breast of Mr. Shelly, with little hope, however, that they will rouse this benumbed and long-forgotten faculty, to any thing like feeling. It is well for mankind that the life of the Atheist is so just a comment upon his creed, and that none can feel a wish to join his standard, but he who has become an alien from virtue, and the enemy of his species.

We had intended to indulge in further observation, and to bring the principles of the declaration of rights more prominently and distinctly before our readers; but for the present we shall forbear. A government founded on

Atheism, or conducted by Atheists, would be the greatest curse the world has ever felt. It was inflicted for a short season, as a visitation, on a neighboring country, and its reign was avowedly and expressly the reign of terror. The declarers of rights, intoxicated by their sudden elevation, and freed from every restraint, became the most ferocious tyrants; and, while they shut up the temples of God, abolished his worship, and proclaimed death to be an eternal sleep, they converted, by their principles and spirit, the most polished people in Europe into a horde of assassins; the seat if voluptuous refinement, of pleasure and of arts, into a theatre of blood.

With an example so recent and so fearfully instructive before our eyes, it is not probable that we shall be deluded by Mr. Shelly or any of his school; the splendours of a poetical imagination may dazzle and delight, and they may prove a mighty engine of mischief to many who have more fancy than judgment; but they will never impose upon the sober and calculating part of the community; they will never efface the impression from our minds, that Atheism is an inhuman, bloody, ferocious system, equally hostile to every useful restraint, and to every virtuous affection; that having nothing above us to excite awe, or around us to awaken tenderness, it wages war with Heaven and with earth: its first object is to dethrone God; its next to destroy man. With such conviction, the enlightened and virtuous inhabitants of Great Britain will not surely be tempted to their fate by such a rhapsody as the following, with which Mr. Shelly concludes his Declaration of Rights, and with which we take our leave of him . . .

—UNSIGNED REVIEW, *The Necessity of Atheism* and *Declaration of Rights*, Brighton Magazine, May 1822, pp. 540–544

ZASTROZZI AND ST. IRVYNE

UNSIGNED (1810)

ZASTROZZI is one of the most savage and improbable demons that ever issued from a diseased brain. His mother, who had been seduced by an Italian nobleman by the name of Verezzi, and left him in wretchedness and want, conjures her son, on her death bed, to revenge her wrongs on Verezzi and his progeny for ever! Zastrozzi fulfills her diabolical injunctions, by assassinating her seducer, and pursues the young Verezzi, his son, with unrelentless and savage cruelty. The first scene which opens this *shameless* and disgusting volume, represents Verezzi in a damp cell, chained to a wall.

'His limbs, which not even a little straw kept from the rock, were fixed by immense staples to the flinty floor; and but one of his hands left at liberty to take the scanty pittance of bread and water which was daily allowed him.'

This beautiful youth (as he is described), is released from his confinement by the roof of the cell falling in during a most terrific storm. He is then conducted, though in a raging fever, by the emissaries of the fiend-like Zastrozzi to the cottage of an old woman, which stands on a lone heath, remote from all human intercourse. From this place he contrives to escape, and we find him at another old woman's cottage near Passau. Here he saves the life of Matilda, La Contessa di Laurentini, who, in a fit of desperation and hopeless love for the Adonis Verezzi, plunges herself into the river. The author does not think proper to account to his readers when and how these two persons had become acquainted, or how Verezzi could know the unbounded and disgusting passion which Matilda entertains for him. It is vaguely intimated, that Verezzi loves and is beloved by Julia Marchesa di Strobazzo, who is as amiable as Matilda is diabolical; but we are left to conjecture how the connection between Zastrozzi and Matilda is brought about. But these inconsistencies need not surprise us, when we reflect that a more discordant, disgusting, and despicable performance has not, we are persuaded, issued from the press for some time. Verezzi accompanies Matilda to Passau, with whom he remains, and by whom he is informed of the death of Julia. This intelligence throws him into another fever; on his recovery, Matilda conveys him to a castellan of her own, situated in the Venetian territory. Here she practices every art and assumes all the amiable appearances and fascinating manners she is mistress of, which she thinks most likely to wean Verezzi from his fondness for the memory of Julia, and to inspire him with an affection for herself. But all her arts prove f[r]iutless, till Zastrozzi suggests the scheme of affecting to assassinate Verezzi, when Matilda is to interpose and make him believe that she saves his life. Verezzi, who is a poor fool, and any thing but a man, falls into the snare, forgets his Julia, indulges a vicious passion for Matilda, which the author denominates love, but which is as far removed from that exalted passion as modesty is from indecency, and deserves a name which we shall not offend the readers by repeating. Revelling in an inordinate and bestial passion, of which the fiend Matilda is the object, he discovers that Julia still lives. This causes momentary regret, but awakens the jealousy of Matilda, which he calms by the most indelicate professions, and whilst he is about to drink a goblet of wine to the happiness of his infamous paramour, Julia glides into the room. Verezzi is instantly seized with a frenzy, and stabs himself. Matilda is rendered furious by this death blow to her criminal gratifications.

> 'Her eyes *scintillated*,' (a favourite word with the author, which he introduces in almost every page), 'with fiend-like expression. She advanced to the lifeless corpse of Verezzi, she plucked the dagger from his bosom, it was stained with his life's blood, which trickled fast from the point to the floor, she raised it on high, and impiously called upon the God of nature to doom her to endless torments should Julia survive her vengeance.'

She is as good as her word; she stabs Julia in a thousand places; and, with exulting pleasure, again and again buries her dagger in the body of the unfortunate victim of her rage. Matilda is seized by the officers of justice, as well as Zastrozzi, who confesses that he had planned the whole business, and made Matilda the tool by which he satiated his revenge.

The story itself, and the style in which it is told, are so truly contemptible, that we should have passed it unnoticed, had not our indignation been excited by the open and barefaced immorality and grossness displayed throughout. Matilda's character is that of a la[s]civious fiend, who dignifies a vicious, unrestrained passion by the appellation of love.

Does the author, whoever he may be, think his gross and wanton pages fit to meet the eye of a modest young woman? Is this the instruction to be instilled under the title of a romance? Such trash, indeed, as this work contains, is fit only for the inmates of a brothel. It is by such means of corruption as this that the tastes of our youth of both sexes become vitiated, their imaginations heated, and a foundation laid for their future misery and dishonor. When a taste for this kind of writing is imbibed, they may bid farewell to innocence, farewell to purity of thought, and all that makes youth and virtue lovely!

We know not when we have felt so much indignation as in the perusal of this execrable production. The author of it cannot be too severely reprobated. Not all, his '*scintillated eyes*,' his '*battling emotions*,' his '*frigorific torpidity* of despair,' not his '*Lethean torpor*,' with the rest of his nonsensical and stupid jargon, ought to save him from infamy, and his volume from the flames.

—Unsigned, *Critical Review*, November 1810, pp. 329–331

John Cordy Jeaffreson (1885)

Jeaffreson (1831–1901) was born in Suffolk. He attended Pembroke College, Oxford, and completed his B.A. in 1852. He turned his back on a career in law in order to become a full-time writer. Jeaffreson contributed regularly to *The Athenaeum* and found success in his *A Book About* ... series. These included *A Book about Doctors* (1860), *A Book about Lawyers* (1866), and *A Book about the Clergy* (1870). Later in his career, he became an archivist of corporation records and several private manuscript collections throughout England. He continued his literary scholarship by publishing biographies of Shelley (1888), Byron (1883), and Lord Nelson and Lady Hamilton (1888, 1889). Jeaffreson published his own memoirs, *A Book of Recollections*, in 1894.

Published in a single duodecimo volume, this tale of horror *(Zastrozzi)* contains about as many words as a single volume of an ordinary three-volume novel. Perhaps more horrors have never been crowded into so short a romance. The tortures endured by Verezzi during his successive imprisonments afflict the memory. Verezzi's father is poniarded to death by his bastard son. Julia's faithful servant, Paulo, dies in the presence of his poisoners, groaning horribly and writhing in hideous convulsions. Matilda makes a futile attempt to throw herself into the Danube. The dagger-scene in the vicinity of the Castella di Laurentini would not have been more terrific had the mock-assailant been a veritable bravo. The Count Verezzi commits suicide. Julia is stabbed in a thousand different spots of her body. Zastrozzi is racked to death. The Contessa di Laurentini is left for execution.

Affording not a single indication of literary taste or wholesome sentiment, the story is badly written, morbid, unnatural, and superlatively foolish, from its first to its last page. To Shelley's reasonable and honest biographers, the performance is of great value and interest on account of the view it gives of the future poet's culture, attainments, and mental condition towards the close of his career at Eton. Allowance should of course be made for the author's youth, his inexperience of human nature and society, and the difficulties besetting every puerile essayist in an arduous department of literature. But when all allowances have been made, the book remains a thing of evidence to the utter discredit of all the fine things that have been written by certain of the poet's adulators about his intellectual precocity. He would not have laboured at this crude tale in his seventeenth year, corrected it for the press, and published it in his eighteenth year, hoping to win fame by it, had he, in his boyhood, acquired the knowledge of English literature, for which several historians of his earlier career have given him credit, or had he been the sincere and strenuous student of natural science the same writers have declared him. Had he perused the works of the higher English writers with critical discernment as well as delight, the Etonian would have written his mother tongue with less inelegance and feebleness. Had his care for natural science exceeded the commonplace curiosity of a youth, given to play tricks with an air-pump, an electrical machine, and a chest of chemical materials, his mind would have been too fully occupied to have a hankering for the miserable distinction that comes to the writers of bad novels.

Though it is not regarded as a faultless performance in the coteries of the Shelleyan enthusiasts, passages of considerable merit and indications of fine feeling have been discovered in this superlatively foolish story, by some of the gentlemen who have in these later years constituted themselves the peculiar guardians of Shelley's honour, and the especial interpreters of his philosophical utterances.

In the superabundance of his veneration for every line written, and every scrap of paper known to have been touched by the poet, Mr. Buxton Forman is educating the English people to regard *Zastrozzi* as a performance that, instead of being perused lightly and laughed over merrily, should be studied with due regard to the various readings of its two different editions,—the original edition of 1810, and the reprint of 1839, in *The Romancist and Novelist's Library*. Wherever those editions differ by an inverted comma, a mark of punctuation, a dropt letter, or a letter too many, Mr. Buxton Forman calls attention to the difference, as though each trivial diversity of the two texts were a matter of high importance. Believing that delicate meanings may be found in the poet's occasional slips of spelling, Mr. Forman calls attention to the remarkable fact, that the word 'ceiling' in the reprint is spelt 'deling' in the original edition; the no less curious and significant circumstance that the word 'escrutoire' of the later edition is spelt *escrutoire* in the edition that passed straight to the world from the author's own hand and eye. In like manner we are invited to notice the difference of a perfectly formed Y between the 'mishapen' of Shelley's own text, and the 'misshapen' of the reprint. Mr. Forman calls attention to an even bolder departure from the original text in the reprint, which may well be regarded with suspicion and mistrust by the Shelleyan specialists. Whilst the original edition contains the sentence, 'The most horrible scheme of vengeance at at this instant glances across Zastrozzi's mind,' the editor of the 1839 edition has the daring (not altogether innocent of irreverence) to omit the second 'at.' From the standpoint and principles of an editor, who regards Shelley as a being who might have been the Saviour of the World, Mr. Buxton Forman is of course right in attaching great importance to these differences of the two editions, of an almost sacred performance. But to the profane mind of the present writer, who, instead of thinking Shelley in any respect comparable with the Saviour of the World, and conceives him to have been a rather foolish schoolboy in the earlier months of 1809, a very foolish Oxford undergraduate in the later months of 1810, and a still more foolish undergraduate in the earlier months of 1811, it appears that these differences of the two editions of *Zastrozzi* are of no more importance than the proverbial difference between 'tweedledum' and 'tweedle-dee.'

—JOHN CORDY JEAFFRESON,
The Real Shelley, 1885,
vol. 1, pp. 117–119

WALTER RALEIGH (1894)

Raleigh (1861–1922) was born in London. After living with his uncle, Adam Gifford (Lord Gifford), in Edinburgh, Raleigh studied at University College School and University College, London. He completed his B.A. in 1881 and then attended King's College, Cambridge, where he served as editor of the *Cambridge Review*. Raleigh taught in India from 1885 to 1887, eventually leaving the country because of poor health. He resumed his teaching career in 1890 at University College, Liverpool (which was also home to A.C. Bradley), where he worked as professor of modern literature. His first book was *The English Novel* (1894). He published numerous studies of leading literary figures, including *Robert Louis Stevenson: An Essay* (1895), *Milton* (1900), *Shakespeare* (1907), and *Six Essays on Johnson* (1910). In 1904, Raleigh was appointed first chair of English literature at Oxford and later became a distinguished lecturer at Cambridge. He also helped to found the English faculty library at Oxford in 1914. Raleigh was knighted in 1911, and after the outbreak of World War I, turned increasingly to political writing. These works include *Might Is Right* (1914), *The War of Ideas* (1916), *The Faith of England* (1917), *Some Gains of the War* (1918), *The War and the Press* (1918), and *England and the War* (1918). Raleigh also accepted the task of writing the official history of the Royal Air Force, titled *The War in the Air* (1922). He was only able to complete the first volume before his death from typhoid fever, which he contracted during a trip to the Middle East while researching the second volume of the history.

With the two romances of the boyhood of Shelley, *Zastrozzi* (1810) and *St. Irvyne; or, The Rosicrucian* (1811), this account of the revivalists may fitly close. The sovereign transmutation that the dull, hard stuff of Godwin's doctrines suffered in the crucible of Shelley's imagination is known to all readers of the poems. In the *Epipsychidion* the nightingale pours forth a song suggested to her by the croaking of the frog. But in his earlier romances Shelley's imagination is wild and crude, so that they combine more than the violence of Maturin's early work with more than the absurdity of Godwin's complacent dogma. It is a strange mixture, and an odd world. These lovers, who regard legal marriage as an impropriety, and these villains, whose mildest feeling is an ecstasy of malignity, are types drawn from different schools. Romance, in these works, has once more reached the extreme of its tether; the world of adjectives is exhausted, raptures fall back into the ineffable, agonies into the indescribable. So monotonous a protest of the inadequacy of language ceases to work its effect, and instead of heightening the situation, serves only to lower the literary art.

—WALTER RALEIGH,
The English Novel,
1894, pp. 251–252

THE REVOLT OF ISLAM

JOHN GIBSON LOCKHART "OBSERVATIONS ON *THE REVOLT OF ISLAM*" (1819)

John Gibson Lockhart (1794–1854) was born in Lanarkshire, Scotland. His father, Dr. John Lockhart, was the parish minister in the family's hometown of Cambusnethan and later served as college minister of Blackfriars in Glasgow. Lockhart began his college career at the University of Glasgow, before entering Balliol College, Oxford. He received a first-class degree in classics there in 1813. While in school, Lockhart excelled in classic and modern languages. In 1817, he began working for *Blackwood's Edinburgh Magazine*, where he excelled in the art of literary criticism. Along with his fellow Tory reviewers, Lockhart gained fame with his negative "Cockney School" reviews of Keats, Shelley, Hazlitt, and Hunt, many of which he wrote under the pseudonym "Z." He befriended both William Blackwood and Walter Scott, marrying Scott's eldest daughter, Sophia, in April 1820. This close relationship with Scott led to one of Lockhart's most famous works, his seven-volume biography of his father-in-law in 1837–38 (which was later expanded to ten volumes in 1839). The *Life of Scott* remains one of the most important examples of biographical writing. During his time at *Blackwood's*, Lockhart traveled to Germany and published his two-volume translation of Frederick Schlegel, *Lectures on the History of Literature* in 1818. He followed this with the three-volume *Peter's Letters to His Kinfolk* (1819), a work told through the point of view of Dr. Peter Morris, a Welshman recording his impressions of the people and places of Scotland. Lockhart also continued his work on translations, publishing *Ancient Spanish Ballads* in 1823. His novels include *Valerius: A Roman Story* (1821); *Some Passages in the Life of Mr. Adam Blair* (1822), with its daring description of a Presbyterian minister's adulterous relationship; *Reginald Dalton* (1823); and *The History of Matthew Wald* (1824). In 1825, Lockhart accepted the editorship of the *Quarterly Review* and was instrumental in the journal's success for nearly thirty years.

We forbear from making any comments on this strange narrative; because we could not do so without entering upon other points which we have already professed our intention of waving for the present. It will easily be seen, indeed, that neither the main interest nor the main merit of the poet at all consists in the conception of his plot or in the arrangement of his incidents. His praise is, in our judgment, that of having poured over his narrative a very rare strength and abundance of poetic imagery and feeling—of having steeped every word in the essence of his inspiration. *The Revolt of Islam* contains no detached passages at all comparable with some which our readers recollect in the works of the great poets our contemporaries; but neither does it contain any such intermixture of prosaic materials as disfigure even the greatest of them. Mr Shelly has displayed his possession of a mind intensely poetical, and of an exuberance of poetic language, perpetually strong and perpetually varied. In spite, moreover, of a certain perversion in all his modes of thinking, which, unless he gets rid of it, will ever prevent him being acceptable to any considerable or respectable body of readers, he has displayed many glimpses of right understanding and generous feeling, which must save him from the unmingled condemnation even of the most rigorous judges. His destiny is entirely in his own hands; if he acts wisely, it cannot fail to be a glorious one; if he continues to pervert his talents, by making them the instruments of a base sophistry, their splendor will only contribute to render his disgrace the more conspicuous. Mr Shelly, whatever his errors may have been, is a scholar, a gentleman, and a poet; and he must therefore despise from his soul the only eulogies to which he has hitherto been accustomed—paragraphs from the *Examiner*, and sonnets from Johnny Keats. He has it in his power to select better companions; and if he does so, he may very securely promise himself abundance of better praise.

—John Gibson Lockhart, "Observations on
*The Revolt of Islam," Blackwood's Edinburgh
Magazine,* January 1819, pp. 475–482

PROMETHEUS UNBOUND

UNSIGNED (1821)

It would be highly absurd to deny, that this gentleman has manifested very extraordinary powers of language and imagination in his treatment of the allegory, however grossly and miserably he may have tried to pervert its purpose and meaning. But of this more anon. In the meantime, what can be more deserving of reprobation than the course which he is allowing his intellect to take, and that too at the very time when he ought to be laying the foundations of a lasting and honourable name. There is no occasion for going round about the bush to hint what the poet himself has so unblushingly and sinfully blazoned forth in every part of his production. With him, it is quite evident that the Jupiter whose downfall has been predicted by Prometheus, means nothing more than Religion in general, that is, every human system of religious belief; and that, with the fall of this, he considers it perfectly necessary (as indeed we also believe, though with far different feelings) that every system of human government also should give way and perish. The patience of the contemplative spirit in Prometheus is to be followed by the daring of the active Demagorgon, at whose touch all "old thrones" are at once and for ever to be cast down into the dust. It appears too plainly, from the luscious pictures with which his play terminates, that Mr Shelly looks forward to an unusual relaxation of all moral *rules*—or rather, indeed, to the extinction of all moral feelings, except that of a certain mysterious indefinable *kindliness,* as the natural and necessary result of the overthrow of all civil government and religious belief. It appears, still more wonderfully, that he contemplates this state of things as the ideal SUMMUM BONUM. In short, it is quite impossible that there should exist a more pestiferous mixture of blasphemy, sedition, and sensuality, than is visible in the whole structure and strain of this poem—which, nevertheless, and notwithstanding all the detestation its principles excite, must and will be considered by all that read it attentively, as abounding in poetical beauties of the highest order—as presenting many specimens not easily to be surpassed, of the moral sublime of eloquence—as overflowing with pathos, and most magnificent in description. Where can be found a spectacle more worthy of sorrow than such a man performing and glorying in the performance of such things? His evil ambition,—from all he has yet written, but most of all, from what he has last and best written, his *Prometheus,*—appears to be no other, than that of attaining the highest place among those poets,— enemies, not friends, of their species,—who, as a great and virtuous poet has well said (putting evil consequence close after evil cause),

Profane the God-given strength, and *mar the lofty* line.

—Unsigned,
Blackwood's Edinburgh Magazine,
September 1820, p. 680

Unsigned (1821)

As Mr. Shelley disdains to draw his materials from nature, it is not wonderful that his subjects should in general be widely remote from every thing that is level with the comprehension, or interesting to the heart of man. He has been pleased to call *Prometheus Unbound* a lyrical drama, though it has neither action nor dramatic dialogue. The subject of it as the transition of Prometheus from a state of suffering to a state of happiness; together with a corresponding change in the situation of mankind. But no distinct account is given of either of these states, nor of the means by which Prometheus and the world pass from the one to the other. The Prometheus of Mr. Shelley is not the Prometheus of ancient mythology. He is a being who is neither a God nor a man, who has conferred supreme power on Jupiter. Jupiter torments him; and Demogorgon, by annihilating Jupiter's power, restores him to happiness. Asia, Panthea, and Ione, are female beings of a nature similar to that of Prometheus. Apollo, Mercury, the Furies, and a faun, make their appearance; but have not much to do in the piece. To fill up the *personal dramatis,* we have voices of the mountains, voices of the air, voices of the springs, voices of the whirlwinds, together with several echos. Then come spirits without end: spirits of the moon, spirits of the earth, spirits of the human mind, spirits of the hours; who all attest their super-human nature by singing and saying things which no human being can comprehend. We do not find fault with this poem, because it is built on notions which no longer possess any influence over the mind, but because its basis and its materials are mere dreaming, shadowy, incoherent abstractions. It would have been quite as absurd and extravagant in the time of Aeschylus, as it is now.

It may seem strange that such a volume should find readers, and still more strange that it should meet with admirers. We were ourselves surprised by the phenomenon: nothing similar to it occurred to us, till we recollected the numerous congregations which the incoherencies of an itinerant Methodist preacher attract. These preachers, without any connected train of thought, and without attempting to reason, or to attach any definite meaning to the terms which they use, pour out a deluge of sonorous words that relate to sacred objects and devout feelings. These words, connected as they are with all that is most venerable in the eyes of man, excite a multitude of pious associations in the hearer, and produce in him a species of mental intoxication. His feelings are awakened, and his heart touched, while his imagination and understanding are bewildered; and he receives temporary pleasure, sometimes even temporary improvement, at the expense of the essential and even permanent depravation of his character. In the same way, poetry like that of Mr. Shelley presents every where glittering constellations of words, which taken separately have a meaning, and either communicate some activity to the imagination, or dazzle it by their brilliance. Many of them relate to beautiful or interesting objects, and are therefore capable of imparting pleasure to us by the associations attached to them. The reader is conscious that his mind is raised from a state of stagnation, and he is willing to believe, that he is astounded and bewildered, not by the absurdity, but by the originality and sublimity of the author.

—UNSIGNED, *Quarterly Review*,
October 1821, pp. 175–176

WILLIAM P. TRENT "APROPOS OF SHELLEY" (1899)

William Peterfield Trent (1862–1939) was born in Richmond, Virginia. He attended the University of Virginia, New York University, and Johns Hopkins University, and went on to teach at the University of the South, Columbia University, and Barnard College. Trent also founded the *Sewanee Review* in 1892 and served as editor from 1893 to 1900. His interest in literature began early in life, and he published *William Gilmore Simms* in 1892. Trent also published studies of John Milton, Robert E. Lee, Henry Wadsworth Longfellow, and Daniel Defoe. Throughout his writing and teaching career, he firmly believed in the power of literature to provide moral instruction to its readers. This idea is at the heart of the majority of his writings, including *The Authority of Criticism* (1899), a book that deals with writers such as Milton, Byron, and Shelley.

With regard to what may be called the intellectual claims put forth for this poem which has been edited for schools and been made the subject of essays by the dozen, I can say only that, however true they may be when applied to special passages, they are by no means true when applied to the drama as a whole. The fourth act, which is a favorite with the Shelleyans, seems to have been an afterthought, and is a most lame and impotent conclusion. The characters are, except for short intervals, vague, misty and devoid of personality. The solution proposed for the problem of human destiny, for the freeing of the Promethean spirit of man is as impossible and ineffectual as if it had been generated in the heated brain of a maniac. This great poem is really like more than a series of wonderful phantasmagoria flashed forth upon the curtain of the reader's mind by a very unsteady hand. When the reader voluntarily shuts off the light, i.e., ceases to think or judge, the effect is dazzling; when he allows the light of reason to play upon his mind, the effect is just the reverse. I admire the *Prometheus Unbound* as the daring and in parts splendid achievement of a brilliant, unbalanced, but noble poetic nature; but I cannot admit that it is worthy of language which would be hyperbolical in the case of any other poet than Shakspere or Milton.

—WILLIAM P. TRENT, "Apropos of Shelley,"
The Authority of Criticism and Other Essays, 1899, pp. 86–87

THE CENCI

WILLIAM GODWIN (1820)

William Godwin (1756–1836), the son and grandson of dissenting ministers, was born in Cambridgeshire. His early writings included *Life of Lord Chatham* (1783), contributions to the *English Review*, *New Annual Register*, and *Political Herald*, as well as novels such as *Italian Letters*, *Damon and Delia*, and *Imogen*. He began his diary in 1788, a written record of his daily affairs, which he kept for the rest of his life. With the publication of *An Enquiry Concerning Political Justice* in 1793, Godwin gained fame among reformers and liberals. The following year, Godwin published his best known novel, *The Adventures of Caleb Williams*. In 1797, Godwin married fellow reformist and author Mary Wollstonecraft. Their daughter, Mary Wollstonecraft Godwin, was born later that year; her birth was followed by Wollstonecraft's death less than two weeks later. This loss left Godwin to care for Mary and Wollstonecraft's daughter from a previous relationship, Fanny Imlay. Despite his grief and financial problems, Godwin composed a controversial memoir of Wollstonecraft in 1798 and another novel, *St. Leon*, in 1799. In 1801, he married Mary Jane Clairmont, who brought two more children into the Godwin household. Their son, William Godwin, Jr., was born in 1803. Despite ongoing financial problems, Godwin continued to write. *Fleetwood* was published in 1805, and that same year, he opened a London bookshop and began writing children's literature. In 1812, Godwin met Percy Shelley for the first time, and Shelley began supporting Godwin financially. Their friendship was strained, however, by Shelley and Mary's elopement to continental Europe against Godwin's wishes. Godwin and Shelley reconciled after Percy and Mary were married, but their communication broke off in 1820. Godwin published additional novels, including *Mandeville* (1817), *Cloudesley* (1830), and *Deloraine* (1833). He also published *History of the Commonwealth of England* in four volumes (1824–28). C. Kegan Paul's *William Godwin: His Friends and Contemporaries* (2 vols., 1876) helped reestablish Godwin's literary reputation; in more recent years, several studies have focused on the relationship between the two families, including Henry Noel Brailsford's *Shelley, Godwin, and Their Circle* (1913) and William St. Clair's *The Godwins and the Shelleys* (1989).

I have read the tragedy of *Cenci*, and am glad to see Shelley at last descending to what really passes among human creatures. The story is certainly an unfortunate one, but the execution gives me a new idea of Shelley's powers. There are passages of great strength, and the character of Beatrice is certainly excellent.

—WILLIAM GODWIN, letter to Mary Shelley,
March 30, 1820, cited in C. Kegan Paul,
*William Godwin: His Friends and
Contemporaries*, 1876, vol. 2, p. 272

R. PICKETT SCOTT (1878)

Robert Pickett Scott (1856–1931) was born in London and later attended King's College, Cambridge. He served as assistant master at the Central Foundation School in London from 1881 to 1882. In the succeeding years, Scott served in several other education-relation administrative posts, including as staff inspector of secondary schools with the Board of Education from 1904 to 1911, member of the Foreign Office Committee on Chinese Education from 1921 to 1922, and honorary secretary of the Headmasters' Association from 1890 to 1903. With R.T. Wallas, he coauthored *The Call of the Homeland*.

Some critics have accused him of selecting the story of the *Cenci* solely on account of its horror. De Quincey on the other hand declares that Shelley found the "whole attraction of this dreadful tale in the angelic nature of

Beatrice as revealed in local traditions and in the portrait of her by Guido." In our idea, neither of these criticisms is entirely true or entirely false, for though it is certain that a nature such as Beatrice's would have special claim for a soul such as Shelley's, yet it is quite certain also—this fact being amply evidenced by some of his favourite words and images—that the horrible possessed for him at times an almost irresistible fascination: both Keats and Coleridge were careful not to linger in this region of Cimmerian darkness, and Shelley's work would have been the better if he had followed their example.

This tragedy has even been denounced by one critic as "a loathsome subject loathsomely treated;" we shall examine the grounds of this accusation. The second count of it cannot, we believe, by any argument be substantiated; granted that the subject is a loathsome one, we challenge the critic to point out the work of any dramatist whatsoever wherein the matters which excite our abhorrence are more delicately touched upon. Shelley has so shrunk away from the hateful crime of Cenci, "as actually to leave it doubtful," writes De Quincey, "whether the murder were in punishment of the last outrage committed or in repulsion of a menace continually repeated." The first count of the accusation is, then, the only one which need occupy us.

We first note that Shelley and his critic choose to look upon the dramatic art from standpoints essentially different, and from thence this tragedy will ever be severally praised and condemned. The ground on which Shelley took his stand was that art is supreme; his critic's was that art devoid of morality is *not* supreme. We take our stand here with the critic.

Shelley entirely separated the ideas of art and morality, this the critic refused to do. Shelley declared that some actions are admirably adapted for poetic and dramatic purposes which are essentially immoral. These are the words in which he justified to Mrs Gisborne the introduction even of the crime of incest into poetry:—"Incest," he writes, "is, like many other incorrect things, a very poetical circumstance. It may be the excess of love or hate. It may be the defiance of everything for the sake of another, which clothes itself in the glory of the highest heroism; or it may be that cynical rage, which, confounding the good and bad in existing opinions, breaks through them for the purpose of rioting in selfishness and antipathy." These are the arguments by which Shelley seeks to justify his practice, but the critic can imagine the glory of a higher heroism than that of a self-sacrificing lust, and turns away with disgust from a picture which, however truly, presents a human being "rioting in selfishness and antipathy." Shelley pleads again that his object, like that of our greatest dramatist, is "to hold the mirror up to nature," to which the critic replies that though his power would be great who could represent things as they are, adding no new beauty, and taking away no old deformity, yet his would be a depraved taste and an impure imagination who should occupy himself in describing, however minutely and to the life, the plague-spots of society, and the desperate ugliness of wicked hearts. "Granted," says the critic, "that the poetry is peerless, that the lyrics are faultless and that the dramatic art here reaches its perfection, I see in the work a moral ugliness which effectually eclipses all its artistic beauty. To me the darkness is more intense because a very source of light has been clouded over." Shelley, indeed, does not deserve so sweeping a condemnation, because he has attempted to draw a strong contrast of light and darkness in the characters of Beatrice and her father; but the fact that such darkness is willingly painted by him is enough to prove that he had not yet attained that purity of motive, that singleness of eye which can alone render the whole heart full of light and leave no corner dark.

It has been said that the figures of Beatrice and of Cenci stand out in too bold relief, compared with the other *dramatis persona;* but it should be remembered, that such characters would always stand out boldly in whatsoever age and under whatsoever circumstances they had been placed: nay, more,— as a matter of fact, these are the characters whose true history after more than two centuries still has power to awaken the strongest feelings of pity and of hatred in all ranks of the people, even in the great and busy city of the Seven Hills. "On my arrival at Rome"—writes Shelley—"I found that the story of the Cenci was a subject not to be mentioned in Italian society without awakening a deep and breathless interest. All ranks of people knew the outlines of this history, and participated in the overwhelming interest which it seems to have the magic of exciting in the human heart. I had a copy of Guido's picture of Beatrice, which is preserved in the Colonna Palace, and my servant instantly recognised it as the portrait of *La Cenci.*" The character of a Beatrice, of a Cenci, would always be prominent, even in an age of strong men, for both had essentially the strong will and the unswerving purpose which mark the leader of men. Surely, his profile should not be lightly sketched, before whom his children cowered as to a tyrant, and the princes of the church and the haughtiest nobles as to a demon: surely, too, the lines of her character should be touched in with no uncertain stroke who was a refuge for her stepmother and brothers, and who, beside daring to denounce her father in his own hall, when the sternest man there crept silently away, had yet the courage to compass the tyrant's death, since by no other means could she escape his persecution.

How pitilessly clear-cut is the figure of old Cenci, boasting of his 'strength and wealth, and pride and lust, and length of days;' vaunting that he has 'no remorse and little fear,' while the action of the play declares this to be no false estimate of himself. It is a picture of the natural man possessing all that the natural heart can desire, absolute in his good fortune, and unchecked by any scruple of conscience, believing only in a God as a means by which he shall himself be respited from hell, and by which he shall be revenged upon those he hates. Through observation of this character we notice in Shelley a capacity for satire of which moreover the poet himself was conscious. Still more certain evidence of this exists in *Peter Bell the Third*,—but though our poet's faculty in this direction was sufficiently marked, yet his efforts were beyond all question immeasurably surpassed by Byron, whose *Vision of Judgment* is, perhaps, the most powerful satire the world has ever seen. The reason that Byron should excel Shelley in this branch of their craft it is not difficult to determine: in satire, Shelley's musical verse did not help him, and his earnestness was absolutely against him, whereas Byron's sneer and the grand force of his intellect were invaluable for the perfection of this weapon.

—R. Pickett Scott,
The Place of Shelley among the Poets of His Time, 1878, pp. 32–36

GENERAL COMMENTARY ON THE POETRY

William Hazlitt "Shelley's Posthumous Poems" (1824)

Mr Shelley's style is to poetry what astrology is to natural science—a passionate dream, a straining after impossibilities, a record of fond conjectures, a confused embodying of vague abstractions,—a fever of the soul, thirsting and craving after what it cannot have, indulging its love of power and novelty at the expense of truth and nature, associating ideas by contraries, and wasting great powers by their application to unattainable objects.

Poetry, we grant, creates a world of its own; but it creates it out of existing materials. Mr Shelley is the maker of his own poetry—out of nothing. Not that he is deficient in the true sources of strength and beauty, if he had given himself fair play (the volume before us, as well as his other productions, contains many proofs to the contrary): But, in him, fancy, will, caprice, predominated over and absorbed, the natural influences of things; and he had no respect for any poetry that did not strain the intellect as well as fire the imagination—and was not sublimed into a high spirit of metaphysical philosophy. Instead of giving a language to thought, or lending the heart a tongue, he utters dark sayings, and deals in allegories and riddles. His Muse offers her services to clothe shadowy doubts and inscrutable difficulties in a robe of glittering words, and to turn nature into a brilliant paradox. We thank him—but we must be excused. Where we see the dazzling beacon-lights streaming over the darkness of the abyss, we dread the quicksands and the rocks below. Mr Shelley's mind was of 'too fiery a quality' to repose (for any continuance) on the probable or the true—it soared 'beyond the visible diurnal sphere,' to the strange, the improbable, and the impossible. He mistook the nature of the poet's calling, which should be guided by involuntary, not by voluntary impulses. He shook off, as an heroic and praise-worthy act, the trammels of sense, custom, and sympathy, and became the creature of his own will. He was 'all air,' disdaining the bars and ties of mortal mould. He ransacked his brain for incongruities, and believed in whatever was incredible. Almost all is effort, almost all is extravagant, almost all is quaint, incomprehensible, and abortive, from aiming to be more than it is. Epithets are applied, because they do not fit: subjects are chosen, because they are repulsive: the colours of his style, for their gaudy, changeful, startling effect, resemble the display of fireworks in the dark, and, like them, have neither durability, nor keeping, nor discriminate form. Yet Mr Shelley, with all his faults, was a man of genius; and we lament that uncontrollable violence of temperament which gave it a forced and false direction. He has single thoughts of great depth and force, single images of rare beauty, detached passages of extreme tenderness; and, in his smaller pieces, where he has attempted little, he has done most. If some casual and interesting idea touched his feelings or struck his fancy, he expressed it in pleasing and unaffected verse: but give him a larger subject, and time to reflect, and he was sure to get entangled in a system. The fumes of vanity rolled volumes of smoke, mixed with sparkles of fire, from the cloudy tabernacle of his thought. The success of his writings is therefore in general in the inverse ratio of the extent of his undertakings; inasmuch as his desire to teach, his ambition to excel, as soon as it was brought into play, encroached upon, and outstripped his powers of execution.

Mr Shelley was a remarkable man. His person was a type and shadow of his genius. His complexion, fair, golden, freckled, seemed transparent with an inward light, and his spirit within him so divinely wrought, That you might almost say his body thought.

He reminded those who saw him of some of Ovid's fables. His form, graceful and slender, drooped like a flower in the breeze. But he was crushed beneath the weight of thought which he aspired to bear, and was withered in the lightning-glare of a ruthless philosophy! He mistook the nature of his own faculties and feelings—the lowly children of the valley, by which the skylark makes its bed, and the bee murmurs, for the proud cedar or the mountain-pine, in which the eagle builds its eyry, 'and dallies with the wind, and scorns the sun.'—He wished to make of idle verse and idler prose the frame-work of the universe, and to bind all possible existence in the visionary chain of intellectual beauty—

> More subtle web Arachne cannot spin,
> Nor the fine nets, which oft we woven see
> Of scorched dew, do not in th' air more lightly flee.

Perhaps some lurking sense of his own deficiencies in the lofty walk which he attempted, irritated his impatience and his desires; and urged him on, with winged hopes, to atone for past failures, by more arduous efforts, and more unavailing struggles.

With all his faults, Mr Shelley was an honest man. His unbelief and his presumption were parts of a disease, which was not combined in him either with indifference to human happiness, or contempt for human infirmities. There was neither selfishness nor malice at the bottom of his illusions. He was sincere in all his professions; and he practised what he preached—to his own sufficient cost. He followed up the letter and the spirit of his theoretical principles in his own person, and was ready to share both the benefit and the penalty with others. He thought and acted logically, and was what he professed to be, a sincere lover of truth, of nature, and of human kind. To all the rage of paradox, he united an unaccountable candour and severity of reasoning: in spite of an aristocratic education, he retained in his manners the simplicity of a primitive apostle. An Epicurean in his sentiments, he lived with the frugality and abstemiousness of an ascetick. His fault was, that he had no deference for the opinions of others, too little sympathy with their feelings (which he thought he had a right to sacrifice, as well as his own, to a grand ethical experiment)—and trusted too implicitly to the light of his own mind, and to the warmth of his own impulses. He was indeed the most striking example we remember of the two extremes described by Lord Bacon as the great impediments to human improvement, the love of Novelty, and the love of Antiquity. 'The first of these (impediments) is an extreme affection of two extremities, the one Antiquity, the other Novelty; wherein it seemeth the children of time do take after the nature and malice of the father. For as he devoureth his children, so one of them seeketh to devour and suppress the other; while Antiquity envieth there should be new additions, and Novelty cannot be content to add, but it may deface. Surely the advice of the Prophet is the true direction in this matter: *Stand upon the old ways, and see which is the right and good way, and walk therein.* Antiquity deserveth that reverence, that men should make a stand thereupon, and discover what is the best way; but when the discovery is well taken, then to take progression. And to speak truly, *Antiquitus seculi Juventas mundi*. These times are the ancient times, when the world is ancient, and not those which we count ancient, *ordine-retrogrado*, by a computation backwards from ourselves.' (*Advancement of Learning*, Book 1. p. 46.)—Such is the text: and Mr Shelley's writings are a splendid commentary on one half of it. Considered in this point of view, his career may not be uninstructive even to those whom it most offended; and might be held up as a beacon and warning no less to the bigot than the sciolist. We wish to speak of the errors of a man of genius with tenderness. His nature was kind, and his sentiments noble; but in him the rage of free inquiry and private judgment amounted to a species of madness. Whatever was new, untried, unheard of, unauthorized, exerted a kind of fascination over his mind. The examples of the world, the opinion of others, instead of acting as a check upon him, served but to impel him forward with double velocity in his wild and hazardous career. Spurning the world of realities, he rushed into the world of nonentities and contingencies, like air into a *vacuum*. If a thing was old and established, this was with him a certain proof of its having no solid foundation to rest upon: if it was new, it was good and right. Every paradox was to him a self-evident truth; every prejudice an undoubted absurdity. The weight of authority, the sanction of ages, the common consent of mankind, were vouchers only for ignorance, error, and imposture. Whatever shocked the feelings of others, conciliated his regard; whatever was light, extravagant, and vain, was to him a proportionable relief from the dullness and stupidity of established opinions. The worst of it however was, that he thus gave great encouragement to those who believe in all received absurdities, and are wedded to all existing abuses: his extravagance seeming to sanction their grossness and selfishness, as theirs were a full justification of his folly and eccentricity. The two extremes in this way often meet, jostle,—and confirm one another. The infirmities of age are a foil to the presumption of youth; and 'there the antics sit,' mocking one another—the ape Sophistry pointing with reckless scorn at 'palsied eld,' and the bed-rid hag, Legitimacy, rattling her chains, counting her beads, dipping her hands in blood, and blessing herself from all change and from every appeal to common sense and reason! Opinion thus alternates in a round of contradictions: the impatience or obstinacy of the human mind takes part with, and flies off to one or other of the two extremes 'of affection' and leaves a horrid gap, a blank sense and feeling in the middle, which seems never likely to be filled up, without a total change in our mode of proceeding. The martello-towers with which we are to repress, if we cannot destroy, the systems of fraud and oppression should not be castles in the air, or clouds in the verge of the horizon, but the enormous and accumulated pile of abuses which have arisen out of their own continuance. The principles of sound morality, liberty and humanity, are not to be found only in a few recent writers, who have discovered the secret of the greatest happiness to the greatest numbers, but are truths as old as the creation. To be convinced of the existence of wrong, we should read history rather than poetry: the levers with which we must work out our regeneration are not the cobwebs of the brain, but the warm, palpitating fibres of the human heart. It is the collision of passions and interests, the petulance of party-spirit, and the perversities of self-will and self-opinion that have been the great obstacles to social improvement—not stupidity or ignorance; and the caricaturing one side of the question and shocking the most pardonable prejudices on the other, is not the way to allay hearts or produce unanimity. By flying to the extremes of scepticism, we make others shrink back, and shut themselves up in the strongholds of bigotry and superstition— by mixing up doubtful or offensive matters with salutary and demonstrable truths, we bring the whole into question, flyblow the cause, risk the principle, and give a handle and a pretext to the enemy to treat all philosophy and all reform as a compost of crude, chaotic, and monstrous absurdities. We thus arm the virtues as well as the vices of the community against us; we trifle with their understandings, and exasperate their self-love; we give to superstition and injustice all their old security and sanctity, as if they were the only alternatives of impiety and profligacy, and league the natural with the selfish prejudices of mankind in hostile array against us. To this consummation, it must be confessed that too many of Mr Shelley's productions pointedly tend. He makes no account of the opinions of others, or the consequences of any of his own; but proceeds—tasking his reason to the utmost to account for every thing, and discarding every thing as mystery and error for which he cannot account by an effort of mere intelligence— measuring man, providence, nature, and even his own heart, by the limits of the understanding—now hallowing high mysteries, now desecrating pure sentiments, according as they fall in with or exceeded those limits; and exalting and purifying, with Promethean heart; whatever he does not confound and debase.

—WILLIAM HAZLITT, from
"Shelley's Posthumous Poems,"
Edinburgh Review, July 1824, pp. 494–498

Unsigned "Review of New Publications" (1824)

Cold is the heart that can visit with severity the moral aberrations of departed genius. Poor Shelley, as well as his noble friend, Lord Byron, might, if he had been spared, have redeemed his fame, and have died the glory of his race and time. In the "Posthumous Poems of Percy Bysshe Shelley," there is much that we cannot possibly praise; much that, under different circumstances, we should feel ourselves called upon most deeply to censure: as it is, our pen shall not inflict an additional wound upon the yet bleeding heart of an affectionate and devoted widow. From the poet's last wreath we will select one little tender blossom of remembrance: —

> The odour from the flower is gone,
> Which like thy kisses breathed on me
> The colour from the flower is flown,
> Which glowed of thee, and only thee.
>
> A shrivelled, lifeless, vacant form,
> It lies on my abandoned breast;
> And mocks the heart, which yet is warm,
> With cold and silent rest.
>
> I weep—my tears revive it not!
> I sign—it breathes no more on me;
> Its mute and uncomplaining lot,
> Is such as mine should be.

—UNSIGNED, "Review of New Publications,"
Belle Assemblée, August 1824, p. 81

ALBANY FONBLANQUE
"LITERARY NOTICES" (1824)

Albany William Fonblanque (1793–1872) was born in London, the youngest son of John Fonblanque, a lawyer and a Member of Parliament representing Camelford. Albany followed in his father's footsteps, choosing a career in law but soon left the profession to become a full-time writer. His early contributions appear in *The Times*, the *Morning Chronicle*, the *Westminster Review*, and *The Atlas*. In 1826, Fonblanque became a political writer and reviewer for Leigh Hunt's *Examiner*. He eventually became owner of the periodical from 1832 to 1865 and continued its radical reputation, supporting reform movements including the 1832 Reform Bill. In 1837, he collected some of his *Examiner* articles in *England under Seven Administrations*.

This interesting volume is a publication, by Mrs. Shelley, of all the posthumous poems remaining in her possession of her highly-gifted husband; one whose benevolent and exalted spirit rendered him as much the idol of those who knew him, as his incompressible mental independence subjected him to the rancorous virulence and calumny of those who knew him not. With the exception of a reprint of all this productions which are scattered in periodical publications, and of "Alastor, or the Spirit of Solitude," which it had become difficult to obtain, this publication (a very sizeable octavo) is made up of original poetry, much of which is altogether finished, and the remainder collected from manuscript books, written impulsively at the time of composition, and never retouched. In reference to the latter, with the candour and earnestness so pleasing in themselves, but doubly interesting from the pen and the relationship, Mrs. Shelley thus observes: —

> "I do not know whether the critics will reprehend the insertion of some of the most imperfect among these; but I frankly own, that I have been more actuated by the fear lest any monument of his genius should escape me, than the wish of presenting nothing but what was complete to the fastidious reader. I feel secure that the lovers of Shelley's Poetry (who knew how more than any other poet of the present day, every line and word he wrote is instinct with peculiar beauty) will pardon and thank me: I consecrate this volume to them."

Thus the present volume contains all the poetry of Mr. Shelley which either previously existed not in a collective or purchaseable form, or which is wholly original; and when followed by the publication of his prose pieces, which Mrs. Shelley also promises, the entire labours of a man of exalted and original genius,—the more than Lycidas of his day, both in his untimely fate and rare endowments,—will assume a form for the library, and take their proper station among the imperishable records of the passing age.

On the character of the lofty imaginings of the muse of Shelley, it is not here intended to expatiate, and still less to dwell on those eminent moral properties of mind by which he was distinguished—his bold independence and fearless enthusiasm. In reference to the former, as in the case of Byron, the virulent sources of injustice, the pestilent slanderers of everything which dares to think and breathe out of their own miry railway, have done him tardy justice; and as to the latter, setting aside the annoyance of a temporary *brutum fulmen*, the spirit of society is too much out of pupillage for its ultimate decision of character to be affected by the vulgar and common-place clamour of interested bigotry. Agreeing to hide "a bright particular star," they shout and hurl dust into the air; but the intellectual light, and the firmament in which it is placed, are beyond their reach; the noise and the dust subside, and the star shi[n]es serenely on, to be seen and admired for ever.

Among the contents of this volume, we have to remark "The Triumph of Life" as one of the most elaborate of the finished poems of Mr. Shelley, being indeed the latest of that description; it is instinct with poetry and thought, as are several more of kindred length and importance, all of which, down to the briefer fragments, abound with elegant and reflective beauty. There are also some most admirable Translations, and among the rest that of the never translated, or at least never translated and published, "Prologue to Faust,"—a task which possibly, from kindred genius and associations, Mr. Shelley was more equal to than any man on earth. This Prologue, and the "May-day Night," at the Brocken—that indescribably grand and wild emanation of original genius, were given up in the translation of Lord Leveson Gower. Mr. Shelley has given a version of them both; and did this volume contain nothing else, the world of thought and imagination would exclaim, "All hail!"

We have already mentioned the preface by Mrs. Shelley, and may further observe, that the slight portion of narrative it conveys is very interesting. Mrs. S. had hoped for a Life of Mr. Shelley, by one with whom his living spirit had mingled in most intimate and affectionate communication. Circumstances prevented the fulfilment of this hope on the present occasion, but the hope is still entertained.

—Albany Fonblanque, "Literary Notices,"
Examiner, June 13, 1824, p. 370

Unsigned (1824)

Posthumous Poems of Percy B. Shelly, 8 vo.—These effusions of the late Mr. Shelly, are published by his widow, with the avowed purpose of doing honour to his memory. But though some of the pieces possess merit, the volume will hardly add to the literary reputation of the author.

—Unsigned, *Ladies' Monthly Museum*,
August 1824, p. 106

Unsigned "Shelley's Posthumous Poems" (1824)

Amidst the crowd of feeble and tawdry writers with which we are surrounded, tantalizing us with a mere shew of power, and rendering their native baldness more disgusting by the exaggerations and distortions with which they attempt to hide it, it is refreshing to meet a work upon which the genuine mark of intellectual greatness is stamped. Here are no misgivings, no chilling doubts, no reasoning with ourselves as to the grounds of our temporary admiration; no comparison of canons, no reference to criterions of beauty. . . . It is a reviving feeling—a sense of deliverance and of exultation; we are emancipated from the minute and narrowing restraints to which an habitual intercourse with petty prejudices almost insensibly subjects us; we breathe freely in the open air of enlarged thought; and we deem ourselves ennobled by our relation to a superior mind, and by the sense of our own capabilities which its grand conceptions awaken in us.

Such were the feelings—mixed, it is true, and alternating with feelings of a different kind—with which we perused the posthumous poems of Percy Shelley. We are aware that this expression of our sentiments will probably astonish some, and scandalize others. We know that public opinion (that opinion to which every one is now required to surrender the independent suggestions of his own reason and conscience, on pain of ridicule and obloquy) has doomed the name of Shelley to unmixed reprobation. We are a review-and-newspaper-ridden people; and while we contend clamorously for the right of thinking for ourselves, we yet guide ourselves unconsciously by the opinion of censors whom we know to be partial and incompetent. Shelley was a leveller in politics—this all knew; and they had been told that Shelley was an Atheist, that he was a man of flagitious character, and that his poems are nothing more than a heap of bombast and verbiage, of immortality and blasphemy. They believe implicitly what they are taught, and he who would disturb the fixed persuasion runs some danger of being himself involved in the obloquy which he would remove from another. We may be excused from ceremony in contradicting the decisions of an authority of which we do not acknowledge the legitimacy. Let it not be supposed that we are standing forth as the panegyrists of Shelley, when we state our personal delinquency, whether literary or moral. It was not merely that he erred, but that his errors (so far as they were such) were unpopular, and that he was incapable of concealing them. Could he have truckled to the time, —could he have refrained from violating the majesty of custom,— could he have avoided collision with established interests,—could he have condescended, as many others have done, to mask his peculiar opinions under a decent guise of conformity, he might have remained undisturbed. Besides this, the extravagant lengths to which he carried his system afforded more than ordinary facilities for attack; his poetical errors, being errors of excess and not of effect, were peculiarly obnoxious to that kind of ridicule in which modern criticism delights to indulge; and, to crown all, he was the friend of Leigh Hunt and Hazlitt. Hence the critics of one party assailed him without mercy; and as the vindication of his fame was not calculated to serve any temporary purpose, the critics of the other party forbore to defend him!

* * *

Even if our respect for truth did not prevent us from insulting its dignity by a shew of deference to such assailants, it would avail little to set the public opinion right on a particular subject, unless we could at the same time eradicate the servile principle which is the endless source of errors on all subjects. Our only aim in these remarks is to impress on the reader the self-evident truth, that the intellectual as well as the moral character of Shelley's writings is to be judged of from the writings themselves.

* * *

We had intended to add something like a delineation of Shelley's poetical character; but we feel that the task would demand many qualifications which we do not possess. It may suffice to say, as a general description, that his element lay in the mixture of passion and imagination—the imagery being, as it were, impregnated with the passion which brooded over it. His extraordinary sensitive power overbalanced his power of reflection; he would otherwise have been even greater than he was. He wants pliancy of genius; no first-rate poet ever possessed less variety of powers; there is not merely a want of thought, but a want of human interest in his productions.† But no words can do justice to the mixed sublimity and sweetness of his images. It is as if the solid grandeur of Milton were combined with the thrilling vividness and overpowering sweetness of Jeremy Taylor. It is like the glory of the noontide sun, and the glory of the lightning, united in one.

† We except that most powerful work, the Tragedy of the Cenci.

—Unsigned, "Shelley's Posthumous Poems,"
Knight's Quarterly Magazine,
August 1824, pp. 182–199

Unsigned (1824)

There is peace, there is pardon, there is tenderness in the grave. That which in life is denominated by crime, is by death almost softened into error, and Pity goes hand in hand with Reprobation. It is with these feelings we take up this last record of Shelley. Like his other productions, in it are blended beauty and blasphemy, trash by the side of some fine poetry; in short, we can but liken his genius to some African river,—there is gold in its waters, but it is imbedded in sand, mud, slime, and filth. The Witch of Atlas is a good specimen of this author's style: wild, imaginative, revelling in dreams of unreal beauty, it is in the author's peculiar manner.

* * *

This volume is edited by the Widow of the poet, and has a Preface of panegyrie, which may perhaps be excused in consideration of her feeling. But surely it is too hyperbolical to be the effusion of genuine sorrow;* and its estimate of the very rubbish which loads almost every page, cannot be listened to without a direct denial. It is almost inconceivable how extremely certain classes of writers delude themselves. Is it Egotism, which as it were sanctifies to themselves every syllable they pen? What but such a feeling could induce any one to write or publish such trash as the following, which is a fair, or rather favourable average of nine-tenths of this publication:

> Arethusa arose
> From her couch of snows
> In the Aeroceraenian mountains, —
> From cloud and from crag,
> With many a jag,
> Shepherding her bright fountains.
>
> – – – – –
>
> Then Alpheus bold,
> On his glacier cold,
> With his trident the mountains strook;
> And opened a chasm
> In the rocks;—with a spasm
> All Erymanthus shook.

But the brain which conceived, and the hand which wrote, shall conceive and write no more: *This* has mitigated and shortened, and now closes our criticisms.

* "The sea, by its restless moaning, seemed to desire to inform us of what we would not learn." p. vi. &c. &c.

—UNSIGNED, *Posthumous Poems of Percy Bysshe Shelley, Literary Gazette*, 17 July 1824, pp. 451–452

Mary Shelley "Preface" (1839)

Obstacles have long existed to my presenting the public with a perfect edition of Shelley's Poems. These being at last happily removed, I hasten to fulfil an important duty,—that of giving the productions of a sublime genius to the world, with all the correctness possible, and of, at the same time, detailing the history of those productions, as they sprung, living and warm, from his heart and brain. I abstain from any remark on the occurrences of his private life; except, inasmuch as the passions which they engendered, inspired his poetry. This is not the time to relate the truth; and I should reject any colouring of the truth. No account of these events has ever been given at all approaching reality in their details, either as regards himself or others; nor shall I further allude to them than to remark, that the errors of action, committed by a man as noble and generous as Shelley, may, as far as he only is concerned, be fearlessly avowed, by those who loved him, in the firm conviction, that were they judged impartially, his character would stand in fairer and brighter light than that of any contemporary. Whatever faults he had, ought to find extenuation among his fellows, since they proved him to be human; without them, the exalted nature of his soul would have raised him into something divine.

The qualities that struck any one newly introduced to Shelley, were, first, a gentle and cordial goodness that animated his intercourse with warm affection, and helpful sympathy. The other, the eagerness and ardour with which he was attached to the cause of human happiness and improvement; and the fervent eloquence with which he discussed such subjects. His conversation was marked by its happy abundance, and the beautiful language in which he clothed his poetic ideas and philosophical notions. To defecate life of its misery and its evil, was the ruling passion of his soul: he dedicated to it every power of his mind, every pulsation of his heart. He looked on political freedom as the direct agent to effect the happiness of mankind; and thus any new-sprung hope of liberty inspired a joy and an exultation more intense and wild than he could have felt for any personal advantage. Those who have never experienced the working of passion on general and unselfish subjects cannot understand this; and it must be difficult of comprehension to the younger generation rising around, since they cannot remember the scorn and hatred with which the partisans of reform were regarded some few years ago, nor the persecutions to which they were exposed. He had been from youth the victim of the state of feeling inspired by the reaction of the French Revolution; and believing firmly in the justice and excellence of his views, it cannot be wondered that a nature as sensitive, as impetuous, and as generous as his, should put its whole force into the attempt to alleviate for others the evils of those systems from which he had himself suffered. Many advantages attended his birth; he spurned them all when balanced with what he considered his duties. He was generous to imprudence, devoted to heroism.

These characteristics breathe throughout his poetry. The struggle for human weal; the resolution firm to martyrdom; the impetuous pursuit; the glad triumph in good; the determination not to despair. Such were the features that marked those of his works which he regarded with most complacency, as sustained by a lofty subject and useful aim.

In addition to these, his poems may be divided into two classes,—the purely imaginative, and those which sprung from the emotions of his heart. Among the former may be classed *The Witch of Atlas, Adonais,* and his latest composition, left imperfect, *The Triumph of Life.* In the first of these particularly, he gave the reins to his fancy, and luxuriated in every idea as it rose; in all, there is that sense of mystery which formed an essential portion of his perception of life—a clinging to the subtler inner spirit, rather than to the outward form—a curious and metaphysical anatomy of human passion and perception.

The second class is, of course, the more popular, as appealing at once to emotions common to us all; some of these rest on the passion of love; others on grief and despondency; others on the sentiments inspired by natural objects. Shelley's conception of love was exalted, absorbing, allied to all that is purest and noblest in our nature, and warmed by earnest passion; such it appears when he gave it a voice in verse. Yet he was usually averse to expressing these feelings, except when highly idealised; and many of his more beautiful effusions he had cast aside, unfinished, and they were never seen by me till after I had lost him. Others, as for instance, *Rosalind and Helen,* and "Lines written among the Euganean Hills," I found among his papers by chance; and with some difficulty urged him to complete them. There are others, such as the "Ode to the Sky Lark," and "The Cloud," which, in the opinion of many critics, bear a purer poetical stamp than any other of his productions. They were written as his mind prompted, listening to the carolling of the bird, aloft in the azure sky of Italy; or marking the cloud as it sped across the heavens, while he floated in his boat on the Thames.

No poet was ever warmed by a more genuine and unforced inspiration. His extreme sensibility gave the intensity of passion to his intellectual pursuits; and rendered his mind keenly alive to every perception of outward objects, as well as to his internal sensations. Such a gift is, among the sad vicissitudes of human life, the disappointments we meet, and the galling sense of our own mistakes and errors, fraught with pain; to escape from such, he delivered up his soul to poetry, and felt happy when he sheltered himself from the influence of human sympathies, in the wildest regions of fancy. His imagination has been termed too brilliant, his thoughts too subtle. He loved to idealise reality; and this is a taste shared by few. We are willing to have our passing whims exalted into passions, for this gratifies our vanity; but few of us understand or sympathise with the endeavour to ally the love of abstract beauty, and adoration of abstract good, the *to agathon to kalon* of the Socratic philosophers, with our sympathies with our kind. In this Shelley resembled Plato; both taking more delight in the abstract and the ideal, than in the special and tangible. This did not result from imitation; for it was not till Shelley resided in Italy that he made Plato his study; he then translated his *Symposium* and his *Ion*; and the English language boasts of no more brilliant composition, than Plato's *Praise of Love*, translated by Shelley. To return to his own poetry. The luxury of imagination, which sought nothing beyond itself, as a child burthens itself with spring flowers, thinking of no use beyond the enjoyment of gathering them, often showed itself in his verses: they will be only appreciated by minds which have resemblance to his own; and the mystic subtlety of many of his thoughts will share the same fate. The metaphysical strain that characterises much of what he has written, was, indeed, the portion of his works to which, apart from those whose scope was to awaken mankind to aspirations for what he considered the true and good, he was himself particularly attached. There is much, however, that speaks to the many. When he would consent to dismiss these huntings after the obscure, which, entwined with his nature as they were, he did with difficulty, no poet ever expressed in sweeter, more heart-reaching or more passionate verse, the gentler or more forcible emotions of the soul.

A wise friend once wrote to Shelley "You are still very young, and in certain essential respects you do not yet sufficiently perceive that you are so." It is seldom that the young know what youth is, till they have got beyond its period; and time was not given him to attain this knowledge. It must be remembered that there is the stamp of such inexperience on all he wrote; he had not completed his nine-and-twentieth year when he died. The calm of middle life did not add the seal of the virtues which adorn maturity to those generated by the vehement spirit of youth. Through life also he was a martyr to ill health, and constant pain wound up his nerves to a pitch of susceptibility that rendered his views of life different from those of a man in the enjoyment of healthy sensations. Perfectly gentle and forbearing in manner, he suffered a good deal of internal irritability, or rather excitement, and his fortitude to bear was almost always on the stretch; and thus, during a short life, had gone through more experience of sensation, than many whose existence is protracted. "If I die tomorrow," he said, on the eve of his unanticipated death, "I have lived to be older than my father." The weight of thought and feeling burdened him heavily; you read his sufferings in his attenuated frame, while you perceived the mastery he held over them in his animated countenance and brilliant eyes.

He died, and the world showed no outward sign; but his influence over mankind, though slow in growth, is fast augmenting, and in the ameliorations that have taken place in the political state of his country, we may trace in part the operation of his arduous struggles. His spirit gathers peace in its new state from the sense that, though late, his exertions were not made in vain, and in the progress of the liberty he so fondly loved.

He died, and his place among those who knew him intimately, has never been filled up. He walked beside them like a spirit of good to comfort and benefit—to enlighten the darkness of life with irradiations of genius, to cheer it with his sympathy and love. Any one, once attached to Shelley, must feel all other affections, however true and fond, as wasted on barren soil in comparison. It is our best consolation to know that such a pure-minded and exalted being was once among us, and now exists where we hope one day to join him;—although the intolerant, in their blindness, poured down anathemas, the Spirit of Good, who can judge the heart, never rejected him.

In the notes appended to the poems, I have endeavoured to narrate the origin and history of each. The loss of nearly all letters and papers which refer to his early life, renders the execution more imperfect than it would otherwise have been. I have, however, the liveliest recollection of all that was done and said during the period of my knowing him. Every impression is as clear as if stamped yesterday, and I have no apprehension of any mistake in my statements as far as they go. In other respects, I am, indeed, incompetent; but I feel the importance of the task, and regard it as my most sacred duty. I endeavour to fulfil it in a manner he would himself approve; and hope in this publication to lay the first stone of a monument due to Shelley's genius, his sufferings, and his virtues:

> S'al seguir son tarda, Forse avverra che 'l bel nome gentile
> Consacrero con questa stanca penna.

—MARY SHELLEY, "Preface," *The Poetical Works of Percy Bysshe Shelley*, 1839

Edwin P. Whipple "English Poets of the Nineteenth Century" (1845)

Edwin Percy Whipple (1819–86) was born in Gloucester, Massachusetts. One of his earliest published works was an 1843 essay on Thomas Babington Macaulay in the *Boston Miscellany*. Whipple sustained this initial success by publishing other lectures and essays, including *Essays and Reviews* (1848–49) and *Lectures on Subjects Connected with Literature and Life* (1850). After briefly taking the post of literary editor for the *Boston Daily Globe* in 1872, Whipple began a lecture tour, which gained him widespread admiration as one of the leading American writers of his day and likewise resulted in the publication of *Character and Characteristic Men* (1866), *Literature in the Age of Elizabeth* (1869), and *Success and Its Conditions* (1871). The following essay, "English Poets of the Nineteenth Century," was published in *Essays and Reviews*.

The life of Percy Bysshe Shelley presents a notable example of the effects of social persecution on a nature peculiarly fitted to bring us "news from the empyrean." This mode of murder was tried upon Shelley; but his spirit was strong, as well as sensitive, and opposed weapons of ethereal temper to the brutality of his adversaries. His writings, however, give evidence of the injurious influence of the conflict upon the direction of his powers. Possessing one of the most richly gifted minds ever framed by Providence to adorn and bless the world, and a heart whose sympathies comprehended all nature and mankind in the broad sphere of its love, he was still the most unpopular poet of his time—although he indicated, perhaps, more than any other, the tendencies of its imaginative literature, and expressed with more fulness, precision, and beauty, the subtle spirituality of its tone of thought. His character and his writings were elaborately misrepresented. Persons infinitely inferior to him, we will not say in genius, but in honesty, in benevolence, in virtue, in the practice of those duties of love and self-sacrifice which religion enjoins, still contrived to experience for him a mingled feeling of pity and aversion, unexampled even in the annals of the Pharisees. The same sympathizing apologists for the infirmities of genius who shed tears and manufactured palliatives for Burns and Byron, fell back on the rigor and ice of their morality when they mentioned the name of Shelley. His adversaries were often in ludicrous moral contrast to himself. Venal politicians, fattening on public plunder, represented themselves as shocked by his theories of government. Roues were apprehensive that his refined notions of marriage would encourage libertinism. Smooth, practical atheists preached morality and religion to him from quarterly reviews, and defamed him with an arrogant stupidity, and a sneaking injustice, unparalleled in the effronteries and fooleries of criticism. That pure and pious poet, Thomas Moore, conceived it incumbent on himself to warn his immaculate friend, Lord Byron, from being led astray by Shelley's principles—a most useless monition! Poetasters and rhyme-stringers without number were published, puffed, patronized, paid, and forgotten, during the period when the *Revolt of Islam* and *Prometheus Unbound* were only known by garbled extracts which gleamed amid the dull malice of unscrupulous reviews. Men who could not write a single sentence unstained with malignity, selfishness, or some other deadly sin, gravely rebuked him for infidelity, and volunteered their advice as to the manner by which he might become a bad Christian and a good hypocrite. But Shelley happened to be an honest man as well as a poet, and was better contented with proscription, however severe, than with infamy, however splendid. This was a peculiarity of his disposition which made his conduct so enigmatical to the majority of his enemies.

The mode of judging Shelley adopted by his contemporaries, and followed by many similar spirits in our own day, seems to us radically unjust and foolish. It gives a factitious influence to everything noxious in his poetry, and subverts its own end by the unscrupulous eagerness with which it seizes on bad means. It is therefore not entitled to the praise of judicious falsehood and politic bigotry. The critic who would educe a moral from his writings and conduct, must not begin with substituting horror for analysis. The most favorable view can be taken of his character, without compromising a single principle of morality and religion. While this is the case, we see no reason why, in the cause of morality and religion, we should echo stale invectives at conscientious error, and join the hoarse roar of calumny and falsehood over his tomb.

In these remarks we do not intend to deny that Shelley had faults. The magnitude of his genius and virtues should not cover these from view. But we believe that for every act of his life which his conscience did not in its most refined perceptions of duty approve, he experienced an intensity of remorse which few are conscientious enough to appreciate. His education, and the unfortunate influences to which he was subjected, account for the defects in his view of life, and the heretical opinions which mastered his understanding. His position was such that he was impelled, by what may be called his Christian virtues, into what must be called his errors. His self-denial, his benevolence, his moral courage, his finest affections, his deepest convictions of duty, were so addressed as to force him into opposition to established opinions relating to government and religion. The sorrowful interest with which we follow the events of his life arises from the feeling that he was, to a remarkable degree, the victim and prey of circumstances. He was made to see and feel the abuses of things before he understood their uses. In the most emphatic sense of the word, he was a poet. This title, we fear, is too often considered to designate merely a maker of verses; to point out a person who can express thought and emotion with the usual variety of pause, swell, and cadence; and who often contrives to write one thing and live another. Not in this sense was Shelley a poet. He was always terribly in earnest. What he felt and thought, he felt and thought with such intensity as to make his life identical with his verse. He was a hero in the epic life of the nineteenth century. Ideas, abstractions, which pass like flakes of snow into other minds, fell upon his heart like sparks of fire. "He was no tongue-hero, no fine virtue prattler." He did not speak from his lungs, but from his soul. And, sooner than betray one honest conviction of his intellect, sooner than award "mouth-honor" to what he hated as cruelty and oppression, he was willing to have his genius derided and his name defamed.

We have said that Shelley was poetical in what he lived as well as in what he wrote. Those realities which his soul did grasp, it held with invincible courage. Hymns to "Intellectual Beauty" came from his actions as well as his pen. He was endowed by nature with an intellect of great depth and exquisite fineness; an imagination marvellously endowed with the power to give shape and hue to the most shadowy abstractions, which his soaring mind clutched on the vanishing points of human intelligence; a fancy quick to discern the most remote analogies, brilliant, excursive, aerial, affluent in graceful and delicate images; and a sensibility acutely alive to the most fleeting shades of joy and pain,—warm, full, and unselfish in its love, deep-toned and mighty in its indignation. This fiery spiritual essence was enclosed in a frame sensitive enough to be its fit embodiment. Both in mind and body Shelley was so constituted as to require, in his culture, the utmost discrimination and the most loving care. He received the exact opposite of these. The balance of his mind was early overthrown. He had boyish doubts about religion, which he himself could not consider permanent, for his opinions at college vacillated between D'Holbach, Hume, and Plato. These doubts were met, first with contempt, then with anathemas, then with expulsion and disgrace. The consequences may be seen in that wilderness of eloquent contradictions—*Queen Mab*. His more mature opinions were visited with proscription, and he was robbed of his children. In every case truth was so presented to him that he could not accept it without moral degradation. A mere lie of the lip, recommended to him by his preceptor, would have saved him from expulsion from Oxford; a mere outward conformity to conventional usage would have given him the first rank as a rich country gentleman, with houses, lands, and a seat in parliament. Society is admirably versed in the art of converting those sent to bless and cheer it into partial evils. Its success in Shelley's case is noteworthy. It saw that, with all his logical power, he was unfitted to reason on the practical concerns of life where abstract right is modified by a thousand conditions of expediency; that when he perceived cruelty and oppression under the forms of liberty and love, and cant trampling reason in the dust, he was too indignant to discriminate, with the cool unconcern of statesmanship, between a theory and its practice; it saw, in short, that he was a true and earnest poet, with a pulse of fire and a mind of light; and, of course, it denounced, and simpered, and lifted its hands, and rolled its eyes, and pointed its finger, and shot out its tongue, and mouthed its commonplace horror, and drove him from its sweet presence and companionship!

From the dispensers of the government and religion of his own country, Shelley met with little but injustice; in the country of his adoption he saw government and religion controlled by chicane and despotism. All the accidents and circumstances of his condition, from his birth to his death, concurred in placing the most naturally religious of poets in a position of antagonism to the outward forms and creeds of revealed truth.

The writings of Shelley are, to a considerable extent, the history of his mind and heart, as they were affected by personal experiences, and the events of his time. His works are an eloquent protest against the gulf which separates, in life, the actual world from the world perceived by thought and imagination. He desired society to be pure, free, unselfish, devoted to the realization of goodness and beauty; and he believed it capable of that exaltation. For the simplicity of this faith he was doomed to encounter all the perverted truth and goodness that society could command. No man ever lived with a deeper and more inextinguishable thirst to promote human liberty and happiness. This master passion of his nature controlled all his other ambitions, personal or literary. His sense of the hatefulness of oppression, in any form, almost amounted to bodily torture. A wrong done to a nation, the triumph of power over right, filled him with as much grief and indignation as would be excited in common men by the murder of a son or a brother.

The consuming intensity, indeed, with which his soul burned within him at the sight and thought of tyranny, amounted almost to madness. It ran along his veins like tingling fire. His bursts of vehement feeling appear occasionally to rend and tear his frame in their passionate utterance. In the reaction from these periods of agony and anguish of heart, his representations of life were necessarily one-sided. To his mind, in this state, where great evil existed, it drew all things into itself. The following lines exhibit the aspect under which a whole nation appeared to his sight, while his thoughts were filled with its corruptions. They have a moody grandeur of expression which acts powerfully on the sensibility, though they only exhibit the diseased phase of Shelley's philanthropy:—

ENGLAND IN 1819.
An old, blind, mad, despised and dying king,
Princes, the dregs of their dull race, who flow
Through public scorn—mud from a muddy spring,—

Rulers, that neither see, nor feel, nor know, But, leech-like, to their fainting country cling, Till they drop, blind in blood, without a blow,—A people, starved and stabbed in the unfilled field,—An army, which liberticide and prey Makes as a two-edged sword to those who wield,—Golden and sanguine laws, which tempt and slay,—Religion, Christless, Godless—a book sealed; A Senate—Time's worst statute unrepealed,—Are graves, from which a glorious Phantom may Burst, to illumine our tempestuous day. His poems have been charged with a lack of human sympathy—a singular charge' against a poet whose miseries sprung from the intensity of his human sympathies. Indeed, Shelley's sympathies were naturally almost universal. Had his mind received a genial development, had it not been sent back upon itself to prey upon its own energies, we believe that it would have displayed as much comprehension as intensity; for in reading Shelley's poetry we are impressed with what may be termed the infinite capability of the man. The direction his genius takes in any composition never seemed to indicate the bounds of his powers. What he has done we feel not to be so great as what he might have done. From the maturity of the young man who wrote *Prometheus Unbound* and *The Cenci*, what might not have been expected? As it is, innumerable passages might be quoted from his writings, to show the baselessness of the objections to his writings, founded on the assertion of their lack of human sympathy. The predominance of his spiritual over his animal nature; the velocity with which his mind, loosed from the "grasp of gravitation," darted upwards into regions whither slower-pacing imaginations could not follow; the amazing fertility with which he poured out crowds of magnificent images, and the profuse flood of dazzling radiance, blinding the eye with excess of light, which they shed over his compositions; his love of idealizing the world of sense, until it became instinct with thought, and infusing into things dull and lifeless to the sight and touch the qualities of individual existence; the marvellous keenness of insight with which he pierced beneath even the refinements of thought, and evolved new materials of wonder and delight from a seemingly exhausted subject;—all these, to a superficial observer, carry with them the appearance of unreality. A close examination, however, will often prove that the unreality is merely in appearance,—is, in fact, the perception of a higher reality than the world is willing to acknowledge. But, waiving this consideration, no reader of Shelley can be ignorant that his genius sympathized readily with the humble as well as the lofty; that some of the most beautiful exhibitions of the tenderest and simplest affections of the heart are to be found in his writings; that he had an ear exquisitely tuned to catch the "still, sad music of humanity;" that human hopes, and fears, and loves, all woke sympathetic echoes in his heart; that the language of human passion kindles and burns along his creations, often with a might and freedom almost Shaksperean. Leigh Hunt finely says of him,—"Whether interrogating Nature in the icy solitudes of Chamouny, or thrilling with the lark in the sunshine, or shedding indignant tears with sorrow and poverty, or pulling flowers like a child in the field, or pitching himself back into the depths of time and space, and discoursing with the first forms and gigantic shadows of creation, he was alike in earnest and alike at home."

The great stigma cast upon Shelley's writings is irreligion. As far as this is well founded, it is most certainly to be regretted, and to be condemned. There are many passages in his works evincing much presumption and arrogance, which we could wish blotted out of existence, were it not for the moral they convey to Christians, and the light they throw upon the history of his mind's development. We suppose it would be difficult to adduce any man of genius, who experienced less Christianity from others, and exercised more towards others, than Shelley. It was but natural that a man with so acute a sensibility should confound his own outward experience of religionists with religion. It is a matter of astonishment to us, that those who rail against Shelley for certain rash and wayward infidelities of expressions in his works, do not ask themselves whether excitable minds are not driven daily into similar infidelities by the same causes which influenced him. The man who sees Christianity only in its unnatural connection with fanaticism and hypocrisy, may be pardoned, at least, for rejecting the latter; and they, at the bottom, were what Shelley rejected.

—Edwin P. Whipple, "English Poets
of the Nineteenth Century," 1845,
Essays and Reviews 1850, vol. 1, pp. 308–317

Robert Browning "Introduction" (1852)

Born in London, Robert Browning (1812–89) stands as one of the preeminent poets of the Victorian era and one of the most gifted practitioners of the dramatic monologue. His poems were collected in works such as *Men and Women* (1855), *Dramatis Personae* (1864), and *The Ring and the Book*

(1868–69), the latter establishing Browning's reputation in literary circles. In 1846, he married fellow writer Elizabeth Barrett, and the couple moved to Florence, where they lived until Elizabeth's death in 1861. Browning spent his remaining years enjoying his literary success and achieved national fame as a poet. He died at his son's house in Venice and is buried in Westminster Abbey. The following introduction was republished by the Shelley Society in 1888 as a separate essay. W.T. Harden gives the following introduction to the essay:

> The Circumstances under which the following "Essay" was first published in 1852 were so far unfortunate as that a speedy limit was put to its circulation by the discovery that the letters which it ushered into the world were a literary fraud. But if ever the doing of evil is to be excused because of some resultant good, here is a case which is eminently entitled to such consideration, for we may fairly conclude, and not without a touch of humour, if not also without a tremor of anxiety, that if the fraud had not been perpetrated the essay might never have been penned. Equally fortunate was the fact that some few copies escaped the control and the recall of the publisher, which however were so few that the book is now one of those *opima spolia* that collectors covet and dealers delight in.
>
> For if the letters were spurious and worthless the essay was genuine and most valuable. It was surely by some occult and happy inspiration that the writer treated his subject both broadly and deeply, not toying with the handful of letters, but passing to their supposed author and taking the opportunity to analyse his genius and to vindicate his character. So ably was this done, with such keen appreciation of intellectual qualities and such generous discernment of moral probabilities, that the essay must always remain essential alike to the students of Browning and of Shelley, and deserves to stand both as a prologue to the writings and as an epilogue to the life of Shelley.
>
> The monograph should at least be in the hands of every member of the Shelley Society, constituting as it does a practical defence of Shelley, and concentrating in itself the spirit of those various testimonies of admiration which occur elsewhere in Mr. Browning's writings. Without in the least disparaging the well-known passages in *Pauline* which refer to Shelley it is obvious that the essay has a much higher value; it was, when published, the well-ripened fruit of Mr. Browning's mature judgment, and it remains the acknowledged expression of his final conclusion.... (pp. 7–8).

An opportunity having presented itself for the acquisition of a series of unedited letters by Shelley, all more or less directly supplementary to and illustrative of the collection already published by Mr. Moxon, that gentleman has decided on securing them. They will prove an acceptable addition to a body of correspondence, the value of which towards a right understanding of its author's purpose and work, may be said to exceed that of any similar contribution exhibiting the worldly relations of a poet whose genius has operated by a different law.

Doubtless we accept gladly the biography of an objective poet, as the phrase now goes; one whose endeavour has been to reproduce things external (whether the phenomena of the scenic universe, or the manifested action of the human heart and brain) with an immediate reference, in every case, to the common eye and apprehension of his fellow men, assumed capable of receiving and profiting by this reproduction. It has been obtained through the poet's double faculty of seeing external objects more clearly, widely, and deeply, than is possible to the average mind, at the same time that he is so acquainted and in sympathy with its narrower comprehension as to be careful to supply it with no other materials than it can combine into an intelligible whole. The auditory of such a poet will include, not only the intelligences which, save for such assistance, would have missed the deeper meaning and enjoyment of the original objects, but also the spirits of a like endowment with his own, who, by means of his abstract, can forthwith pass to the reality it was made from, and either corroborate their impressions of things known already, or supply themselves with new from whatever shows in the inexhaustible variety of existence may have hitherto escaped their knowledge. Such a poet is properly the ποιητής, the fashioner; and the thing fashioned, his poetry, will of necessity be substantive, projected from himself and distinct. We are ignorant what the inventor of *Othello* conceived of that fact as he beheld it in completeness, how he accounted for it, under what known law he registered its nature, or to what unknown law he traced its coincidence. We learn only what he intended we should learn by that particular exercise of his power,—the fact itself,—which, with its infinite significances, each of us receives for the first time as a creation, and is hereafter left to deal with, as, in proportion to his own intelligence, he best may. We are ignorant, and would fain be otherwise.

Doubtless, with respect to such a poet, we covet his biography. We desire to look back upon the process of gathering together in a lifetime, the materials of the work we behold entire; of elaborating, perhaps under difficulty and with hindrance, all that is familiar to our admiration in the apparent facility of success. And the inner impulse of this effort and operation, what induced it? Did a soul's delight in its own extended sphere of vision set it, for the gratification of an insuppressible power, on labour, as other men are set on rest? Or did a sense of duty or of love lead it to communicate its own sensations to mankind? Did an irresistible sympathy with men compel it to bring down and suit its own provision of knowledge and beauty to their narrow scope? Did the personality of such an one stand like an open watch-tower in the midst of the territory it is erected to gaze on, and were the storms and calms, the stars and meteors, its watchman was wont to report of, the habitual variegation of his every-day life, as they glanced across its open roof or lay reflected on its four-square parapet? Or did some sunken and darkened chamber of imagery witness, in the artificial illumination of every storied compartment we are permitted to contemplate, how rare and precious were the outlooks through here and there an embrasure upon a world beyond, and how blankly would have pressed on the artificer the boundary of his daily life, except for the amorous diligence with which he had rendered permanent by art whatever came to diversify the gloom? Still, fraught with instruction and interest as such details undoubtedly are, we can, if needs be, dispense with them. The man passes, the work remains. The work speaks for itself, as we say: and the biography of the worker is no more necessary to an understanding or enjoyment of it, than is a model or anatomy of some tropical tree, to the right tasting of the fruit we are familiar with on the market-stall,—or a geologist's map and stratification, to the prompt recognition of the hill-top, our land-mark of every day.

We turn with stronger needs to the genius of an opposite tendency—the subjective poet of modern classification. He, gifted like the objective poet with the fuller perception of nature and man, is impelled to embody the thing he perceives, not so much with reference to the many below as to the one above him, the supreme Intelligence which apprehends all things in their absolute truth,—an ultimate view ever aspired to, if but partially attained, by the poet's own soul. Not what man sees, but what God sees—the *Ideas* of Plato, seeds of creation lying burningly on the Divine Hand—it is toward these that he struggles. Not with the combination of humanity in action, but with the primal elements of humanity he has to do; and he digs where he stands,—preferring to seek them in his own soul as the nearest reflex of that absolute Mind, according to the intuitions of which he desires to perceive and speak. Such a poet does not deal habitually with the picturesque groupings and tempestuous tossings of the forest-trees, but with their roots and fibres naked to the chalk and stone. He does not paint pictures and hang them on the walls, but rather carries them on the retina of his own eyes: we must look deep into his human eyes, to see those pictures on them. He is rather a seer, accordingly, than a fashioner, and what he produces will be less a work than an effluence. That effluence cannot be easily considered in abstraction from his personality,—being indeed the very radiance and aroma of his personality, projected from it but not separated. Therefore, in our approach to the poetry, we necessarily approach the personality of the poet; in apprehending it we apprehend him, and certainly we cannot love it without loving him. Both for love's and for understanding's sake we desire to know him, and as readers of his poetry must be readers of his biography also.

I shall observe, in passing, that it seems not so much from any essential distinction in the faculty of the two poets or in the nature of the objects contemplated by either, as in the more immediate adaptability of these objects to the distinct purpose of each, that the objective poet, in his appeal to the aggregate human mind, chooses to deal with the doings of men, (the result of which dealing, in its pure form, when even description, as suggesting a describer, is dispensed with, is what we call dramatic poetry), while the subjective poet, whose study has been himself, appealing through himself to the absolute Divine mind, prefers to dwell upon those external scenic appearances which strike out most abundantly and uninterruptedly his inner light and power, selects that silence of the earth and sea in which he can best hear the beating of his individual heart, and leaves the noisy, complex, yet imperfect exhibitions of nature in the manifold experience of man around him, which serve only to distract and suppress the working of his brain. These opposite tendencies of genius will be more readily descried in their artistic effect than in their moral spring and cause. Pushed to an extreme and manifested as a deformity, they will be seen plainest of all in the fault of either artist, when subsidiarily to the human interest of his work his occasional illustrations from scenic nature are introduced as in the earlier works of the originative painters—men and women filling the foreground with consummate mastery, while mountain, grove and rivulet show like an anticipatory revenge on that succeeding race of landscape-painters whose "figures" disturb the perfection of their earth and sky. It would be idle to inquire, of these two kinds of poetic faculty in operation, which is the higher or even rarer endowment. If the subjective might seem to be the ultimate requirement of every age, the objective, in the strictest state, must still retain its original value. For it is with this world, as starting point and basis alike, that we shall always have to concern ourselves: the world is not to be learned and thrown aside, but reverted to and relearned. The spiritual comprehension may be infinitely subtilised, but the raw material it operates upon, must remain. There may be no end of the poets who communicate to us what they see in an object with reference to their own individuality; what it was before they saw it, in reference to the aggregate human mind, will be as desirable to know as ever. Nor is there any reason why these two modes of poetic faculty may not issue hereafter from the same poet in successive perfect works, examples of which, according to what are now considered the exigences of art, we have hitherto possessed in distinct individuals only. A mere running-in of the one faculty upon the other, is, of course, the ordinary circumstance. Far more rarely it happens that either is found so decidedly prominent and superior, as to be pronounced comparatively pure: while of the perfect shield, with the gold and the silver side set up for all comers to challenge, there has yet been no instance. Either faculty in its eminent state is doubtless conceded by Providence as a best gift to men, according to their especial want. There is a time when the general eye has, so to speak, absorbed its fill of the pheonomena around it, whether spiritual or material, and desires rather to learn the exacter significance of what it possesses, than to receive any augmentation of what is possessed. Then is the opportunity for the poet of loftier vision, to lift his fellows, with their half-apprehensions, up to his own sphere, by intensifying the import of details and rounding the universal meaning. The influence of such an achievement will not soon die out. A tribe of successors (Homerides) working more or less in the same spirit, dwell on his discoveries and reinforce his doctrine; till, at unawares, the world is found to be subsisting wholly on the shadow of a reality, on sentiments diluted from passions, on the tradition of a fact, the convention of a moral, the straw of last year's harvest. Then is the imperative call for the appearance of another sort of poet, who shall at once replace this intellectual rumination of food swallowed long ago, by a supply of the fresh and living swathe; getting at new substance by breaking up the assumed wholes into parts of independent and unclassed value, careless of the unknown laws for recombining them (it will be the business of yet another poet to suggest those hereafter), prodigal of objects for men's outer and not inner sight, shaping for their uses a new and different creation from the last, which it displaces by the right of life over death,—to endure until, in the inevitable process, its very sufficiency to itself shall require, at length, an exposition of its affinity to something higher,— when the positive yet conflicting facts shall again precipitate themselves under a harmonising law, and one more degree will be apparent for a poet to climb in that mighty ladder, of which, however cloud-involved and undefined may glimmer the topmost step, the world dares no longer doubt that its gradations ascend.

Such being the two kinds of artists, it is naturally, as I have shown, with the biography of the subjective poet that we have the deeper concern. Apart from his recorded life altogether, we might fail to determine with satisfactory precision to what class his productions belong, and what amount of praise is assignable to the producer. Certainly, in the face of any conspicuous achievement of genius, philosophy, no less than sympathetic instinct, warrants our belief in a great moral purpose having mainly inspired even where it does not visibly look out of the same. Greatness in a work suggests an adequate instrumentality; and none of the lower incitements, however they may avail to initiate or even effect many considerable displays of power, simulating the nobler inspiration to which they are mistakenly referred, have been found able, under the ordinary conditions of humanity, to task themselves to the end of so exacting a performance as a poet's complete work. As soon will the galvanism that provokes to violent action the muscles of a corpse, induce it to cross the chamber steadily: sooner. The love of displaying power for the display's sake, the love of riches, of distinction, of notoriety,—the desire of a triumph over rivals, and the vanity in the applause of friends,—each and all of such whetted appetites grow intenser by exercise and increasingly sagacious as to the best and readiest means of self-appeasement,—while for any of their ends, whether the money or the pointed finger of the crowd, or the flattery and hate to heart's content, there are cheaper prices to pay, they will all find soon enough, than the bestowment of a life upon a labour, hard, slow, and not sure. Also, assuming the proper moral aim to have produced a work, there are many and various states of an aim: it may be more intense than clearsighted, or too easily satisfied with a lower field of activity than a steadier aspiration would reach. All the bad poetry in the world (accounted poetry, that is, by its affinities) will be found to result from some one of the infinite degrees of discrepancy between the attributes of the poet's soul, occasioning a want of correspondency between his work and the verities of nature,— issuing in poetry, false under whatever form, which shows a thing not as it is to mankind generally, nor as it is to the particular describer, but as it is supposed to be for some unreal neutral mood, midway between both and of value to neither, and living its brief minute simply through the indolence of whoever accepts it or his incapacity to denounce a cheat. Although of such depths of failure there can be no question here we must in every case betake ourselves to the review of a poet's life ere we determine some of the nicer questions concerning his poetry,—more especially if the performance we seek to estimate aright, has been obstructed and cut short of completion by circumstances,—a disastrous youth or a premature death. We may learn from the biography whether his spirit invariably saw and spoke from the last height to which it had attained. An absolute vision is not for this world, but we are permitted a continual approximation to it, every degree of which in the individual, provided it exceed the attainment of the masses, must procure him a clear advantage. Did the poet ever attain to a higher platform than where he rested and exhibited a result? Did he know more than he spoke of?

I concede however, in respect to the subject of our study as well as some few other illustrious examples, that the unmistakeable quality of the verse would be evidence enough, under usual circumstances, not only of the kind and degree of the intellectual but of the moral constitution of Shelley: the whole personality of the poet shining forward from the poems, without much need of going further to seek it. The Remains—produced within a period of ten years, and at a season of life when other men of at all comparable genius have hardly done more than prepare the eye for future sight and the tongue for speech—present us with the complete enginery of a poet, as signal in the excellence of its several adaptitudes as transcendent in the combination of effects,—examples, in fact, of the whole poet's function of beholding with an understanding keenness the universe, nature and man, in their actual state of perfection in imperfection,—of the whole poet's virtue of being untempted by the manifold partial developments of beauty and good on every side, into leaving them the ultimates he found them,—induced by the facility of the gratification of his own sense of those qualities, or by the pleasure of acquiescence in the short-comings of his predecessors in art, and the pain of disturbing their conventionalisms,—the whole poet's virtue, I repeat, of looking higher than any manifestation yet made of both beauty and good, in order to suggest from the utmost actual realisation of the one a corresponding capability in the other, and out of the calm, purity and energy of nature, to reconstitute and store up for the forthcoming stage of man's being, a gift in repayment of that former gift, in which man's own thought and passion had been lavished by the poet on the else-incompleted magnificence of the sunrise, the else-uninterpreted mystery of the lake,—so drawing out, lifting up, and assimilating this ideal of a future man, thus descried as possible, to the present reality of the poet's soul already arrived at the higher state of development, and still aspirant to elevate and extend itself in conformity with its still-improving perceptions of, no longer the eventual Human, but the actual Divine. In conjunction with which noble and rare powers, came the subordinate power of delivering these attained results to the world in an embodiment of verse more closely answering to and indicative of the process of the informing spirit, (failing as it occasionally does, in art, only to succeed in highest art),—with a diction more adequate to the task in its natural and acquired richness, its material colour and spiritual transparency,—the whole being moved by and suffused with a music at once of the soul and the sense, expressive both of an external might of sincere passion and an internal fitness and consonancy,—than can be attributed to any other writer whose record is among us. Such was the spheric poetical faculty of Shelley, as its own self-sufficing central light, radiating equally through immaturity and accomplishment, through many fragments and occasional completion, reveals it to a competent judgment.

But the acceptance of this truth by the public, has been retarded by certain objections which cast us back on the evidence of biography, even with Shelley's poetry in our hands. Except for the particular character of these objections, indeed, the non-appreciation of his contemporaries would simply class, now that it is over, with a series of experiences which have necessarily happened and needlessly been wondered at, ever since the world began, and concerning which any present anger may well be moderated, no less in justice to our forerunners than in policy to ourselves. For the misapprehensiveness of his age is exactly what a poet is sent to remedy; and the interval between his operation and the generally perceptible effect of it, is no greater, less indeed, than in many other departments of the great human effort. The "E pur si muove" of the astronomer was as bitter a word as any uttered before or since by a poet over his rejected living work, in that depth of conviction which is so like despair.

But in this respect was the experience of Shelley peculiarly unfortunate—that the disbelief in him as a man, even preceded the disbelief in him as a writer; the misconstruction of his moral nature preparing the way for the misappreciation of his intellectual labours. There existed from the beginning,—simultaneous with, indeed anterior to his earliest noticeable works, and not brought forward to counteract any impression they had succeeded in making,—certain charges against his private character and life, which, if substantiated to their whole breadth, would materially disturb, I do not attempt to deny, our reception and enjoyment of his works, however wonderful the artistic qualities of these. For we are not sufficiently supplied with instances of genius of his order, to be able to pronounce certainly how many of its constituent parts have been tasked and strained to the production of a given lie, and how high and pure a mood of the creative mind may be dramatically simulated as the poet's habitual and exclusive one. The doubts, therefore, arising from such a question, required to be set at rest, as they were effectually, by those early authentic notices of Shelley's career and the corroborative accompaniment of his letters, in which not only the main tenor and principal result of his life, but the purity and beauty of many of the processes which had conduced to them, were made apparent enough for the general reader's purpose,—whoever lightly condemned Shelley first, on the evidence of reviews and gossip, as lightly acquitting him now, on that of memoirs and correspondence. Still, it is advisable to lose no opportunity of strengthening and completing the chain of biographical testimony; much more, of course, for the sake of the poet's original lovers, whose volunteered sacrifice of particular principle in favour of absorbing sympathy we might desire to dispense with, than for the sake of his foolish haters, who have long since diverted upon other objects their obtuseness or malignancy. A full life of Shelley should be written at once, while the materials for it continue in reach; not to minister to the curiosity of the public, but to obliterate the last stain of that false life which was forced on the public's attention before it had any curiosity on the matter,—a biography, composed in harmony with the present general disposition to have faith in him, yet not shrinking from a candid statement of all ambiguous passages, through a reasonable confidence that the most doubtful of them will be found consistent with a belief in the eventual perfection of his character, according to the poor limits of our humanity. Nor will men persist in confounding, any more than God confounds, with genuine infidelity and an atheism of the heart, those passionate, impatient struggles of a boy towards distant truth and love, made in the dark, and ended by one sweep of the natural seas before the full moral sunrise could shine out on him. Crude convictions of boyhood, conveyed in imperfect and inapt forms of speech,—for such things all boys have been pardoned. There are growing-pains, accompanied by temporary distortion, of the soul also. And it would be hard indeed upon this young Titan of genius, murmuring in divine music his human ignorances, through his very thirst for knowledge, and his rebellion, in mere aspiration to law, if the melody itself substantiated the error, and the tragic cutting short of life perpetuated into sins, such faults as, under happier circumstances, would have been left behind by the consent of the most arrogant moralist, forgotten on the lowest steps of youth. The responsibility of presenting to the public a biography of Shelley, does not, however lie with me: I have only to make it a little easier by arranging these few supplementary letters, with a recognition of the value of the whole collection. This value I take to consist in a most truthful conformity of the Correspondence, in its limited degree, with the moral and intellectual character of the writer as displayed in the highest manifestations of his genius. Letters and poems are obviously an act of the same mind, produced by the same law, only differing in the application to the individual or collective understanding. Letters and poems may be used indifferently as the basement of our opinion upon the writer's character; the finished expression of a sentiment in the poems, giving light and significance to the rudiments of the same in the letters, and these, again, in their incipiency and unripeness, authenticating the exalted mood and reattaching it to the personality of the writer. The musician speaks on the note he sings with; there is no change in the scale, as he diminishes the volume into familiar intercourse. There is nothing of that jarring between the man and the author, which has been found so amusing or so melancholy; no dropping of the tragic mask, as the crowd melts away; no mean discovery of the real motives of a life's achievement, often, in other lives, laid bare as pitifully as when, at the close of a holiday, we catch sight of the internal lead-pipes and wood-valves, to which, and not to the ostensible conch and dominant Triton of the fountain, we have owed our admired waterwork. No breaking out, in household privacy, of hatred anger and scorn, incongruous with the higher mood and suppressed artistically in the book: no brutal return to self-delighting, when the audience of philanthropic schemes is out of hearing: no indecent stripping off the grander feeling and rule of life as too costly and cumbrous for every-day wear. Whatever Shelley was, he was with an admirable sincerity. It was not always truth that he thought and spoke; but in the purity of truth he spoke and thought always. Everywhere is apparent his belief in the existence of Good, to which Evil is an accident; his faithful holding by what he assumed to be the former, going everywhere in company with the tenderest pity for those acting or suffering on the opposite hypothesis. For he was tender, though tenderness is not always the characteristic of very sincere natures; he was eminently both tender and sincere. And not only do the same affection and yearning after the well-being of his kind, appear in the letters as in the poems, but they express themselves by the same theories and plans, however crude and unsound. There is no reservation of a subtler, less costly, more serviceable remedy for his own ill, than he has proposed for the general one; nor does he ever contemplate an object on his own account, from a less elevation than he uses in exhibiting it to the world. How shall we help believing Shelley to have been, in his ultimate attainment, the splendid spirit of his own best poetry, when we find even his carnal speech to agree faithfully, at faintest as at strongest, with the tone and rhythm of his most oracular utterances?

For the rest, these new letters are not offered as presenting any new feature of the poet's character. Regarded in themselves, and as the substantive productions of a man, their importance would be slight. But they possess interest beyond their limits, in confirming the evidence just dwelt on, of the poetical mood of Shelley being only the intensification of his habitual mood; the same tongue only speaking, for want of the special excitement to sing. The very first letter, as one instance for all, strikes the key-note of the predominating sentiment of Shelley throughout his whole life—his sympathy with the oppressed. And when we see him at so early an age, casting out, under the influence of such a sympathy, letters and pamphlets on every side, we accept it as the simple exemplification of the sincerity, with which, at the close of his life, he spoke of himself, as—

> One whose heart a stranger's tear might wear
> As water-drops the sandy fountain stone;
> Who loved and pitied all things, and could moan
> For woes which others hear not, and could see
> The absent with the glass of phantasy,
> And near the poor and trampled sit and weep,
> Following the captive to his dungeon deep—
> One who was as a nerve o'er which do creep
> The else-unfelt oppressions of this earth.

Such sympathy with his kind was evidently developed in him to an extraordinary and even morbid degree, at a period when the general intellectual powers it was impatient to put in motion, were immature or deficient.

I conjecture, from a review of the various publications of Shelley's youth, that one of the causes of his failure at the outset, was the peculiar *practicalness* of his mind, which was not without a determinate effect on his progress in theorising. An ordinary youth, who turns his attention to similar subjects, discovers falsities, incongruities, and various points for amendment, and, in the natural advance of the purely critical spirit unchecked by considerations of remedy, keeps up before his young eyes so many instances of the same error and wrong, that he finds himself unawares arrived at the startling conclusion, that all must be changed—or nothing: in the face of which plainly impossible achievement, he is apt (looking perhaps a little more serious by the time he touches at the decisive issue), to feel, either carelessly or considerately, that his own attempting a single piece of service would be worse than useless even, and to refer the whole task to another age and person—safe in proportion to his incapacity. Wanting words to speak, he has never made a fool of himself by speaking. But, in Shelley's case, the early fervour and power to *see*, was accompanied by as precocious a fertility to *contrive*: he endeavoured to realise as he went on idealising; every wrong had simultaneously its remedy, and, out of the strength of his hatred for the former, he took the strength of his confidence in the latter—till suddenly he stood pledged to the defence of a set of miserable little expedients, just as if they represented great principles, and to an attack upon various great principles, really so, without leaving himself time to examine whether, because they were antagonistical to the remedy he had suggested, they must therefore be identical or even essentially connected with the wrong he sought to cure,—playing with blind passion into the hands of his enemies, and dashing at whatever red cloak was held forth to him, as the cause of the fireball he had last been stung with— mistaking Churchdom for Christianity, and for marriage, "the sale of love" and the law of sexual oppression.

Gradually, however, he was leaving behind him this low practical dexterity, unable to keep up with his widening intellectual perception; and, in exact proportion as he did so, his true power strengthened and proved itself. Gradually he was raised above the contemplation of spots and the attempt at effacing them, to the great Abstract Light, and, through the discrepancy of the creation, to the sufficiency of the First Cause. Gradually he was learning that the best way of removing abuses is to stand fast by truth. Truth is one, as they are manifold; and innumerable negative effects are produced by the upholding of one positive principle. I shall say what I think,—had Shelley lived he would have finally ranged himself with the Christians; his very instinct for helping the weaker side (if numbers make strength), his very "hate of hate," which at first mistranslated itself into delirious Queen Mab notes and the like, would have got clearer-sighted by exercise. The preliminary step to following Christ, is the leaving the dead to bury their dead—not clamouring on His doctrine for an especial solution of difficulties which are referable to the general problem of the universe. Already he had attained to a profession of "a worship to the Spirit of good within, which requires (before it sends that inspiration forth, which impresses its likeness upon all it creates) devoted and disinterested homage, *as Coleridge says,*"—and Paul likewise. And we find in one of his last exquisite fragments, avowedly a record of one of his own mornings and its experience, as it dawned on him at his soul and body's best in his boat on the Serchio—that as surely as

> The stars burnt out in the pale blue air, And the thin white moon lay withering there— Day had kindled the dewy woods, And the rocks above, and the stream below, And the vapours in their multitudes, And the Apennine's shroud of summer snow— Day had awakened all things that be;

just so surely, he tells us (stepping forward from this delicious dance-music, choragus-like, into the grander measure befitting the final enunciation),

> All rose to do the task He set to each, Who shaped us to his ends and not our own; The million rose to learn, and One to teach What none yet ever knew or can be known. No more difference than this, from David's pregnant conclusion so long ago!

Meantime, as I call Shelley a moral man, because he was true, simple-hearted, and brave, and because what he acted corresponded to what he knew, so I call him a man of religious mind, because every audacious negative cast up by him against the Divine, was interpenetrated with a mood of reverence and adoration,—and because I find him everywhere taking for granted some of the capital dogmas of Christianity, while most vehemently denying their historical basement. There is such a thing as an efficacious knowledge of and belief in the politics of Junius, or the poetry of Rowley, though a man should at the same time dispute the title of Chatterton to the one, and consider the author of the other, as Byron wittily did, "really, truly, nobody at all."[1] There is even such a thing, we come to learn wonderingly in these very letters, as a profound sensibility and adaptitude for art, while the science of the percipient is so little advanced as to admit of his stronger admiration for Guido (and Carlo Dolce!) than for Michael Angelo. A Divine Being has Himself said, that "a word against the Son of man shall be forgiven to a man," while "a word against the Spirit of God" (implying a general deliberate preference of perceived evil to perceived good) "shall not be forgiven to a man." Also, in religion, one earnest and unextorted assertion of belief should outweigh, as a matter of testimony, many assertions of unbelief. The fact that there is a gold-region is established by the finding of one lump, though you miss the vein never so often.

He died before his youth ended. In taking the measure of him as a man, he must be considered on the whole and at his ultimate spiritual stature, and not be judged of at the immaturity and by the mistakes of ten years before: that, indeed, would be to judge of the author of *Julian and Maddalo* by *Zastrozzi*. Let the whole truth be told of his worst mistake. I believe, for my own part, that if anything could now shame or grieve Shelley, it would be an attempt to vindicate him at the expense of another.

In forming a judgment, I would, however, press on the reader the simple justice of considering tenderly his constitution of body as well as mind, and how unfavourable it was to the steady symmetries of conventional life; the body, in the torture of incurable disease, refusing to give repose to the bewildered soul, tossing in its hot fever of the fancy,—and the laudanum-bottle making but a perilous and pitiful truce between these two. He was constantly subject to "that state of mind" (I quote his own note to *Hellas*) "in which ideas may be supposed to assume the force of sensation, through the confusion of thought with the objects of thought, and excess of passion animating the creations of the imagination:" in other words, he was liable to remarkable delusions and hallucinations. The nocturnal attack in Wales, for instance, was assuredly a delusion; and I venture to express my own conviction, derived from a little attention to the circumstances of either story, that the idea of the enamoured lady following him to Naples, and of the "man in the cloak" who struck him at the Pisan post-office, were equally illusory,—the mere projection, in fact, from himself, of the image of his own love and hate.

> To thirst and find no fill—to wail and wander With short unsteady steps—to pause and ponder— To feel the blood run through the

> veins and tingle When busy thought and blind sensation mingle,—
> To nurse the image of *unfelt caresses* Till dim imagination just possesses The half-created shadow—

of unfelt caresses,—and of unfelt blows as well: to such conditions was his genius subject. It was not at Rome only (where he heard a mystic voice exclaiming, "Cenci, Cenci," in reference to the tragic theme which occupied him at the time),—it was not at Rome only that he mistook the cry of "old rags." The habit of somnambulism is said to have extended to the very last days of his life.

Let me conclude with a thought of Shelley as a poet. In the hierarchy of creative minds, it is the presence of the highest faculty that gives first rank, in virtue of its kind, not degree; no pretension of a lower nature, whatever the completeness of development or variety of effect, impeding the precedency of the rarer endowment though only in the germ. The contrary is sometimes maintained; it is attempted to make the lower gifts (which are potentially included in the higher faculty) of independent value, and equal to some exercise of the special function. For instance, should not a poet possess common sense? Then the possession of abundant common sense implies a step towards becoming a poet. Yes; such a step as the lapidary's, when, strong in the fact of carbon entering largely into the composition of the diamond, he heaps up a sack of charcoal in order to compete with the Koh-i-noor. I pass at once, therefore, from Shelley's minor excellencies to his noblest and predominating characteristic.

This I call his simultaneous perception of Power and Love in the absolute, and of Beauty and Good in the concrete, while he throws, from his poet's station between both, swifter, subtler, and more numerous films for the connexion of each with each, than have been thrown by any modern artificer of whom I have knowledge; proving how, as he says,

> The spirit of the worm within the sod,
> In love and worship blends itself with God.

I would rather consider Shelley's poetry as a sublime fragmentary essay towards a presentment of the correspondency of the universe to Deity, of the natural to the spiritual, and of the actual to the ideal, than I would isolate and separately appraise the worth of many detachable portions which might be acknowledged as utterly perfect in a lower moral point of view, under the mere conditions of art. It would be easy to take my stand on successful instances of objectivity in Shelley: there is the unrivalled *Cenci;* there is the *Julian and Maddalo* too; there is the magnificent "Ode to Naples:" why not regard, it may be said, the less organised matter as the radiant elemental foam and solution, out of which would have been evolved, eventually, creations as perfect even as those? But I prefer to look for the highest attainment, not simply the high,—and, seeing it, I hold by it. There is surely enough of the work "Shelley" to be known enduringly among men, and, I believe, to be accepted of God, as human work may; and around the imperfect proportions of such, the most elaborated productions of ordinary art must arrange themselves as inferior illustrations.

It is because I have long held these opinions in assurance and gratitude, that I catch at the opportunity offered to me of expressing them here; knowing that the alacrity to fulfil an humble office conveys more love than the acceptance of the honour of a higher one, and that better, therefore, than the signal service it was the dream of my boyhood to render to his fame and memory, may be the saying of a few, inadequate words upon these scarcely more important supplementary letters of SHELLEY.

Notes

1. Or, to take our illustrations from the writings of Shelley himself, there is such a thing as admirably appreciating a work by Andrea Verocchio,—and fancifully characterising the Pisan Torre Guelfa by the Ponte a Mare, black against the sunsets,—and consummately painting the islet of San Clemente with its penitentiary for rebellious priests, to the west between Venice and the Lido—while you believe the first to be a fragment of an antique sarcophagus,— the second, Ugolino's Tower of Famine (the vestiges of which should be sought for in the Piazza de'Cavalieri)—and the third (as I convinced myself last summer at Venice), San Servolo with its madhouse—which, far from being "windowless," is as full of windows as a barrack.

—ROBERT BROWNING, "Introduction,"
Letters of Percy Bysshe Shelley, 1852

ALGERNON CHARLES SWINBURNE
"NOTES ON THE TEXT OF SHELLEY" (1869)

It is no bad way of testing an opinion held vaguely but sincerely to take it up and rub it, as it were, against the opinion of some one else, who is clearly worth agreeing with or disagreeing. Mr. (Matthew) Arnold, with whose clear and critical spirit it is always good to come in contact, as disciple or as dissenter, has twice spoken of Shelley, each time, as I think, putting forth a brilliant error, shot through and spotted with glimpses of truth. Byron and Shelley, he says, "two members of the aristocratic class," alone in their day, strove "to apply the modern spirit" to English literature. "Aristocracies are, as such, naturally impenetrable by ideas; but their individual members have a high courage and a turn for breaking bounds; and a man of genius, who is the born child of the idea, happening to be born in the aristocratic ranks, chafes against the obstacles which prevent him from freely developing it." To the truth of this he might have cited a third witness; for of the English poets then living, three only were children of the social or political idea, strong enough to breathe and work in the air of revolution, to wrestle with change and hold fast the new liberty, to believe at all in the godhead of people or peoples, in the absolute right and want of the world, equality of justice, of work and truth and life; and these three came all out of the same rank, were all born into one social sect, men of historic blood and name, having nothing to ask of revolution, nothing (as the phrase is now) to gain by freedom, but leave to love and serve the light for the light's sake. Landor, who died last, was eldest, and Shelley, who died first, was youngest of the three. Each stood alike apart from the rest, far unlike as each was to the other two; not, like Coleridge, blind to the things of the time, nor, like Keats, alien to all things but art; and leaving to Southey or Wordsworth the official laurels and loyalties of courtly content and satisfied compliance. Out of their rank the Georges could raise no recruits to beat the drum of prose or blow the bagpipes of verse in any royal and constitutional procession towards nuptial or funereal goal.[1]

So far we must go with Mr. Arnold; but I cannot follow him when he adds that Byron and Shelley failed in their attempt; that the best "literary creation" of their time, work "far more solid and complete than theirs," was due to men in whom the new spirit was dead or was unborn; that, therefore, "their names will be greater than their writings." First, I protest against the bracketing of the two names. With all reserve of reverence for the noble genius and memory of Byron, I can no more accept him as a poet equal or even akin to Shelley on any side but one, than I could imagine Shelley endowed with the various, fearless, keen-eyed, and triumphant energy which makes the greatest of Byron's works so great. With all his glory of ardour and vigour and humour, Byron was a singer who could not sing; Shelley outsang all poets on record but some two or three throughout all time; his depths and heights of inner and outer music are as divine as nature's, and not sooner exhaustible. He was alone the perfect singing-god; his thoughts, words, deeds, all sang together. This between two singing-men is a distinction of some significance; and must be, until the inarticulate poets and their articulate outriders have put down singing-men altogether as unrealities, inexpedient if not afflictive in the commonwealth of M. Proudhon and Mr. Carlyle. Till the dawn of that "most desired hour, more loved and lovely than all its sisters," these unblessed generations will continue to note the difference, and take some account of it. Again, though in some sense a "child of the idea," Byron is but a foundling or bastard child; Shelley is born heir, and has it by birthright; to Byron it is a charitable nurse, to Shelley a natural mother. All the more praise, it may be said, to Byron for having seen so much as he did and served so loyally. Be it so then; but let not his imperfect and intermittent service, noble and helpful now, and now alloyed with baser temper or broken short through sloth or spite or habit, be set beside the flawless work and perfect service of Shelley. His whole heart and mind, his whole soul and strength, Byron could not give to the idea at all; neither to art, nor freedom, nor any faith whatever. His life's work therefore falls as short of the standard of Shelley's as of Goethe's work. To compare *Cain* with *Prometheus*, the *Prophecy of Dante* with the "Ode to Naples," is to compare *Manfred* with *Faust*. Shelley was born a son and soldier of light, an archangel winged and weaponed for angels' work. Byron, with a noble admixture of brighter and purer blood, had in him a cross of the true Philistine breed....

Of the relation between Shelley and Byron I have here no more to say; but before ending these notes I find yet another point or so to touch upon. Perhaps to every student of any one among the greater poets there seems to be something in his work not yet recognised by other students, some secret power or beauty reserved for his research. I do not think that justice has yet been done to Shelley as to some among his peers, in all details and from every side. Mr. Arnold, in my view, misconceives and misjudges him not less when set against Keats than when bracketed with Byron. Keats has indeed a divine magic of language applied to nature; here he is unapproachable; this is his throne, and he may bid all kings of song come bow to it. But his ground is not Shelley's ground; they do not run in the same race at all. The "Ode to Autumn," among other such poems of Keats, renders nature as no man but Keats ever could. Such poems as the "Lines Written among the Euganean Hills" cannot compete with it. But do they compete with it? The poem of Keats, Mr. Arnold says, "*renders* Nature;" the poem of Shelley "tries *to render* her." It is this that I deny. What Shelley tries to do he does; and he does not try to do the same thing as Keats. The comparison is as empty and profitless as one between the sonnets of Shakespeare and the sonnets of Milton. Shelley never in his life wrote a poem of that exquisite contraction and completeness, within that round and perfect limit. This poem of the Euganean Hills is no piece of spiritual sculpture or painting after the life of natural things. I do not pretend to assign it a higher or a lower place; I say simply that its place is not the same. It is a rhapsody of thought and feeling coloured by contact with nature, but not born of the contact; and such as it is all Shelley's work is, even when most vague and vast in its elemental scope of labour and of aim. A soul as great as the world lays hold on the things of the world; on all life of plants, and beasts, and men; on all likeness of time, and death, and good things and evil. His aim is rather to render the effect of a thing than a thing itself; the soul and spirit of life rather than the living form, the growth rather than the thing grown. And herein he too is unapproachable.

Notes

1. The one kindly attempt of Landor to fill Southey's place for him when disabled could scarcely have proved acceptable to his friend's official employers.
But since thou liest sick at heart And worn with years, some little part Of thy hard office let me try, Tho' inexpert was always I To *toss the litter of Westphalian swine From under human to above divine.*
(*Works*, vol. ii. p. 654.)
"Call you that backing of your friends"—when they happen to be laureates?

—Algernon Charles Swinburne,
from "Notes on the Text of Shelley,"
Fortnightly Review, May 1869, pp. 555–559

Leslie Stephen
"Godwin and Shelley" (1879)

Born in London, Stephen (1832–1904) published in the fields of history, philosophy, and literature. His first major publication was the two-volume *English Thought in the Eighteenth Century* (1876). Several years later, Stephen attempted to merge science and ethics in his *The Science of Ethics* (1882). The work was influenced by Stephen's own struggle with his early evangelical upbringing and his later agnostic beliefs. The work proved unpopular, however, and drove Stephen into other avenues of scholarship. He served as literary editor for the *Cornhill Magazine* from 1878 to 1882 and of the *Dictionary of National Biography* from 1882 to 1889. Stephen also completed a study of *The English Utilitarians* (1900) and continued his early interest in the preceding century with *English Literature and Society in the Eighteenth Century* (1904). Perhaps his most significant claims to fame, however, are his daughters, Virginia Woolf and Vanessa Bell. The following excerpt comes from the essay "Godwin and Shelley," from his series *Hours in a Library* (1874–79), in which Stephen stressed the importance of reading for moral improvement.

The Godwinism (of Shelley) is strongest in the crude poetry of *Queen Mab*, where many passages read like the *Political justice* done into verse.... After pointing to some of the miseries which afflict unfortunate mankind, and observing that they are not due to man's "evil nature," which, it seems, is merely a figment invented to excuse crimes, the question naturally suggests itself, to what, then, can all this mischief be due? Nature has made everything perfect and harmonious, except man. On man alone she has, it seems, heaped "ruin, vice, and slavery." But the indignant answer is given:

> Nature! No!
> Kings, priests, and statesmen blast the human flower Even in its tender bud; their influence darts Like subtle poison through the bloodless veins Of desolate society.

According to this ingenious view, "kings, priests, and statesmen" are something outside of, and logically opposed to, Nature. They represent the evil principle in this strange dualism. Whence this influence arises, how George III and Paley and Lord Eldon came to possess an existence independent of Nature, and acquired the power of turning all her good purpose to nought, is one of those questions which we can hardly refrain from asking, but which it would be obviously unkind to press. Still less would it be to the purpose to ask how this beneficent Nature is related to the purely neutral Necessity, which is "the mother of the world," or how, between the two, such a monstrous birth as the "prolific fiend" Religion came into existence. The crude incoherence of the whole system is too obvious to require exposition; and yet it is simply an explicit statement of Godwin's theories put forth with inconvenient excess of candour. The absurdities slurred over by the philosopher are thrown into brilliant relief by the poet. Shelley improved as a poet, and in a degree rarely exemplified in poetry, between *Queen Mab* and the *Prometheus;* but even in the *Prometheus* and his last writings we find a continued reflection of Godwin's characteristic views. Everywhere as much a prophet as a poet, Shelley is always announcing, sometimes in exquisite poetry, the advent of the millennium. His conception of the millennium, if we try to examine precisely what it is, always embodies the same thought, that man is to be made perfect by the complete dissolution of all the traditional ties by which the race is at present bound together. In the passage which originally formed the conclusion to the *Prometheus,* the "Spirit of the Hour" reveals the approaching consummation. The whole passage is a fine one, and it is almost a shame to quote fragments; but we may briefly observe that in the coming world everybody is to say exactly what he thinks; women are to be—

> gentle radiant forms,
> From custom's evil taint exempt and pure; Speaking the wisdom once they could not think, Looking emotions once they feared to feel.

Thrones, altars, judgment seats, and prisons are to be abolished when reason is absolute; and when

> The loathsome mask has fallen, the man remains Sceptreless, free, uncircumscribed, but man Equal, unclassed, tribeless, and nationless, Exempt from awe, worship, degree, the king Over himself.

To be "unclassed, tribeless, and nationless," and we may add, without marriage, is to be in the lowest depths of barbarism. It is so, at least, in the world of realities. But the description will fit that "state of nature" of which philosophers of the time delighted to talk. The best comment is to be found in Godwin. The great mistake of Rousseau, says that writer, was that whilst truly recognising government to be the source of all evil, he chose to praise the state which preceded government, instead of the state which, we may hope, will succeed its abolition. When we are perfect, we shall get rid of all laws of every kind, and thus, in some sense, the ultimate goal of all progress is to attain precisely to that state of nature which Rousseau regretted as a thing of the past and which is described in Shelley's glowing rhetoric.

The difficulty of making this view coherent is curiously reflected in the mechanism of Shelley's great poem; great it is, for the marvel of its lyrical excellence is fortunately independent of the conceptions of life and human nature which it is intended to set forth. If all the complex organisation which has slowly evolved itself in the course of history, the expression of which is civilisation, order, coherence, and co-operation in the different departments of life, is to be set down as an unmitigated evil, the fruit of downright imposture, all history becomes unintelligible. Man, potentially perfectible, has always been the sport of what seems to be a malignant and dark power of utterly inexplicable origin and character. Shelley, we are told, could not bear to read history. The explanation offered is that he was too much shocked by the perpetual record of misery, tyranny, and crime. A man who can see nothing else in history is obviously a very inefficient historian. Godwin tells us that he had learnt from Swift's bitter misanthropy the truth that all political institutions are hopelessly corrupt. A fusion of the satirist's view, that all which is is bad, with the enthusiast's view, that all which will be will be perfect, just expresses Shelley's peculiar mixture of optimism and pessimism. When we try to translate this into a philosophical view or a poetical representation of the world, the consequence is inevitably perplexing.

Thus Shelley tells us in the preface to the *Prometheus* that he could not accept the view, adopted by Aeschylus, of a final reconciliation between Jupiter and his victim. He was "averse from a catastrophe so feeble as that of reconciling the champion with the oppressor of mankind." He cannot be content with the intimate mixture of good and evil which is presented in the world as we know it. He must have absolute good on one side, contrasted with absolute evil on the other. But it would seem—as far as one is justified in attaching any precise meaning to poetical symbols—that the fitting catastrophe to the world's drama must be in some sense a reconciliation between Prometheus and Jupiter; or, in other words, between the reason and the blind forces by which it is opposed. The ultimate good must be not the annihilation of all the conditions of human life, but the slow conquest of nature by the adaptation of the life to its conditions. We learn to rule nature, as it is generally expressed, by learning to obey it. Any such view, however, is uncongenial to Shelley, though he might have derived it from Bacon, one of the professed objects of his veneration. The result of his own view is that the catastrophe of the drama is utterly inexplicable and mysterious. Who are Jupiter and Demogorgon? Why, when Demogorgon appears in the car of the Hours, and tells Jupiter that the time is come, and that they are both to dwell together in darkness henceforth, does Jupiter immediately give up with a cry of Ai! Ai! and descend (as one cannot help irreverently suggesting) as through a theatrical trapdoor? Dealing with such high matters, and penetrating to the very ultimate mystery of the universe, we must of course be prepared for surprising inversions. A mysterious blind destiny is at the bottom of everything, according to Shelley, and of course it may at any moment crush the whole existing order in utter annihilation. And yet, it is impossible not to feel that here, too, we have still the same incoherence which was shown more crudely in *Queen Mab*. The absolute destruction of all law, and of law not merely in the sense of human law, but of the laws in virtue of which the stars run their course and the frame of the universe is bound together, is the end to which we are to look forward. It will come when it will come; for it is impossible to join on such a catastrophe to any of the phenomenal series of events, of which alone we can obtain any kind of knowledge. The actual world, it is plain, is regarded as a hideous nightmare. The evil dream will dissolve and break up when something awakes us from our mysterious sleep; but that something, whatever it may be, must of course be outside the dream, and not a consummation worked out by the dream itself. We expect a catastrophe, not an evolution. And, finally, when the dream dissolves, when the "painted veil" called life is drawn aside, what will be left?

Some answer—and a remarkable answer—is given by Shelley. But first we may say one word in reference to a point already touched. The entire dissolution of all existing laws was part of Shelley's, as of Godwin's, programme. The amazing calmness with which the philosopher summarily disposes of marriage in a cursory paragraph or two, as (in the words of the old story) a fond thing, foolishly invented and repugnant to the plain teaching of reason, is one of the most grotesque crudities of his book. This doctrine has to be taken into account both in judging of Shelley's character and considering some of his poetical work. It is, of course, frequently noticed in extenuation or aggravation of the most serious imputation upon his character. We are told that Shelley can be entirely cleared by revelations which have not as yet been made. That is satisfactory, and would be still more satisfactory if we were sure that his apologists fully appreciated the charge. According to the story as hitherto published, we can only say that his conduct seems to indicate a flightiness and impulsiveness inconsistent with real depth of sentiment. The complaint is that he behaved ill to the first Mrs. Shelley, considered not as a wife, but as a human being, and as a human being then possessing a peculiar and special claim upon his utmost tenderness. This is only worth saying in order to suggest the answer to a casuistical problem which seems to puzzle his biographers. Is a man the better or the worse because, when he breaks a moral law, he denies it to be moral? Is he to be more or less condemned because, whilst committing a murder, he proceeds to assert that everybody ought to commit murder when he chooses? Without seeking to untwist all the strands of a very pretty problem, I will simply say that, to my mind, the question must in the last resort be simply one of fact. What we have to ask is the quality implied by his indifference to the law? If a man acts wrongly from benevolent feeling, misguided by some dexterous fallacy, his error affords no presumption that he is otherwise intrinsically bad. If, on the other hand, his indifference to the law arises from malice, or sensuality, it must of course lower our esteem for him in proportion, under whatever code of morality he may please to shelter his misdoings.

In Shelley's particular case we should probably be disposed to ascribe his moral deficiencies to the effect of crude but specious theory upon a singularly philanthropic but abnormally impulsive mind. No one would accuse him of any want of purity or generosity; but we might regard him as wanting in depth and intensity of sentiment. Allied to this moral weakness is his incapacity for either feeling in himself or appreciating in others the force of ordinary human passions directed to a concrete object. The only apology that can be made for his selection of the singularly loathsome motive for his drama is in the fact that in his hands the chief character becomes simply an incarnation of purely intellectual wickedness; he is a new avatar of the mysterious principle of evil which generally appears as a priest or king; he represents the hatred to good in the abstract rather than subservience to the lower passions. It is easy to understand how Shelley's temperament should lead him to undervalue the importance of the restraints which are rightly regarded as essential to social welfare, and fall in with Godwin's tranquil abolition of marriage as an uncomfortable fetter upon the perfect liberty of choice. But it is also undeniable that the defect not only makes his poetry rather unsatisfying to those coarser natures which cannot support themselves on the chameleon's diet, but occasionally leads to unpleasant discords. Thus, for example, the worshippers of Shelley generally regard the *Epipsychidion* as one of his finest poems, and are inclined to warn off the profane vulgar as unfitted to appreciate its beauties. It is, perhaps, less difficult to understand than to sympathise very heartily with the sentiment by which it is inspired. There are abundant precedents, both in religious and purely imaginative literature, for regarding a human passion as in some sense typifying, or identical with, the passion for ideal perfection. So far a want of sympathy may imply a deficiency in poetic sensibility. But I cannot believe that the *Vita Nuova* (to which we are referred) would have been the better if Dante had been careful to explain that there was another lady besides Beatrice for whom he had an almost equal devotion; nor do I think that it is the prosaic part of us which protests when Shelley thinks it necessary to expound his anti-matrimonial theory in the *Epipsychidion*. Why should he tell us that—

> I never was attached to that great sect,
> Whose doctrine is that each one should select
> Out of the crowd a mistress or a friend,

and so on; in short, that he despises the "modern morals" which distinctly approve of monogamy? Human love, one would say, becomes a fitting type of a loftier emotion, in so far as it implies exclusive devotion to its object. During this uncomfortable intrusion of a discordant theory, we seem to be listening less to the passionate utterance of a true poet than to the shrill tones of a conceited propagator of flimsy crotchets, proclaiming his tenets without regard to truth or propriety. Mrs. Shelley does not seem to have entered into the spirit of the composition; and we can hardly wonder if she found this little bit of argument rather a stumbling-block to her comprehension.

To return, however, from these moral deductions to the more general principles. It is scarcely necessary to insist at length upon the peculiar idealism implied in Shelley's poetry. It is, of course, the first characteristic upon which every critic must fasten. The materials with which he works are impalpable abstractions where other poets use concrete images. His poetry is like the subtle veil woven by the witch of Atlas from "threads of fleecy mists," "long lines of light," such as are kindled by the dawn and "star-beams." When he speaks of natural scenery the solid earth seems to be dissolved, and we are in presence of nothing but the shifting phantasmagoria of cloudland, the glow of moonlight on eternal snow, or the "golden lightning of the setting sun." The only earthly scenery which recalls Shelley to a more material mind is that which one sees from a high peak at sunrise, when the rising vapours tinged with prismatic colours shut out all signs of human life, and we are alone with the sky and the shadowy billows of the sea of mountains. Only in such vague regions can Shelley find fitting symbolism for those faint emotions suggested by the most abstract speculations, from which he alone is able to extract an unearthly music. To insist upon this would be waste of time. Nobody, one may say briefly, has ever expanded into an astonishing variety of interpretation the familiar text of Shakespeare—

> We are such stuff
> As dreams are made on, and our little lives
> Are rounded with a sleep.

The doctrine is expressed in a passage in *Hellas,* where Ahasuerus states this as the final result of European thought. The passage, like so many in Shelley, shows that he had Shakespeare in his mind without exactly copying him. The Shakesperian reference to the "cloud-capped towers" and "gorgeous palaces" is echoed in the verses which conclude with the words:—

> This whole
> Of suns and worlds, and men and beasts, and flower
> With all the violent and tempestuous working
> By which they have been, are, or cease to be,
> Is but a vision: *all that it inherits*
> Are motes of a sick eye, bubbles and dreams;
> Thought is its cradle and its grave, nor less
> The future and the past are idle shadows
> Of thought's eternal flight—they have no being.
> Nought is but that it feels itself to be.

The italicised words point to the original in the *Tempest;* but Shelley proceeds to expound his theory more dogmatically than Prospero, and we are not quite surprised when Mahmoud is puzzled and declares that the words "stream like a tempest of dazzling mist through his brain." The words represent the most characteristic effect of Shelley as accurately as the aspect of consistent idealism to a prosaic mind.

It need not be said how frequently the thought occurs in Shelley. We might fix him to a metaphysical system if we interpreted him prosaically. When in *Prometheus* Panthea describes to Asia a mysterious dream, suddenly Asia sees another shape pass between her and the "golden dew" which gleams through its substance. "What is it?" she asks. "It is mine other dream," replies Panthea. "It disappears," exclaims Asia. "It passes now into my mind," replies Panthea. We are, that is, in a region where dreams walk as visible as the dreamers, and pass into or out of a mind which is indeed only a collection of dreams. The archaic mind regarded dreams as substantial or objective realities. In Shelley the reality is reduced to the unsubstantiality of a dream. To the ordinary thinker, the spirit is (to speak in materialist language) the receptacle of ideas. With Shelley, a little further on, we find that the relation is inverted; spirits themselves inhabit ideas; they live in the mind as in an ocean. Thought is the ultimate reality which contains spirits and ideas and dreams, if, rather, it is not simpler to say that everything is a dream.

The Faery-land of Spenser might be classified in our inadequate phraseology as equally "ideal" with Shelley's impalpable scenery. But Spenser's allegorical figures are as visible as the actors in a masque; and, in fact, the *Faery Queen* is a masque in words. His pages are a gallery of pictures, and may supply innumerable subjects for the artist. To illustrate Shelley would be as impossible as to paint a strain of music, unless, indeed, some of Turner's cloud scenery may be taken as representative of his incidental descriptions.

This language frequently reminds us of metaphysical doctrines which were unknown to Shelley in their modern shape. Nobody, perhaps, is capable of thinking in this fashion in ordinary life; and Shelley, with all his singular visions and hallucinations, probably took the common-sense view of ordinary mortals in his dealings with commonplace or facts. It is surprising enough that, even for purely poetical purposes, he could continue this to the ordinary conceptions of object and subject. But his familiarity with this point of view may help to explain some of the problems as to his ultimate belief. It is plain that he was in some sense dissatisfied with the simple scepticism of Godwin. But he found no successor to guide his speculations. Coleridge once regretted that Shelley had not applied to him instead of Southey, who, in truth, was as ill qualified as a man could well be to help a young enthusiast through the mazes of metaphysical entanglement. It is idle to speculate upon the possible result. Shelley, if we may judge from a passage in his epistle to Mrs. Gisborne, had no very high opinion of Coleridge's capacity as a spiritual guide. Shelley, in fact, in spite of his so-called mysticism, was an ardent lover of clearness, and would have been disgusted by the haze in which Coleridge enwrapped his revelations to mankind. But Coleridge might possibly have introduced him to a sphere of thought in which he could have found something congenial. One parallel may be suggested which will perhaps help to illustrate this position.

Various passages have been quoted from Shelley's poetry to prove that he was a theist and a believer in immortality. His real belief, it would seem, will hardly run into any of the orthodox moulds. It is understood as clearly as may be in the conclusion to the "Sensitive Plant":

> —in this life
> Of error, ignorance, and strife, Where nothing is, but all things seem, And we see the shadows of the dream. It is a modest creed, and yet Pleasant if one considers it, To own that death itself must be Like all the rest, a mockery. That garden sweet, that lady fair, And all sweet shapes and odours there In truth have never passed away;
> 'T is we, 't is ours have changed; not they.

A fuller exposition of the thought is given in the *Adonais;* and some of the phrases suggest the parallel to which I refer.... one of the popular works of Fichte, the *Vocation of Man,* (is) a vigorous description of that state of utter scepticism, which seems at one point to be the final goal of his idealism, as it was that of the less elaborate form of the same doctrine which Godwin had learnt from Berkeley. Godwin ... was content to leave the difficulty without solution. Fichte escaped, or thought that he escaped, by a solution which restores a meaning to much of the orthodox language. Whether his mode of escape was satisfactory or his final position intelligible, is of course another question. But it is interesting to observe how closely the language in which his final doctrine is set forth to popular readers resembles some passages in the *Adonais*. I will quote a few phrases which may be sufficiently significant.

Shelley, after denouncing the unlucky *Quarterly Reviewer* who had the credit of extinguishing poor Keats, proceeds to find consolation in the thought that Keats has now become

> A portion of the eternal, which must glow
> Through time and change, unquenchably the same
> Whilst thy cold embers choke the sordid hearth of shame.
> Peace, peace! he is not dead, he doth not sleep—
> He hath awakened from the dream of life;
> 'T is we who, lost in stormy visions, keep
> With phantoms an unprofitable strife,
> And, in mad trance, strike with our spirit's knife
> Invulnerable nothings—*we* decay
> Like corpses in a charnel, fear and grief
> Convulse and consume us day by day,
> And cold hopes swarm like worms within our living clay.

So, when Fichte has achieved his deliverance from scepticism, his mind is closed for ever against embarrassment and perplexity, doubt, uncertainty, grief, repentance, and desire. "All that happens belongs to the plan of the eternal world and is good in itself." If there are beings perverse enough to resist reason, he cannot be angry with them, for they are not free agents. They are what they are, and it is useless to be angry with "blind and unconscious nature." "What they actually are does not deserve my anger; what might deserve it they are not, and they would not deserve it if they were. My displeasure would strike an impalpable nonentity," an "invulnerable nothing," as Shelley puts it. They are, in short, parts of the unreal dream to which belong grief, and hope, and fear, and desire. Death is the last of evils, he goes on; for the hour of death is the hour of birth to a new and more excellent life. It is, as Shelley says, waking from a dream. And now, when we have no longer desire for earthly things, or any sense for the transitory and perishable, the universe appears clothed in a more glorious form. The dead heavy mass, which did but stop up space, has perished; and in its place there flows onward, with the rushing music of mighty waves, an eternal stream of life, and power, and action, which issues from the original source of all life—from thy life, O Infinite One! for all life is thy life, and only the religious eye penetrates to the realm of true Beauty. In all the forms that surround me I behold the reflection of my own being, broken up into countless diversified shapes, as the morning sun, broken in a thousand dewdrops, sparkles towards itself, a phrase which recalls Shelley's famous passage a little further on:—

Life, like a dome of many coloured glass, Stains the white radiance of eternity. The application, indeed, is there a little different; but Shelley has just the same thought of the disappearance of the "dead heavy mass" of the world of space and time. Keats, too, is translated to the "realm of true beauty."

> He is a portion of the loveliness Which once he made more lovely; he doth bear The part, while the one spirit's plastic stress Sweeps through the dull dense world, compelling
>   there
> All new successions to the forms they wear! Torturing the unwilling dross that checks its flight To its own likeness, as each mass may bear; And bursting in its beauty and its might From trees, and beasts, and men, into the heaven's
>   light.

There are important differences, as the metaphysician would point out, between the two conceptions, and language of a similar kind might be found in innumerable writers before and since. I only infer that the two minds are proceeding, if one may say so, upon parallel lines. Fichte, like Shelley, was accused of atheism, and his language would, like Shelley's, be regarded by mere readers as an unfair appropriation of old words to new meanings. Shelley had of course no definite metaphysical system to set beside that of the German philosopher; and had learnt what system he had rather from Plato than from Kant. It may also be called significant that Fichte finds the ultimate point of support in conscience or duty; whereas, in Shelley's theory, duty seems to vanish, and the one ultimate reality to be rather love or the beautiful. But it would be pedantic to attempt the discovery of a definite system of opinion when there is really nothing but a certain intellectual tendency. One can only say that, somehow or other, Shelley sought comfort under his general sense that everything is but the baseless fabric of a vision, and moreover a very uncomfortable vision, made up of pain, grief, and the "unrest which men miscall delight," in the belief, or, if belief is too strong a word, the imagination of a transcendental and eternal world of absolute perfection, entirely beyond the influence of "chance, and death, and mutability." Intellectual beauty, to which he addresses one of his finest poems, is the most distinct name of the power which he worships. Thy light alone, he exclaims—

> Thy light alone, like mist on mountains driven,
> Or music by the night wind sent
> Through strings of some still instrument,
> Or moonlight on a midnight stream,
> Gives peace and truth to life's unquiet dream.

In presence of such speculations, the ordinary mass of mankind will be content with declaring that the doctrine, if it can be called a doctrine, is totally unintelligible. The ideal world is upon this vein so hopelessly dissevered from the real, that it can give us no consolation. If life is a dream, the dream is the basis of all we know, and it is small comfort to proclaim its unreality. A truth existing all by itself in a transcendental vacuum entirely unrelated to all that we call fact, is a truth in which we can find very small comfort. And upon this matter I have no desire to differ from the ordinary mass of mankind. In truth, Shelley's creed means only a vague longing, and must be passed through some more philosophical brain before it can become a fit topic for discussion.

But the fact of this unintelligibility is by itself an explanation of much of Shelley's poetical significance. When the excellent Godwin talked about perfectibility and the ultimate triumph of truth and justice, he was in no sort of hurry about it. He was a good deal annoyed when Malthus crushed his dreams, by recalling him to certain very essential conditions of earthly life. Godwin, he said in substance, had forgotten that human beings have got to find food and standing-room on a very limited planet, and to rear children to succeed them. Remove all restraints after the fashion proposed by Godwin, and they will be very soon brought to their senses by the hard pressure of starvation, misery, and vice. Godwin made a feeble ostensible reply, but, in practice, he was content to adjourn the realisation of his hopes for an indefinite period. Reason, he reflected, might be omnipotent, but he could not deny that it would take a long time to put forth its power. He had the strongest possible objections to any of those rough-and-ready modes of forcing men to be reasonable which had culminated in the revolution. So he gave up the trade of philosophising, and devoted himself to historical pursuits, and the preparation of wholesome literature for the infantile mind. To Shelley, no such calm abnegation of his old aims was possible. He continued to assert passionately his belief in the creed of his early youth; but it became daily more difficult to see how it was to be applied to the actual men of existence. He might hold in his poetic raptures that the dreams were the only realities, and the reality nothing but a dream; but he, like other people, was forced to become sensible to the ordinary conditions of mundane existence.

The really exquisite strain in Shelley's poetry is precisely that which corresponds to his dissatisfaction with his master's teaching. So long as Shelley is speaking simply as a disciple of Godwin, we may admire the melodious versification, the purity and fineness of his language, and the unfailing and, in its way, unrivalled beauty of his aerial pictures. But it is impossible to find much real satisfaction in the informing sentiment. The enthusiasm rings hollow, not as suggestive of insincerity, but of deficient substance and reality. Shelley was, in one aspect, a typical though a superlative example of a race of human beings, which has, it may be, no fault except the fault of being intolerable. Had he not been a poet (rather a bold hypothesis, it must be admitted), he would have been a most insufferable bore. He had a terrible affinity for the race of crotchet-mongers, the people who believe that the world is to be saved out of hand by vegetarianism, or female suffrage, or representation of minorities, the one-sided, one-ideaed, shrill-voiced and irrepressible revolutionists. I say nothing against these particular nostrums, and still less against their advocates. I believe that bores are often the very salt of the earth, though I confess that the undiluted salt has for me a disagreeable and acrid savour. The devotees of some of Shelley's pet theories have become much noisier than they were when the excellent Godwin ruled his little clique. It is impossible not to catch in Shelley's earlier poetry, in *Queen Mab* and in the *Revolt of Mam,* the apparent echo of much inexpressibly dreary rant which has deafened us from a thousand platforms. The language may be better; the substance is much the same.

This, which to some readers is annoyance, is to others a topic of extravagant eulogy. Not content with urging the undeniable truth that Shelley was a man of wide and generous sympathy, a detester of tyranny and a condemner of superstition, they speak of him as though he were both a leader of thought and a practical philanthropist. To make such a claim is virtually to expose him to an unfair test. It is simply ridiculous to demand for Shelley the kind of praise which we bestow upon the apostles of great principles in active life. What are we to say upon this hypothesis to the young gentleman who is amazed because vice and misery survive the revelations of Godwin, and whose reforming ardours are quenched—so far as any practical application goes—by the surprising experience that animosities fostered by the wrongs of centuries are not to be pacified by publishing a pamphlet or two about Equality, Justice, and Freedom, or by a month's speechification in Dublin? If these were Shelley's claims upon our admiration, we should be justified in rejecting them with simple contempt, or we should have to give the sacred name of philanthropist to any reckless impulsive schoolboy who thinks his elders fools and proclaims as a discovery the most vapid rant of his time. Admit that Shelley's zeal was as pure as you please, and that he cared less than nothing for money or vulgar comfort; but it is absurd to bestow upon him the praise properly reserved for men whose whole lives have been a continuous sacrifice for the good of their fellows. Nor can I recognise anything really elevating in those portions of Shelley's poetry which embody this shallow declamation. It is not the passionate war-cry of a combatant in a deadly grapple with the forces of evil, but the wail of a dreamer who has never troubled himself to translate the phrases into the language of fact. Measured by this—utterly inappropriate—standard, we should be apt to call Shelley a slight and feverish rebel against the inevitable, whose wrath is little more than the futile, though strangely melodious, crackling of thorns.

To judge of Shelley in this mode would be to leave out of account precisely those qualities in which his unique excellence is most strikingly manifested. Shelley speaks, it is true, as a prophet; but when he has reached his Pisgah, it turns out that the land of promise is by no means to be found upon this solid earth of ours, or definable by degrees of latitude and longitude, but is an unsubstantial phantasmagoria in the clouds. It is in vain, too, that he declares that it is the true reality, and that what we call a reality is a dream. The transcendental world is—if we may say so—not really the world of archetypal ideas, but a fabric spun from empty phrases. The more we look at it the more clearly we recognise its origin; it is the refracted vision of Godwin's prosaic system seen through an imaginative atmosphere. But that which is really admirable is, not the vision itself but the pathetic sentiment caused by Shelley's faint recognition of its obstinate unsubstantiality. It is with this emotion that every man must sympathise in proportion as his intellectual aspirations dominate his lower passions. Forgetting all tiresome crotchets and vapid platitudes, we may be touched, almost in proportion to our own elevation of mind, by the unsatisfied yearning for which Shelley has found such manifold and harmonious utterance. There are moods in which every sensitive and philanthropic nature groans under the

> heavy and the weary weight Of all this unintelligible world.

Whatever our ideal may be, whatever the goal to which we hope to see mankind approximate, our spirits must often flag with a sense of our personal insignificance, and of the appalling dead weight of multiform impediments which crushes the vital energies of the world, like Etna lying upon the Titan. This despair of finding any embodiment for his own ideal, of bridging over the great gulf fixed between the actual world of sin, and sorrow, and stupidity, and the transcendental world of joy, love, and pure reason, represents the final outcome of Shelley's imperfect philosophy, and gives the theme of his most exquisite poetry. The doctrine symbolised in the *Alastor* by the history of the poet who has seen in vision a form of perfect beauty, and dies in despair of ever finding it upon earth (he seems, poor man! to have looked for it somewhere in the neighbourhood of Afghanistan), is the clue to the history of his own intellectual life. He is happiest when he can get away from the world altogether into a vague region, having no particular relation to time or space; to the valleys haunted by the nymphs in the *Prometheus*; or the mystic island in the *Epipsychidion*, where all sights and sounds are as the background of a happy dream, fitting symbols of sentiments too impalpable to be fairly grasped in language: or that "calm and blooming cove" of the lines in the Euganean hills.

The lyrics which we all know more or less by heart are but so many different modes of giving utterance to— The desire of the moth for the star,

> Of the night for the morrow, The devotion to something afar
> From the sphere of our sorrow.

He is always dwelling upon the melancholy doctrine expressed in his last poem by the phrase that God has made good and the means of good irreconcilable. The song of the skylark suggests to him that we are doomed to "look before and after," and to "pine for what is not." Our sweetest songs (how should it be otherwise?) are those which tell of saddest thought. The wild commotion in sea, sky, and earth, which heralds the approach of the south-west wind, harmonises with his dispirited restlessness, and he has to seek refuge in the vague hope that his thoughts, cast abroad at random like the leaves and clouds, may somehow be prophetic of a magical transformation of the world. His most enduring poetry is, in one way or other, a continuous comment upon the famous saying in *Julian and Maddalo*, suggested by the sight of his fellow-Utopian, whose mind has been driven into madness by an uncongenial world.

> Most wretched men Are cradled into poetry by wrong; They learn in suffering what they teach in song.

Some poets suffer under evils of a more tangible kind than those which tormented Shelley; and some find a more satisfactory mode of escape from the sorrows which beset a sensitive nature. But the special beauty of Shelley's poetry is so far due to the fact that we feel it to be the voice of a pure and lofty nature, however crude may have been the form taken by some of his unreal inspiration.

—LESLIE STEPHEN, "Godwin and Shelley," 1879, *Hours in a Library*, 1874–79, 1904, vol. 3, pp. 377–406

STOPFORD A. BROOKE
"SOME THOUGHTS ON SHELLEY" (1880)

Stopford Augustus Brooke (1832–1916) was born near Letterkenny, County Donegal, Ireland. Though he spent his life and gained fame as a minister, Brooke was also an accomplished writer and literary critic. He attended Trinity College, Dublin (completing his B.A. in 1856 and M.A. in 1858), and while there, earned prizes in divinity and English verse. He moved to London, where he taught at Queen's College and became curate at St. Mary Abbots in Kensington. In the years that followed, Brooke worked as a chaplain for the court of Queen Victoria and, in 1876, became curate of Bedford Chapel in Bloomsbury, a post he would hold until 1895. During his years as a preacher, in which he also gravitated to Unitarian beliefs, Brooke indulged his love of poetry through several publications dealing with English poets from Old English to contemporary times. These include *The Theology of the English Poets* (1873) and *English Literature* (1877). In the context of Shelley studies, Brooke is also notable for being the first speaker at the inaugural meeting of the Shelley Society on March 10, 1886. In *The Shelley Society's Note-book*, Part I (published in London in 1888), Brooke's speech is summarized in the opening pages:

> The Rev. Stopford Brooke, in stating the objects of the Society, said that the humour of about a hundred persons might alone be considered a good reason for the existence of any Society whatever, but the founders of the Shelley Society desired to connect together all that would throw light on the poet's personality and his work, to ascertain the truth about him, to issue reprints, and above all to do something to further the objects of Shelley's life and work, and perhaps to better understand and love a genius which was ignored and abused in his own time, but which had risen from the grave into which the critics had trampled it to live in the hearts of men. (p. 2)

When the sea gave up its dead, all of Shelley's body that was rescued from flood and fire was laid where the rise of the ground ends in a dark nook of the Aurelian wall. So deep is that resting-place in shadow that the violets blossom later there than on "the slope of green access" where, seen from Shelley's grave, the flowers grow over the dust of Adonais. It is well that both were buried in Italy rather than in England, for, though no Italian could have written their poetry, yet it was,—in all things else different,—of that spirit which Italy awakens in Englishmen who love her, rather than of the purely English spirit. The Italian air, the sentiment of Italy, fled and dreamed through their poems, but most through those of Shelley. It was but fitting, then, that Shelley, whose fame was England's, should be buried in the city which is the heart of Italy. But he was born far away from this peaceful and melancholy spot, and grew up to manhood under the grey skies of England, until its Universities, its Church, its Society, its Law and its dominant policy became inhospitable to him, nay, even his own father cast him out. They all had, in the opinion of sober men of that time, good cause to make him a stranger, for he attacked them all, and it would be neither wise nor true, nor grateful to Shelley himself, were he to be put forward as a genius unjustly treated, or as one who deserved or asked for pity. Those who separate themselves from society, and war against its dearest maxims, if they are as resolute in their choice, and as firm in their beliefs as Shelley, count the cost, and do not or rarely complain when the penalty is exacted. He was exiled, and it was no wonder. The opinion of the world did not trouble him, nor was that a wonder. But as this exile is the most prominent fact of his life, its influence is sure to underlie his work. One of the questions that any one who writes of Shelley has to ask, is, How did this exile from the Education, Law, Religion, and Society of his country, and from the soil of his country itself, affect his poetry?

It had a very great influence, partly for good and partly for evil. The good it did is clear. It deepened his individuality and the power which issued from that source. It set him free from the poetic conventions to which his art might have yielded too much obedience in England—a good which the obscurity of Keats also procured for him—it prevented him from being worried too much by the blind worms of criticism, it enabled him to develop himself more freely, and it placed him in contact with a natural scenery, fuller and sunnier than he could ever have had in England, in which his love of beauty found so happy and healthy a food that it came to perfect flower. In Italy also, where impulse even more than reason urges intelligence and inspires genius, lyrical poetry, which is born of impulse, is more natural and easy, though not better, than elsewhere, and the very inmost spirit of Shelley, deeper than his metaphysics or his love of Man and inspiring both, deeper even than any personal passion, was the lyrical longing of his whole body, soul, and spirit—"O that I had wings like a dove; then would I flee away, and be at rest."

But the good this exile did his art was largely counterbalanced by its harm. Shelley's individuality, unchecked by that of others, grew too great, and tended not only to isolate him from men, but to prevent his art from becoming conversant enough with human life. The absence of critical sympathy of a good kind, such as that which flows from one poet to another in a large society, left some of his work, as it left some of Keats', more formless, more intemperate, more impalpable, more careless, more apart from the realities of life, than it ought to have been in the most poetical of poets since the days of Elizabeth. Even in his lyric work, the impassioned impulse would have failed less often to fulfil its form perfectly; there would not have been so many fragments thrown aside for want of patience or power to complete them, had he been less personal, less subject to individual freakishness, more subject to the unexpressed criticism which floats, as it were, in the air of a large literary society, and constrains the art of the poet into measured act and power. And as to Nature, we should perhaps have had, with his genius, a much wider and less ideal representation of her, had he not been so enthralled by the vastness and homelessness of Swiss, and by the ideality of Italian scenery. Even when he did write in England itself, the recollected love of Switzerland and the Rhine mingled with the impressions he received from the Thames, and produced a scenery, as in certain passages in *Alastor* and the *Revolt of Islam*, which is not directly studied from anything in heaven or earth. It is none the worse for that, but it is not Nature, it is Art.

These are general considerations, but there were some more particular results, partly good and partly evil, of this separation of Shelley from the ordinary religious and political views of English society.

A good deal of his poetry became polemical, and polemical, like satiric poetry, is apart from pure art. It attacks evil directly, and the poet, his mind being then fixed not on the beautiful but on the base, writes prosaically. Or it embodies a creed in verse, and, being concerned with doctrine, becomes dull. In both cases the poet misses, as Shelley did, that inspiration of the beautiful which arises from the seeing of truth, not from the seeing of a lie; from the love of true ideas, not from their intellectual perception. The verses, for example, in the "Ode to Liberty," which directly attack kingcraft and priestcraft, however gladly one would see their sentiments in prose, are inferior as poetry to all the rest; and it is the same throughout all Shelley's poetry of direct attack on evil. This polemical element in the *Revolt of Islam*, and the endeavour to lay down in it his revolutionary creed, are additional causes of the wastes of prosaic poetry which make it so unreadable. The very splendour and passion of the passages devoted to Nature and Love contrast so sharply, like burning spaces of sunlight on a grey sea, with wearisome whole, that they lose half their value, and disturb, like so much else, the unity of the poem. The same things seem true of *Rosalind and Helen*, and of those political poems which are direct attacks on abuses in England. On the other hand, when Shelley wrote on these evils indirectly inspired by the opposing truths, concerned with their beauty, and borne upwards by delight in them, his work entered the realm of art, and his poetry became magnificent. There is no finer example of this than *Prometheus Unbound*. The subject is at root the same as that of the *Revolt of Islam*, the things opposed are the same, the doctrine is the same, but the whole method of approaching his idea and fulfilling its form is changed, and all the questions are brought into that artistic representation which stirs around them inspiring and enduring emotion.

The good Shelley did in this way was very great. At a time when England, still influenced by its abhorrence of the Reign of Terror, by its fear of France and Napoleon, was most dead to the political ideas that had taken form in 1789, Shelley gave voice, through art, to these ideas, and encouraged that hope of a golden age which, however vague, does so much for human progress. He threw around these things imaginative emotion, and added all its power to the struggle for freedom.

Still greater is the unrecognised work he did in the same way for theology in England. That theology was no better than all theology had become under the influence of the imperial and feudal ideas of Europe. Its notion of God, and of man in relation to God, partly Hebraic, and therefore sacerdotal and sacrificial, partly deeply dyed with asceticism and other elements derived from the Oriental notion of the evil of matter, was further modified by the political views of the Roman Empire, transferred to God by the Roman Church. And when the universal ideas regarding mankind, and a return to nature, were put forth by France, they clashed instantly with this limited, sacerdotal, ascetic, aristocratic, and feudal theology. The sovereign right of God, because He was omnipotent, to destroy the greater part of His subjects, the right of a caste of priests to impose their doctrines on all, and to exile from religion all who did not agree with them; the view that whatever God was represented to do was right, though it might directly contradict the nature, the conscience, and the heart of Man; these, and other related views had been brought to the bar of humanity, and condemned from the intellectual point of view by a whole tribe of thinkers. But if a veteran theology is to be disarmed and slain, it needs to be brought not only into the arena of thought and argument, but into the arena of poetic emotion. A great part of that latter work was done in England by Shelley. He indirectly made, as time went on, an ever-increasing number of men feel that the will of God could not be in antagonism to the universal ideas concerning Man, that His character could not be in contradiction to the moralities of the heart, and that the destiny He willed for mankind must be as universal and as just and loving as Himself. There are more clergymen, and more religious laymen than we imagine, who trace to the emotion Shelley awakened in them when they were young, their wider and better views of God. Many men, also, who were quite careless of religion, yet cared for poetry, were led, and are still led, to think concerning the grounds of a true worship, by the moral enthusiasm which Shelley applied to theology. He made emotion burn around it, and we owe to him a great deal of its nearer advance to the teaching of Christ. But we owe it, not to those portions of his poetry which denounced what was false and evil, but to those which represented and revealed, in delight in its beauty, what was good and true. Had he remained in England, I do not think he would have worked on this matter in the ideal way of *Prometheus Unbound,* because continual contact with the reigning theology would have driven his easily wrought anger into direct violence. In Italy, in exile, it was different. The polemical temper in which he wrote the *Revolt of Islam* changed into the poetical temper in which he wrote *Prometheus Unbound.*

Connected with this, but not with his exile, is the question, in what way his belief as to a Source of Nature influenced his art. He was not an atheist or a materialist. If he may be said to have occupied any theoretical position, it was that of an Ideal Pantheist; the position which, with regard to Nature, a modern poet who cares for the subject, naturally— whatever may be his personal view—adopts in the realm of his art. Wordsworth, a plain Christian at home, wrote about Nature as a Pantheist: the artist loves to conceive of the Universe, not as dead, but as alive. Into that belief Shelley, in hours of inspiration, continually rose, and his work is seldom more impassioned and beautiful than in the passages where he feels and believes in this manner. The finest example is towards the close of the *Adonais*. In his mind, however, the living spirit which, in its living, made the Universe, was not conceived of as Thought, as Wordsworth conceived it, but as Love operating into Beauty; and there is a passage on this idea in the fragment of the "Coliseum," which is as beautiful in prose as that in *Adonais* is in verse. But it is only in higher poetic hours that Shelley seems or cares to realise this belief. In the quieter realms of poetry, in daily life, he confessed no such creed plainly; he had little or no belief in a thinking or loving existence behind the phenomenal universe. It is infinitely improbable, he says, that the cause of mind is similar to mind.

Nothing can be more characteristic of him—and he has the same temper in other matters—than that he should have a faith with regard to a Source of Nature, into which he could soar when he pleased, in which he could live for a time, but which he did not choose to live in, to define, or to realise, continuously. When, in the *Prometheus Unbound*, he is forced, as it were, to realise a central cause, he creates Demogorgon, the dullest of all his impersonations. It is scarcely an impersonation. Once he calls it a "living spirit," but it has neither form nor outline in his mind. He keeps it before him as an "awful Shape."

The truth is, the indefinite was a beloved element of his life. "Lift not the painted veil," he cries, "which those who live call Life." His worst pain was when he thought he had lifted it, and seemed to know the reality. But he did not always believe that he had done so, or he preferred to deny his conclusion. Not as a thinker in prose, but as a poet, he frequently loved the vague with an intensity which raised it almost into an object of worship. The speech of the Third Spirit, in the "Ode to Heaven," is a wonderful instance of what I may call the rapture in indefiniteness. But this rapture had its other side, and when he was depressed by ill-health, the sense of a voiceless, boundless abyss, which for ever held its secret, and in which he floated, deepened his depression. The horror of a homeless and centreless heart which then beset him, is passionately expressed in the *Cenci*. Beatrice is speaking—

> Sweet Heaven, forgive weak thoughts, if there should be
> No God, no Heaven, no Earth, in the void world; The wide, grey, lampless, deep, unpeopled world.

But, on the whole, whether it brought him pain or joy, he preferred to be without a fixed belief with regard to a source of Nature. Could he have done otherwise, could he have given continuous substance in his thoughts to the great conception of ideal Pantheism in which Wordsworth rested, Shelley's whole work on Nature and his description of her would have been more direct, palpable, and homely. He would have loved Nature more, and made us love it more.

The result of all this is that a great deal of his poetry of Nature has no ground in thought, and consequently wants power. It is not that he could not have had this foundation and its strength. Both are his when he chooses. But, for the most part, he did not choose. Such was his temperament that he liked better to live with Nature and be without a centre for her. He would be

Dizzy, lost—but unbewailing.

But I am not sure whether the love of the undefined did not, in the first instance, arise out of his love of the constantly changing, and that itself out of the very character of his intellect, and the temper of his heart. His intellect, incessantly shaken into movement by his imagination, continually threw into new shapes the constant ideas he possessed. His heart, out of which are the issues of imagination, loved deeply a few great conceptions, but wearied almost immediately of any special form in which he embodied them, and changed it for another. In the matter of human love, he was uncontent with all the earthly images he formed of the ideal he had loved and continued to love in his own soul, and he could not but tend to change the images. In the ordinary life of feeling, the moment any emotion arose in his heart, a hundred others came rushing from every quarter into the original feeling, and mingled with it, and changed its outward expression. Sometimes they all clamoured for expression, and we see that Shelley often tried to answer their call. It is when he does this that he is most obscure—obscure through abundance of feelings and their forms. His intellect, heart, and imagination were in a kind of Heraclitean flux, perpetually evolving fresh images, and the new, in swift succession, clouding the old; and then, impatient weariness of rest or of any one thing whatever, driving forward within him this incessant movement, he sank, at last and for the time, exhausted—"As summer clouds disburthened of their rain."

There is no need to illustrate this from his poetry. The huddling rush of images, the changeful crowd of thoughts are found on almost every page. It is often only the oneness of the larger underlying emotion or idea which makes the work clear. We strive to grasp a Proteus as we read. In an instant the thought or the feeling Shelley is expressing becomes impalpable, vanishes, reappears in another form, and then in a multitude of other forms, each in turn eluding the grasp of the intellect, until at last we seize the god himself, and know what Shelley meant, or Shelley felt. In all this he resembles, at a great distance, Shakspere; and has, at that distance, and in this aspect of his art, a strength and a weakness similar to, but not identical with, that which Shakspere possessed,—the strength of changeful activity of imagination, the weakness of being unable, through eagerness, to omit, to select, to coordinate his images. Yet, at his highest, when the full force of genius is urged by full and dominant emotion, what poetry it is! How magnificent is the impassioned unity of the whole in spite of the diversity of the parts! But this lofty height is reached in only a few of Shelley's lyrics, and in a few passages in his longer poems.

At almost every point, the scenery of the sky he drew so fondly images this temper of Shelley's mind, this incessant building and unbuilding, this cloud-changefulness of his imagination.

> I silently laugh at my own cenotaph, And out of the caverns of rain,
> Like a child from the womb, like a ghost from the
> tomb,
> I arise and unbuild it again.

That is a picture of Shelley himself at work on a feeling or on a thought. "I change, but I cannot die."

I might illustrate this love of "the changing" from the history of his life, of his affections, of his theories; from his varied nature, and way of work, as the prose thinker and the poet; from the variety of the subjects on which he wrote, and which he half attempted—for he naturally fell into the fragmentary—from the eagerness with which he searched for new thought, new experiences of feeling, new literatures, even from his love of the strange and sometimes of the horrible; from that uncontent he had in the doctrines of others, until he had added to them, as he did to Plato's doctrine of Love, something of his own in order to make them new,—were there any necessity to enlarge on that which stands so clear. In all these things, what was said of Shelley's movements to and fro in the house at Lerici is true of his movement through the house of thought or of feeling. "Oh, he comes and goes like a spirit, no one knows when or where." But it remains to be said, that all through this secondary changefulness, he held fast to certain primary ideas of life, of morality, and of his art, which no one who cares for him can fail to discover.

There was, then, in Shelley this love of indefiniteness, and this love of changefulness. Which of the two was the cause of the other I cannot tell, but I am inclined to think that the latter was the first. It is better, however, to keep them both equally in view in the study of Shelley's art, and they are both well illustrated in his poetry of Nature.

I have said that his love of the indefinite with regard to a source of Nature weakened his work on Nature. His love of changefulness also weakened it by luring the imagination away from a direct sight of the thing into the sight of a multitude of images suggested by the thing.

But in the case of those who have great genius, that which enfeebles one part of their work often gives strength to another, and in three several ways these elements in Shelley's mind made his work on Nature of great value.

1. His love of that which is indefinite and changeful made him enjoy and describe better than any other English poet that scenery of the clouds and sky which is indefinite owing to infinite change of appearance. The incessant forming and unforming of the vapours which he describes in the last verse of "The Cloud," is that which he most cared to paint. Wordsworth often draws, and with great force, the aspect of the sky, and twice with great elaboration in the *Excursion*; but it is only a momentary aspect, and it is mixed up with illustrations taken from the works of men, with the landscape of the earth below where men are moving, with his own feelings about the scene, and with moral or imaginative lessons. Shelley, when he is at work on the sky, troubles it with none of these human matters, and he describes not only the momentary aspect, but also the change and progress of the sunset or the storm. And he does this with the greatest care, and with a characteristic attention to those delicate tones and halftones of colour which resemble the subtle imaginations and feelings he liked to discover in human Nature, and to which he gave form in poetry.

In his very first poem, in *Queen Mab* (Part II.), there is one of these studies of Sunset. It is splendidly eclipsed by that in the beginning *of Julian and Maddalo,* where the Euganean Hills are lifted away from the earth and made a portion of the scenery of the sky. A special moment of sunset, with the moon and the evening-star in a sky reddened with tempest, is given in *Hellas,* but here, being in a drama, it is mingled with the fate of an empire. The Dawns are drawn with the same care as the sunsets, but with less passion. There are many of them, but the most beautiful perhaps is that in the beginning of the second act of the *Prometheus.* The changes of colour, as the light increases in the spaces of pure sky and in the clouds, are watched and described with precise truth; the slow progress of the dawn, during a long time, is noted down line by line, and all the movement of the mists and of the clouds "shepherded by the slow unwilling wind." Nor is that minuteness of observation wanting which is the proof of careful love. Shelley's imaginative study of beauty is revealed in the way the growth of the dawn is set before us by the waxing and waning of the light of the star, as the vapours rise and melt before the morn.

The Storms are even better than the sunsets and dawns. The finest is at the beginning of the *Revolt of Islam*. It might be a description of one of Turner's storm-skies. The long trains of tremulous mist that precede the tempest, the cleft in the storm-clouds, and seen through it, high above, the space of blue sky fretted with fair clouds, the pallid semicircle of the moon with mist on its upper horn, the flying rack of clouds below the serene spot—all are as Turner saw them; but painting cannot give what Shelley gives—the growth and changes of the storm.

There is another description at the beginning of the eleventh canto of the same poem, in which the vast wall of blue cloud before which grey mists are flying is cloven by the wind, and the sunbeams, like a river of fire flowing between lofty banks, pour through the chasm across the sea, while the shattered vapours which the coming storm has driven forth to make the opening, are tossed, all crimson, into the sky. This is a favourite picture of Shelley's. In the "Vision of the Sea" it is transferred from sunset to sunrise. The fierce wind coming from the west rushes like a flooded river upon the dense clouds which are piled in the east, and rends them asunder, and through the gorge thus cleft

> the beams of the sunrise flow in, Unimpeded, keen, golden and crystalline, Banded armies of light and air.

The description is a little over-wrought, but criticism has no voice when it thinks that no other poet has ever attempted to render, with the same absolute loss of himself, the successive changes, minute by minute, of such an hour of tempest and of sunrise. We are alone with Nature; I might even say, we see Nature alone with herself. Still greater, more poetic, less sensational, is the approach of the gale in the "Ode to the West Wind," where the wind itself is the river on which the forest of the sky shakes down its foliage of clouds, and these are tossed upwards like a Msenad's "uplifted hair," or trail downwards, like the "locks" of Typhon, the vanguard of the tempest. In gathered mass behind, the congregated might of vapours is rising to vault the heaven like a sepulchral dome. Nothing can be closer than the absolute truth to the working of the clouds that fly before the main body of a storm, which is here kept in the midst of these daring comparisons of the imagination.

The same delight in the indefinite and changeful aspects of Nature appears in Shelley's power of describing vast landscapes, such as that seen at noontide from the Euganean Hills, or that which the poet in *Alastor* looks upon from the edge of the mountain precipice. Both swim in the kind of light that makes all objects undefined, deep noon, and sunset light.

Kindred to this is Shelley's pleasure in the intricate, changeful, and incessant weaving and unweaving of nature's life in a great forest. In the "Recollection" it is the Pisan Pineta he describes, and that is a painting directly after Nature. But he has his own ideal forest, of which he tells in *Alastor*, in *Rosalind and Helen*, in the *Triumph of Life*, and again and again in the *Prometheus*. It is no narrow wood, but a universe of forest; full of all trees and flowers, in which are streams, and pools, and lakes, and lawny glades, and hills, and caverns; and in whose multitudinous scenery Shelley's imagination could lose and find itself without an end. The special love of caverns, with their dim recesses, adds another characteristic touch. These then,—the scenery of the sky, of the forest, of the vast plain,—are the aspects of nature Shelley loved the most, and out of the weakness that elsewhere made him too indefinite, and too uncertain through desire of change, for Wordsworth's special kind of descriptive power, arose the force with which he realised them.

2. Again, just because Shelley had no wish to conceive of Nature as involved in one definite thought, he had the power of conceiving the life of separate things in Nature with astonishing individuality. When he wrote of the Cloud, or of Arethusa, or of the Moon, or of the Earth, as distinct existences, he was not led away from their solitary personality by any universal existence in which they were merged, or by the necessity of adding to these any tinge of humanity, any elements of thought or love, such as the Pantheist is almost sure to add. His imagination was free to realise pure Nature, and the power by which he does this, as well as the work done, are quite unique in modern poetry. Theology, with its one Creator of the Universe; Pantheism, with its "one spirit's plastic stress;" Science, with its one Energy, forbid the modern poet, whose mind is settled into any one of these three views, to see anything in Nature as having a separate life of its own. He cannot, as a Greek could do, divide the life of the Air from that of the Earth, of the cloud from that of the stream. But Shelley, able to loosen himself from all these modern conceptions which unite the various universe, could and did, when he pleased, divide and subdivide the life of Nature in the same way as a Greek—and this is the cause why even in the midst of wholly modern imagery and a modern manner, one is conscious of a Greek note in many passages of his poetry of Nature. The following little poem on the Dawn might be conceived by a primitive Aryan. It is a Nature myth:—

> The pale stars are gone!
> For the sun, their swift shepherd,
> To their folds them compelling,
> In the depths of the dawn,

Hastes, in meteor-eclipsing array, and they flee
Beyond his blue dwelling
As fawns flee the leopard.

But Shelley's conceptions of the life of these natural things are less human than even the Homeric Greek or early Indian poet would have made them. They described the work of Nature in terms of human act. Shelley's spirits of the Earth and Moon are utterly apart from our world of thought and from our life. Of this class of poems "The Cloud" is the most perfect example. It describes the life of the Cloud as it might have been a million years before man came on earth. The "sanguine Sunrise" and the "orbed Maiden," the moon, who are the playmates of the cloud, are pure elemental beings.

The same observation is true if we take a poem on a living thing in Nature, like "The Skylark," into which human sentiment is introduced. The sentiment belongs to Shelley, not to the lark. The bird has joy, but it is not our joy. It is "unbodied joy," nor "can we come near it." Wordsworth's "Skylark" is truer, perhaps, to the every-day life of the bird, and the poet remembers, because he loves his own home, that the singer will return to its nest; but Shelley sees and hears the bird who, in its hour of inspired singing, will not recollect that it has a home. Wordsworth humanises the whole spirit of "the pilgrim of the sky"—"True to the kindred points of heaven and home." Shelley never brings the bird into contact with us at all. It is left in the sky, singing; it will never leave the sky. It is the archetype of the lark we seem to listen to, and yet we cannot conceive it, we have no power—"What thou art we know not." The flowers in the "Sensitive Plant" have the same apartness from humanity, and are wholly different beings and in a different world from the Daisy or the Celandine of Wordsworth. It is only the Sensitive Plant, and that is Shelley himself, which has an inner sympathy with the Lady of the garden.

Shelley, then, could isolate and perceive distinct existences in Nature as if he were himself one of these existences. It was a strange power, and we naturally cannot love with a human love things so represented. In Wordsworth's poems we touch the human heart of flowers and birds. In Shelley's we touch "Shapes that haunt Thought's wildernesses." Yet it is quite possible, though we cannot feel affection for Shelley's Cloud or Bird, that they are both truer to the actual fact of things than Wordsworth made his birds and clouds. Strip off the imaginative clothing from "The Cloud," and Science will support every word of it. Let the Skylark sing, let the flowers grow, for their own joy alone. In truth, what sympathy have they, what sympathy has Nature with Man? We may not like to think of Nature in this way; we are left quite cold by "The Cloud," and by the spirits of the Earth and Moon in the

Prometheus; and if we are not left as cold by "The Skylark," it is because we are made to think of our own sorrow, not because we care for the bird. But whether we like or no to see Nature in this fashion, we should be grateful for these unique representations, and to the poet who was able to make them. In this matter also Shelley's want of a central and uniting Thought in Nature made his strength.

The other side of Shelley's relation to Nature is a remarkable contrast to this statement. When he was absorbed in his own being, and writing poems which concerned himself alone, he makes Nature the mere image of his own feelings, the creature of his mood. In his "life alone doth Nature live." This was the natural result, at these times, of his intellectual rejection of such Pantheism as enabled Wordsworth always to distinguish between himself and the Nature he perceived. The Nature Wordsworth saw we can love well, because it is not ourselves—never a reflection of ourselves. The Nature such as Shelley saw in *Alastor* is not easy to love, because it is ourselves in other form. For this reason also we are not able to love Nature, when thus represented by Shelley, so well as we love her in Wordsworth.

Shelley's love of the undefined and changing is still further illustrated by the fact that we see Nature in his poetry in these three ways—on all of which I have dwelt. We sometimes look on her as the ideal Pantheist beholds her; we look on her again as the mere reflection of the poet's moods; we look on her often as she may be in herself, apart from theories about her, apart from man.

3. Lastly, on this subject, the vagueness and changefulness of Shelley's feeling and view of Nature, except in the instances mentioned, the dreams and shadows of it in his poetry that incessantly form and dissolve like the upper clouds of the sky, each fleeting while its successor is being born, and few living long enough to be outlined, are the only images we possess in art, save perhaps in music, of the many hours we ourselves pass with Nature when we neither think nor feel, but drift and dream incessantly from one impression to another, enjoying, but never defining our enjoyment, receiving moment by moment, but never caring to say to any single impression, "Stay and keep me company." In this thing also, Shelley's weakness made his power.

This want of definite belief and of its force belongs also to his conception of the ideal state of mankind. He does not see quite clearly what he desires for man, and describes the golden age chiefly by negatives of wrong. At times he rises into a passionate realisation of his Utopia, as he rises into Pantheism, but he cannot long remain in it. The high-wrought prophecy, too weak to keep the height it has gained, sinks down again and again into an abyss of seeming hopelessness. The last stanza of the "Ode to Liberty" is the type of many an hour of his life, and of the close of many a poem. But he never let hopelessness or depression master him. Shelley is full of resurrection power, and the fall from the peak of prophecy is more the result of reaction after impassioned excitement, than the result of any unbelief in his hopes for men, or in that on which they were grounded.

These hopes, that belief, had their strong foundation. There was one thing at least that Shelley grasped and realised with force in poetry—the moralities of the heart in their relation to the progress of Mankind. Love and its eternity; mercy, forgiveness, and endurance, as forms of love; joy and freedom, justice and truth as the results of love; the sovereign right of Love to be the ruler of the Universe, and the certainty of its victory,—these were the deepest realities, the only absolute certainty, the only centre in Shelley's mind; and whenever, in behalf of the whole Race, he speaks of them, and of the duties and hopes that follow from them, strength is then instinctive and vital in his imagination. Neither now nor hereafter can men lose this powerful and profound impression. It is Shelley's great contribution to the progress of humanity.

But he could not combine with this large view and this large sympathy with the interests of Man, personal sympathy with personal human life. That is absent from his poetry, and his want of it was confirmed by his exile. Confined to a small circle of which he was the centre, among foreigners, feeling himself repudiated by the society of his own country, and incapable of such quiet association with the lives of men and women as Wordsworth loved and enjoyed, it is no wonder that large spaces of human life are entirely unreflected and unidealised in his poetry. The common human heart was not his theme, nor did he care to write of it. And, so far, he is less universal than Wordsworth, and less the great poet. But on the other hand he did two things, in his work on human nature, that Wordsworth could not do. First, he realised in song, so far as it was possible, the impalpable dreams of the poetic temperament, those which, when they arise in happiness, he expresses in the little poem, "On a poet's lips I slept," and others also less joyous—the lonely wanderings of regretful thought, the imagination in its hours of childlike play with images, the moments when we are on the edge where emotion and thought incessantly change into one another, the visions of Nature which we compose but which are not Nature, the sorrows and depressions which have no name and to which we allot no cause, the depths of passionate fancy when we have not only no relation to mankind, but hate to feel that relation. Of all this Wordsworth gives us nothing; and though what he does give us is of more use and worth to us as men who have to do with men, yet Shelley's work in this is dear to our personal life, and has in fact as much to do with one realm of humanity as the sorrow of Michael, or the daily life of the dalesmen have with another. English poetry needed the expression of these things; Shelley's expression of them is unique, but I doubt whether he would ever have expressed them in so complete a way had he not been thrown into isolation.

Secondly, there is an element almost altogether wanting in Wordsworth, the absence of which forbids us to class him as a poet who has touched all the important sides of human life—the element of passionate love. A few of his poems, such as "Barbara," or in another kind, "Laodameia," solemnly glide into it and retreat, but on the whole, this, the most universal subject of lyric poetry, was not felt by Wordsworth. It was felt by Shelley, but not quite naturally, not as Burns, or even Byron felt it. Love, in his poetry, sometimes dies into dreams, sometimes likes its imagery better than itself. It is troubled with a philosophy; it seems now and again to be even bored, if I may be allowed the word, by its own ideality. As Shelley soared but rarely into definite Pantheism, so he rose but rarely into definite passion, nor does he often care to realise it. It was frequently his deliberate choice to celebrate the love which did not "deal with flesh and blood," and as frequently, when he writes directly of love, he prefers to touch the lip of the cup, but not to drink, lest in the reality he should lose the charm of indefiniteness, of ignorance, of pursuit. Of course he was therefore fickle.

For this very reason, however, two realms in this aspect of his art belong to him. Neither of them is the realm of joyous passion, but one is the realm of its ideal approaches, and the other the realm of its ideal regret. No one has expressed so well the hopes, and fears, and fancies, and dreams, which the heart creates for its own pleasure and sorrow, when it plays with love which it realises within itself, but which it never means to realise without; and this is a realm which is so much lived in by many that they ought to be grateful to Shelley for his expression of it. No one else has done it, and it is perfectly done.

But still more perfect, and perhaps more beautiful than any other work of his, are the poems written in the realm of ideal Regret. Whenever he came close to earthly love, touched it, and then of his own will passed it by, it became, as he looked back upon it, ideal, and a part of that indefinite world he loved. The ineffable regret of having lost that which one did not choose to take, is most marvellously, most passionately expressed by Shelley. Song after song records it. The music changes from air to air, but the theme is the same, and so is the character of the music. And, like all the rest of his work, it is unique.

But in this matter, a change passed over Shelley before he died. It is impossible not to feel that the poems written for Mrs. Williams, a whole chain of which exist, are different from the other love poems. They have the same imaginative qualities as the previous songs, and they belong also to the two realms of which I have written above, but there is a new note in them, the beginning of the unmistakable directness of passion. It is, of course, modified by the circumstances, but there it is. And it is from the threshold of this actual world that he looks back on *Epipsychidion* and feels that it belonged to "a part of him that was already dead." The philosophy which made Emilia the shadow of a spiritual Beauty is conspicuous by its total absence from all these later love poems. Moreover, they are not, like the others, all written in the same atmosphere. The atmosphere of ideal love, however varied its cloud-imagery, is always the same thin ether. But these poems breathe in the changing atmosphere of the Earth, and they one and all possess reality. Every one feels that "Ariel to Miranda," "The Invitation," "The Recollection," have the variety of true passion. But none of them reach the natural joy of Burns in passionate love. Two exceptions, however, exist both dating from this time, and both written away from his own life—the "Bridal Song," and the song "To Night." These seem to prove that, had Shelley lived, we might have had from him vivid, fresh, and natural songs of passion.

Had he lived! Had not the sea been too envious, what might we not have possessed and loved! It were too curious perhaps to speculate, but Shelley seems to have been recovering the power of working on subjects beyond himself, in the quiet of those last days at Lerici. He was always capable of rising again, and the extreme clearness and positive element of his intellect acted, like a sharp physician, on his passion-haunted heart and freed it, when it was out-wearied with its own feeling, from self-slavery.

While still at Pisa, at the beginning of 1822, Shelley set to work on a Drama, *Charles I.*, the motive of which was to be the ruin of the king through pride and its weakness, the same motive as *Coriolanus*. It was to be "the birth of severe and high feelings," but severe feeling was not then the temper of his mind, nor could he at that time lose himself enough to create an external world. He laid the play aside, saying that he had not sufficient interest in English history to continue it. Yet it is plain, even from the fragments we possess, how great was the effort Shelley then made to realise, even more than in the *Cenci*, other characters than his own. There is not a trace in it of his own self. It is full of steady power, power more at its ease than in the *Cenci*, and it is quite plain that it cannot be said of the artist who did this piece of work that he had exhausted his vein.

It becomes still more clear that Shelley would have done far more for us when we consider the *Triumph of Life*, to write which he threw aside *Charles I.* It is the gravest poem he ever wrote, and it has a deep interest for this generation. Its personal value as a revelation of his view of life, of the change of some of his views on moral matters and of his retention of youthful theories can scarcely be overestimated, but to analyse it here would take up too much space. It is enough to say here that its interest for humanity is as great as its personal interest. Had he lived then, he would have once more appeared as the Singer of Man and in the cause of men. But the swift wind and the mysterious sea, the things he loved, slew their lover—a common fate—and we hear no more his singing. His work was done, and its twofold nature, as the Poet of Man, and the Poet of his own lonely heart, may well be imaged by the Sea that received him into its breast, for while its central depths know only solitude, over its surface are always passing to and fro the life and fortunes of Humanity.

—STOPFORD A. BROOKE, "Some Thoughts
on Shelley," *Macmillan's Magazine*,
June 1880, pp. 124–135

EDWARD DOWDEN
"LAST WORDS ON SHELLEY" (1887)

Born in Mentenotte, Ireland, Edward Dowden (1843–1913) served as professor of English literature at Alexandra College, Dublin, and the University of Dublin before settling at Trinity College, where he served as department chairperson and professor of English from 1867 to 1913. His biographical subjects include Spenser, Shakespeare, Southey, and Browning. In 1886, Dowden published the two-volume *Life of Percy Bysshe Shelley*, commissioned by Sir Percy and Lady Jane Shelley. Its publication brought Dowden both fame and censure. Some criticized his lack of comment on Shelley's first marriage and the poet's abandonment of Harriet Westbrook. However, the work was considered the first "official" scholarly biography of Shelley. Dowden's interest in Shelley also gave rise to *Letters about Shelley, Interchanged by Three Friends—Edward Dowden, Richard Garnett, and W. Michael Rossetti* (1917).

Idealist as he was, Shelley lived in some important respects in closer and more fruitful relation with the real world than did his great contemporary, Scott. Because he lived with ideas, he apprehended with something like prophetic insight those great forces which have been altering the face of the world during this nineteenth century, and which we sum up under the names of democracy and science; and he apprehended them not from the merely material point of view, but from that of a spiritual being, uniting in his vision with democracy and science a third element not easy to name or to define, an element of spirituality which has been most potent in the higher thought and feeling of our time. Many strange phantasies had Shelley, but no phantasy quite so remote from reality as that of building himself a mock feudal castle, or of living in a world half made up of the modern pseudo-antique. Many strange phantasies he had, but none so strange as phantasies of later date; none so strange as that of reviving the faith of the twelfth century in English brains to-day; none so feebly wild as that of drawing a curtain of worn-out shreds to hide the risen sun of science. As regards science, it is obvious enough that Shelley possessed in no degree the scientific intellect. He was far from being able to contribute to science such anticipations of imaginative genius as those which make the name of Goethe illustrious in botany and comparative anatomy. But Shelley expressed a poet's faith in science and a poet's hopes. Wordsworth, incomparably a greater thinker than Shelley, expressed a poet's fears—fears by no means wholly unjustified—that the pursuit of analytic investigation in things material might dull the eye for what is vital and spiritual in nature and in man. Wordsworth recognised a part of the fact, but Shelley's feelings attached themselves to the more important side of truth in this matter. "Beautiful and ineffectual angel beating in the void his luminous wings in vain." No, not in the void, but amid the prime forces of the modern world; and this ineffectual angel was one of the heralds of the dawn—dawn portentous, it may be, but assuredly real. I recognise in Shelley all the illusions and sophisms of the revolutionary epoch. I recognise the vagueness of much of his humanitarian rhetoric. But humanitarian rhetoric sometimes may be practical beneficence in a nebulous state; let it condense and solidify, and the luminous mist becomes an orb of love—the stout heart of one who would serve the needy and the downcast of our race. If love, justice, hope, freedom, fraternity, be real, then so is the wiser part of the inspiration of Shelley's radiant song.

If, then, Shelley did not hover ineffectually in the void, may we not attempt to define his historical position in our literature? Perhaps it is not rash to assert that when this century of ours is viewed from a point in the distant future, it will be seen that among many great facts of the century the largest and most important are those expressed by the words democracy and science. And what, if we should sum it up in one word, is the leading idea given by democracy to literature? In mediaeval times the heroes on whom imagination fixed its gaze were two, the chivalric knight and the ascetic saint—great and admirable figures. With us the hero is, if you please, no hero at all, but simply the average man. For him we think and toil; our most earnest hopes and wishes are for him. Or, since after all we want a hero, let us say, instead of "the average man," the race, or humanity. And now what, expressed in a single word, is the ruling idea of science? What word can we choose except the great and venerable word, law? Here, then, are two eminent words of our century—the race, or humanity, given to us by democracy; law, given to us by science. Let us connect the two, and we obtain the expression "humanity subject to law"—that is, we have the conception of human progress or evolution. Hence this phrase "progress of humanity," however it may have been spoilt for dainty lips by cheap and vulgar trumpeting, and however we may recognise the fact that for a day or a year, or a group of years, the world's advance may halt on palsied feet—this word, this idea, this faith has had for our age something like the force of a religion. And as the inspiring faith of our century has been this faith in the progress or evolution of man, so its heresy has been the heresy of pessimism; and in literature, side by side with the stronger poetry of hope, there has been a feebler poetry of despair.

In the earlier years of our century the democratic movement concerned itself too exclusively with the individual and his rights, and regarded too little his duties, affections, and privileges as a member of society. It is greatly to the advantage of Shelley's work as a poet, and greatly to his credit as a man, that he assigns to love, that which links us to our fellows, some of the power and authority which Godwin ascribes to reason alone.[1] The French Revolution had been in a great measure a destruction of the ancient order of society, and such poetry as that of Byron, sympathising with the revolution, is too reckless an assertion of individual freedom. Shelley was deeply infected with the same errors. But it is part of the glory of his poetry that in some degree he anticipated the sentiment of this second half of our century, when we desire more to construct or reconstruct than to destroy. Shelley's ideas of a reconstruction of society are indeed often vague or visionary; but there is always present in his poetry the sentiment or feeling which tends to reconstruction, the feeling of love; and the word "fraternity" is for him at least as potent as the word "liberty." In Byron we find an expression of the revolution on its negative side; in Shelley we find this, but also an expression of the revolution on its positive side. As the wave of revolution rolls onward, driven forth from the vast volcanic upheaval in France, and as it becomes a portion of the literary movement of Great Britain, its dark and hissing crest may be the poetry of Byron; but over the tumultuous wave hangs an iris of beauty and promise, and that foam-bow of hope, flashing and failing, and ever reappearing as the wave sweeps on, is the poetry of Shelley.

There is a kind of wisdom, and a very precious kind, of which we find singularly little in Shelley's poetry. The wisdom of common sense, which enables us to steer our way amid the rocks and shoals of life; the wisdom of large benignant humour; the wisdom of the ripened fruits of experience; the wisdom of those axioms intermediate between first principles and practical details—those *axiomata media* which in science Bacon regards as so important; to utter such wisdom in verse was not Shelley's province. But there is another wisdom which the world sometimes counts as folly—that which consists in devotion at all hazards to an ideal, to what stands with us for highest truth, sacred justice, purest love. And assuredly the tendency of Shelley's poetry, however we may venerate ideals other than his, is to quicken the sense that there is such an exalted wisdom as this, and to stimulate us to its pursuit.

Whether we speak of the poet as an inspirer of wisdom, or as one who enlarges and purifies our feelings, or as one who widens the scope of our imagination, we dare not claim for him the title of a great poet unless he has enriched human life and aided men in some way to become better or less incomplete and fragmentary creatures. If Shelley has done this, we may disregard such words as those of Mr Ruskin and Principal Shairp. Let us ask then, "How has Shelley made life better for each of us?" bearing in mind, while we put the question, that only a small part of the full and true answer can find its way into a definite statement. At least we can say this, he helps us to conceive more nobly of nature and more nobly of man. We come through his poetry to feel more vividly the quick influencings which pass from the beauty and splendour and terror of the external world into these spirits of ours. He helps us to lie more open to the joy and sadness of the earth and skies. Who has felt the breath of the autumnal west wind without a sense of its large life and strength and purity, made ampler and more vivid by what Shelley's great ode has contributed to his imagination? Or who has heard the song of the lark in mid-heaven, and not felt how that atom of intense joy above us rebukes our distrust of nature and of life and all our dull despondencies, and feeling this has not remembered that Shelley once helped to interpret for him the rapture of the bird? And though no words of man can make more glorious the spectacle of the midnight heavens, who does not feel in such stanzas as those which begin with the lines—

> Palace-roof of cloudless nights, Paradise of golden lights Deep, immeasurable, vast,

a clarion-cry rousing the imagination and inspiring it with *elan* for that advance which is needful before we can apprehend the splendour and the awful beauty that encircle us?

So Shelley has helped us to feel the glory of external nature, making life a better thing for each of us; and in like manner he quickens within us a sense of the possibilities of greatness and goodness hidden in man and woman. Let us recognise to the full the philosophical errors in the doctrine which lies behind the poetry of the "Prometheus Unbound," the false conception of evil as residing in external powers rather than in man's heart and will, the false ideal of the human society of the future; and recognising these, let us acknowledge that the poem has helped us to conceive more truly and more nobly of the possibilities of man's life, its possibilities of fortitude, endurance, pitying sympathy, heroic martyrdom, aspiration, joy, freedom, love. No poet has more truly conceived, or more vividly presented in words a sense of the measureless importance of one human spirit to another—of the master to the disciple, of the spiritual leader to his followers, of man to woman and woman to man. With a quickened sense of the infinite significance of the relations possible with our fellows, our entire feeling for life and for the virtues which hide in it, more marvellous than the occult virtues of gems, is purified and exalted. Especially has Shelley taught us to recognise the blessedness—blessedness in joy or in anguish—of the higher rule imposed on dedicated spirits, who live for a cause or an idea, a charity or a hope, and for its sake are willing to endure shame and reproach, and a death of martyrdom. But this higher rule, as conceived by Shelley, is not one of voluntary self-mortification or ignoble asceticism; he does honour in verse and prose to music, sculpture, painting, poetry, and quickens our sense of the spiritual power of each. Yet he never settles down to browse with Epicurean satisfaction in any paddock of beauty or pleasure. We are touched through his poetry with a certain divine discontent, so that not music, nor sculpture, nor picture, nor song, can wholly satisfy our spirits; but in and through these we reach after some higher beauty, some divine goodness, which we may not attain, yet towards which we must perpetually aspire. And who has heartened us more than Shelley, amid all his errors, to love freedom, to hope all things, to endure all things, and even while the gloom gathers to have faith in the dawn of light? Who has done more to quicken and refine our sympathy with suffering creatures? To assure us that among the despised and rejected things of the world true goodness dwells, so that even the snake may be in truth a defeated angel in disguise? And who has more powerfully impressed us with the conviction that revenge and reprisals are bitter fruits of the spirit of wrath and pride, and that evil can best be overcome by returning good for evil? From whom do we learn more effectively the duty of loyalty to our convictions, and the duty, imposed upon us at times, to fling out our highest belief as a factor to do its work in the world? And if Shelley rouses within us the spirit that makes us nonconformist (and I, for my part, have a deep reverence for reasonable nonconformity), who has given us a more graceful example than he, in his happier moments, of that rare thing, a nonconformist who is not sour-faced but amiable and gentle?

But in many respects the truth seen by Shelley was seen as broken lights in an imperfect vision. His ideals were in part false ideals. He never quite escaped from the individualism of Godwin's system of thought. When we say, then, that Shelley possessed the wisdom of devotion to the ideal, we must qualify the statement by adding that his so-called ideal was in part no true ideal, but the spurious ideal of a phantast. For what is the good of using the splendid words "truth," "justice," "charity," if the words are used to describe something other than the realities they ought to stand for? These exalted words are wrested in revolutionary times away from their honest sense; they are made a specious veil behind which acts of injustice and cruelty are freely perpetrated. I do not believe that Shelley could ever have been guilty of such acts; but it is the Girondin with his fine phrases who prepares the way for the Jacobin with his atrocious deeds. Shelley's notion, expressed in the *Prometheus Unbound*, that naked manhood,

> Equal, unclassed, tribeless, and nationless,

will remain, when all has been stripped off which humanity has painfully acquired during the ages, is the pseudo-ideal of a Rousseau turned topsy-turvy, or rather of a Rousseau who has turned right-about face, and who sees the fantastic golden age of simplicity, innocence, and freedom not in the past but in the future.

Here, however, I would insist on an important fact, which has never received due attention from the students of Shelley's writings, and which goes far towards establishing his sanity as a thinker, although it indicates a weakness in his poetry. While in such poems as *The Revolt of Islam* and *Prometheus Unbound*, he has imagined an ideal of the future state of society which never can be realised and which we ought not to desire, in his prose writings he often exhibits a justness of view and a moderation which have hardly obtained the recognition they deserve. The contrast between his dreams and visions as a poet, and his very moderate expectations as a practical reformer, is indeed remarkable. "Before the restraints of government are lessened, it is fit that we should lessen the necessity for them," so wrote Shelley in his "Address to the Irish People," and boy as he was, he showed himself by such words to be wiser or honester than some grey-haired counsellors of to-day. "With respect to Universal Suffrage," he wrote, "I confess I consider its adoption in the present unprepared state of public knowledge and feeling a measure fraught with peril." And again, to Leigh Hunt: "The great thing to do is to hold the balance between popular impatience and tyrannical obstinacy . . . I am one of those whom nothing will satisfy, but who are ready to be partially satisfied in all that is practicable." Examples of Shelley's moderation in practical politics could be drawn from every period of his life, evidencing that this was the habit of his mind. His poetry is often vaporous and unreal, although the man himself had a clear perception of reality. Unfortunately the two sides of his mind—the poetical and the practical— seldom worked together. In his verse he set forth his ideals and his visions of the remote future; he reserved his prose for dealing with what was practicable and near. It would have been better for his poetry if he could have put his whole mind into verse, as did Wordsworth. We could readily excuse a prosaic paragraph for the sake of the gain in wisdom and intellectual and moral breadth. And in truth this would, in some degree, have saved his poetry from what is most prosaic in the longer pieces, that doctrinaire background of Godwinian abstractions to which nothing will give reality or life. In one, and only one, of Shelley's longer non-dramatic poems do the two sides of his mind work harmoniously together, in *The Mask of Anarchy*. Although this poem may not rank with his highest work, it enables us to understand how greatly that work would have gained had it been possible for the Shelley who saw visions to have taken counsel with the Shelley who observed and meditated on affairs. But, as he conceived, his ideals of a remote future were not without a practical use for the toilers of the day and hour. We work among petty details in a larger, wiser spirit, and with more of hope and valour and patience, if now and again we lift our eyes and behold the land that is very far off. "We derive tranquillity, and courage, and grandeur of soul," he writes, "from contemplating an object which is because we will it, and may be because we hope and desire it, and must be if succeeding generations of the enlightened, sincerely and earnestly seek it." . . .

Shelley's poetry, says Mr Hutton, is "the poetry of desire." Yes, but here is something to go along with desire and be its counterpoise.

It is, however, fortitude in the presence of pain, and the constancy of a self-sufficing heroic soul in the midst of vicissitude which Shelley honours rather than temperance in the acceptance of delights. Of temperance we find little in his verse. He is always pining for a joy that is gone, or hungering for a rapture that is to come. Only in the last fragment written by Shelley, the admirable *Triumph of Life*, does he appear fully to recognise the danger of yielding the heart intemperately to even the purest passion. In that poem Rousseau and Plato appear as victims of their own hearts: Rousseau, a ruin of manhood; Plato, who had loved more nobly, punished less cruelly, yet a captive to the triumphal car; both suffering the inevitable doom of those who are intemperate in desire and delight. But Socrates, who had known himself and tempered his heart to its object, is no chained victim in that wild career of Life the Triumpher. Thus, through desire Shelley was reaching to a calmer and saner atmosphere as his life drew towards a close.

And perhaps the influence of his lyrical poems upon his readers may be to lift them towards a like calm of mind attained through passion, or at least to purify desire and delight from all grossness, and so to lighten the task of self-control. Aristotle, in a famous passage, speaks of the effect of tragedy in purifying through terror and pity the like passions. In a similar manner the lyrical poetry of delight and desire should purify delight and desire. From the gross throng of conflicting passions the finer are selected by the poet, are given predominance, and are themselves raised to their highest and fairest life. It is the imagination which elevates the gross passion of grief and terror caused by death into the lofty sorrow which is the most human as well as the highest grief, with all of the brute purged away. It is the imagination which elevates the passion of love between man and woman into its nobler forms, where the senses have been taken up into the spirit. So every emotion of pleasure or of pain may be made rarer, finer, more exquisite by the energy of the imagination, and to effect this is one of the highest functions of lyrical poetry. The poet feels more exquisitely than other men, and receives more impulses and intimations from the spiritual side of things. When he sings he not only relieves his own heart, he not only widens our sympathy with human emotion; he chastens and purifies our feelings, rendering them finer and more sane and permanent. The lyrical poetry of Shelley plays thus upon our feelings of delight—delight in external nature, delight in human beauty, delight in art, delight in the beauty of character and action; it plays with its refining influence still more often upon our feelings of desire and of regret. There is a rapture at once calm and impassioned which is admirably expressed in Wordsworth's earlier poetry, a rapture of which Shelley knew little. He does not train us to sober certainties of waking bliss as does Wordsworth. He is in endless pursuit of unattainable ideals, ever at the heels of the flying perfect. Although the man is poor indeed who has not something of Wordsworth's art of sinking profoundly into the joy and peace of things, and drinking a portion of their strength and repose, I am not sure that the fittest attitude of a human creature in this our mortal life is not Shelley's attitude—the attitude of aspiration and desire. Joy is not a thing for us to rest in; joy should rather open into higher joy, light should pass into purer light; from any height or deep at which we may arrive we should still cry, "O altitudo!" to the height or deep beyond. To intensify and to purify desire is, perhaps, no less important for us than to deepen and purify satisfaction. And no one can live for a time in the lyrical poetry of Shelley without an exaltation and purification of desire.

"I can conceive Shelley if he had lived to the present time," wrote Peacock, "passing his days like Volney, looking on the world from his windows without taking part in its turmoils; and perhaps like the same or some other great apostle of liberty (for I cannot at this moment verify the quotation), desiring that nothing should be inscribed on his tomb but his name, the dates of his birth and death, and the single word *Desillusionne*." But it is he who would lie down and rest in some earthly satisfaction who will be disillusioned, not he who forever passes from desire to delight and from delight to desire, with a foot upon the ladder whose top reaches to heaven. Even in respect to political affairs I do not think that Shelley would have looked forth from his window disillusioned. A series of great events would probably have engaged his interest and aroused his imaginative ardour: first, the liberation of Greece, then the emancipation of Catholics in Ireland, then the French Revolution of 1830, then the Reform Bill of 1832; and in 1832 Shelley would have reached his fortieth year, and his character would have gained the enduring ardour of midmanhood. But however this may have been, I cannot conceive Shelley insensible to hope, untouched by desire, incapable of new delights, possessing only the sorry wisdom of a man disillusioned. Rather, I think, he would have continued to live by admiration, hope, and love; and as these were directed to worthier objects and yet more worthy, he would have ascended in dignity of being.

In life and in literature there are three kinds of men to whom we give peculiar honour. The first are the craftsmen, who put true and exact work into all they offer to the world, and find their happiness in such faithful service. Such a craftsman has been described with affectionate reverence by George Eliot in her poem "Stradivarius":—

> That plain, white-aproned man who stood at work, Patient and accurate, full fourscore years, Cherished his sight and touch by temperance; And since keen sense is love of perfectness Made perfect violins, the needed paths For inspiration and high mastery.

We do not reckon Shelley among the craftsmen. The second class is small in numbers; we call these men conquerors, of whom, as seen in literature, the most eminent representatives in modern times have been Shakspere and Goethe. These are the masters of life; and having known joy and anguish, and labour and pleasure, and the mysteries of love and death, of evil and of good, they attain at last a lofty serenity upon heights from which they gaze down, with an interest that has in it something of exalted pity, on the turmoil and strife below. It is their part to bring into actual union, as far as our mortal life permits, what is real and what is ideal. They are at home in both worlds. Shakspere retires to Stratford, and enjoys the dignity and ease and happy activity of the life of an English country gentleman; yet it was he who had wandered with Lear in the tempest, and meditated with Hamlet on the question of self-slaughter. Goethe, councillor to his noble master, the Grand-Duke of Weimar, in that house adorned with treasures of art and science, presides as an acknowledged chief over the intellectual life of a whole generation; yet he had known the storm and stress, had interpreted the feverish heart of his age in *Werther,* and all its spiritual doubts and desires and aspirations in his *Faust.* Such men may well be named conquerors, and Shelley was not one of these. But how shall we name the third class of men, who live for the ideal alone, and yet are betrayed into weakness and error, and deeds which demand an atonement of remorse; men who can never quite reconcile the two worlds in which we have our being, the world of material fact and the spiritual world above and beyond it; who give themselves away for love or give themselves away for light, yet sometimes mistake bitter for sweet, and darkness for light; children who stumble on the sharp stones and bruise their hands and feet, yet who can wing their way with angelic ease through spaces of the upper air. These are they whom we say the gods love, and who seldom reach the fourscore years of Goethe's majestic old age. They are dearer perhaps than any others to the heart of humanity, for they symbolise in a pathetic way, both its weakness and its strength. We cannot class them with the exact and patient craftsmen; they are ever half defeated and can have no claim to take their seats beside the conquerors. Let us name them lovers; and if at any time they have wandered far astray, let us remember their errors with gentleness, because they have loved much. It is in this third class of those who serve mankind that Shelley has found a place.

Notes
1. Godwin, however, it may be noted, desired to banish from philosophy the phrase, "rights of man." *Claims* he would allow, but never *rights.*

—Edward Dowden,
"Last Words on Shelley," 1887,
Transcripts and Studies, 1888, pp. 93–111

W.B. Yeats "The Philosophy of Shelley's Poetry" (1900)

William Butler Yeats (1865–1939) was born in Dublin, Ireland, to John Butler Yeats and Susan Mary Pollexfen. He attended the Metropolitan School of Art from 1883–85 and published his first poems in the *Dublin University Review*. He, along with Edwin Ellis, published a study of William Blake's poetry in 1893, and in his writings, Yeats was influenced by the works of the romantic writers, particularly Shelley. His book *Ideas of Good and Evil* (1903) examines the mysticism of William Blake and Percy Shelley.

The most important, the most precise of all Shelley's symbols, the one he uses with the fullest knowledge of its meaning, is the Morning and Evening Star. It rises and sets for ever over the towers and rivers, and is the throne of his genius. Personified as a woman it leads Rousseau, the typical poet of *The Triumph of Life*, under the power of the destroying hunger of life, under the power of the sun that we shall find presently as a symbol of life, and it is the Morning Star that wars against the principle of evil in *Laon and Cythna*, at first as a star with a red comet, here a symbol of all evil as it is of disorder in *Epipsychidion*, and then as a serpent with an eagle—symbols in Blake too and in the Alchemists; and it is the Morning Star that appears as a winged youth to a woman, who typifies humanity amid its sorrows, in the first canto of *Laon and Cythna*; and it is evoked by the wailing women of *Hellas*, who call it 'lamp of the free' and 'beacon of love' and would go where it hides flying from the deepening night among those 'kingless continents sinless as Eden,' and 'mountains and islands' 'prankt on the sapphire sea' that are but the opposing hemispheres to the senses but, as I think, the ideal world, the world of the dead, to the imagination; and in the 'Ode to Liberty,' Liberty is bid lead wisdom out of the inmost cave of man's mind as the Morning Star leads the sun out of the waves. We know too that had *Prince Athanase* been finished it would have described the finding of Pandemus, the stars' lower genius, and the growing weary of her, and the coming to its true genius Urania at the coming of death, as the day finds the Star at evening. There is hardly indeed a poem of any length in which one does not find it as a symbol of love, or liberty, or wisdom, or beauty, or of some other expression of that Intellectual Beauty, which was to Shelley's mind the central power of the world; and to its faint and fleeting light he offers up all desires, that are as

> The desire of the Moth for the star, The desire for something afar,
> From the sphere of our sorrow.

When its genius comes to Rousseau, shedding dew with one hand, and treading out the stars with her feet, for she is also the genius of the dawn, she brings him a cup full of oblivion and love. He drinks and his mind becomes like sand 'on desert Labrador' marked by the feet of deer and a wolf. And then the new vision, life, the cold light of day moves before him, and the first vision becomes an invisible presence. The same image was in his mind too when he wrote

> Hesperus flies from awakening night
> And pants in its beauty with speed and light,
> Fast fleeting, soft and bright.

Though I do not think that Shelley needed to go to Porphyry's account of the cold intoxicating cup, given to the souls in the constellation of the Cup near the constellation Cancer, for so obvious a symbol as the cup, or that he could not have found the wolf and the deer and the continual flight of his Star in his own mind, his poetry becomes the richer, the more emotional, and loses something of its appearance of idle phantasy when I remember that these are ancient symbols, and still come to visionaries in their dreams. Because the wolf is but a more violent symbol of longing and desire than the hound, his wolf and deer remind me of the hound and deer that Usheen saw in the Gaelic poem chasing one another on the water before he saw the young man following the woman with the golden apple; and of a Galway tale that tells how Niam, whose name means brightness or beauty, came to Usheen as a deer; and of a vision that a friend of mine saw when gazing at a dark-blue curtain. I was with a number of Hermetists, and one of them said to another, 'Do you see something in the curtain?' The other gazed at the curtain for a while and saw presently a man led through a wood by a black hound, and then the hound lay dead at a place the seer knew was called, without knowing why, 'the Meeting of the Suns,' and the man followed a red hound, and then the red hound was pierced by a spear. A white fawn watched the man out of the wood, but he did not look at it, for a white hound came and he followed it trembling, but the seer knew that he would follow the fawn at last, and that it would lead him among the gods. The most learned of the Hermetists said, 'I cannot tell the meaning of the hounds or where the Meeting of the Suns is, but I think the fawn is the Morning and Evening Star.' I have little doubt that when the man saw the white fawn he was coming out of the darkness and passion of the world into some day of partial regeneration, and that it was the Morning Star and would be the Evening Star at its second coming. I have little doubt that it was but the story of Prince Athanase and what may have been the story of Rousseau in *The Triumph of Life*, thrown outward once again from that great memory, which is still the mother of the Muses, though men no longer believe in it.

It may have been this memory, or it may have been some impulse of his nature too subtle for his mind to follow, that made Keats, with his love of embodied things, of precision of form and colouring, of emotions made sleepy by the flesh, see Intellectual Beauty in the Moon; and Blake, who lived in that energy he called eternal delight, see it in the Sun, where his personification of poetic genius labours at a furnace. I think there was certainly some reason why these men took so deep a pleasure in lights, that Shelley thought of with weariness and trouble. The Moon is the most changeable of symbols, and not merely because it is the symbol of change. As mistress of the waters she governs the life of instinct and the generation of things, for as Porphyry says, even 'the apparition of images' in the 'imagination' is through 'an excess of moisture'; and, as a cold and changeable fire set in the bare heavens, she governs alike chastity and the joyless idle drifting hither and thither of generated things. She may give God a body and have Gabriel to bear her messages, or she may come to men in their happy moments as she came to Endymion, or she may deny life and shoot her arrows; but because she only becomes beautiful in giving herself, and is no flying ideal, she is not loved by the children of desire. Shelley could not help but see her with unfriendly eyes. He is believed to have described Mary Shelley at a time when she had come to seem cold in his eyes, in that passage of *Epipsychidion* which tells how a woman like the Moon led him to her cave and made 'frost' creep over the sea of his mind, and so bewitched life and death with 'her silver voice' that they ran from him crying, 'Away, he is not of our crew.' When he describes the Moon as part of some beautiful scene he can call her beautiful, but when he personifies, when his words come under the influence of that great memory or of some mysterious tide in the depth of our being, he grows unfriendly or not truly friendly or at the most pitiful. The Moon's lips 'are pale and waning,' it is 'the cold Moon,' or 'the frozen and inconstant Moon,' or it is 'forgotten' and 'waning,' or it 'wanders' and is 'weary,' or it is 'pale and grey,' or it is 'pale for weariness,' and 'wandering companionless' and 'ever changing,' and finding 'no object worth' its 'constancy,' or it is like a 'dying lady' who 'totters' 'out of her chamber led by the insane and feeble wanderings of her fading brain,' and even when it is no more than a star, it casts an evil influence that makes the lips of lovers 'lurid' or pale. It only becomes a thing of delight when Time is being borne to his tomb in eternity, for then the spirit of the Earth, man's procreant mind, fills it with his own joyousness. He describes the spirit of the Earth and of the Moon, moving above the rivulet of their lives in a passage which reads like a half-understood vision. Man has become 'one harmonious soul of many a soul' and 'all things flow to all' and 'familiar acts are beautiful through love,' and an 'animation of delight' at this change flows from spirit to spirit till the snow 'is loosened from the Moon's lifeless mountains.'

Some old magical writer, I forget who, says if you wish to be melancholy hold in your left hand an image of the Moon made out of silver, and if you wish to be happy hold in your right hand an image of the Sun made out of gold. The Sun is the symbol of sensitive life, and of belief and joy and pride and energy, of indeed the whole life of the will, and of that beauty which neither lures from far off, nor becomes beautiful in giving itself, but makes all glad because it is beauty. Taylor quotes Proclus as calling it 'the Demiurgos of everything sensible.' It was therefore natural that Blake, who was always praising energy, and all exalted overflowing of oneself, and who thought art an impassioned labour to keep men from doubt and despondency, and woman's love an evil, when it would trammel the man's will, should see the poetic genius not in a woman star but in the Sun, and should rejoice throughout his poetry in 'the Sun in his strength.' Shelley, however, except when he uses it to describe the peculiar beauty of Emelia Viviani, who was 'like an incarnation of the Sun when light is changed to love,' saw it with less friendly eyes. He seems to have seen it with perfect happiness only when veiled in mist, or glimmering upon water, or when faint enough to do no more than veil the brightness of his own Star; and in *The Triumph of Life*, the one poem in which it is part of the avowed symbolism, its power is the being and the source of all tyrannies. When the woman personifying the Morning Star has faded from before his eyes, Rousseau sees a 'new vision' in 'a cold bright car' with a rainbow hovering over her, and as she comes the shadow passes from 'leaf and stone,' and the souls she has enslaved seem in 'that light like atomies to dance within a sunbeam,' or they dance among the flowers that grow up newly 'in the grassy verdure of the desert,' unmindful of the misery that is to come upon them. 'These are the great, the unforgotten,' all who have worn 'mitres and helms and crowns or wreaths of light,' and yet have not known themselves. Even 'great Plato' is there because he knew joy and sorrow, because life that could not subdue him by gold or pain, by 'age or sloth or slavery,' subdued him by love. All who have ever lived are there except Christ and Socrates and 'the sacred few' who put away all life could give, being doubtless followers throughout their lives of the forms borne by the flying ideal, or who, 'as soon as they had touched the world with living flame, flew back like eagles to their native noon.'

In ancient times, it seems to me that Blake, who for all his protest was glad to be alive, and ever spoke of his gladness, would have worshipped in some chapel of the Sun, and that Keats, who accepted life gladly though 'with a delicious diligent indolence,' would have worshipped in some chapel of the Moon, but that Shelley, who hated life because he sought 'more in life than any understood,' would have wandered, lost in a ceaseless reverie, in some chapel of the Star of infinite desire.

I think too that as he knelt before an altar, where a thin flame burnt in a lamp made of green agate, a single vision would have come to him again and again, a vision of a boat drifting down a broad river between high hills where there were caves and towers, and following the light of one Star; and that voices would have told him how there is for every man some one scene, some one adventure, some one picture that is the image of his secret life, for wisdom first speaks in images, and that this one image, if he would but brood over it his life long, would lead his soul, disentangled from unmeaning circumstance and the ebb and flow of the world, into that far household, where the undying gods await all whose souls have become simple as flame, whose bodies have become quiet as an agate lamp.

But he was born in a day when the old wisdom had vanished and was content merely to write verses, and often with little thought of more than verses.

—W.B. YEATS, from "The Philosophy of Shelley's Poetry," 1900, *Ideas of Good and Evil*, 1903, pp. 128–141

A.C. BRADLEY "SHELLEY'S VIEW OF POETRY" (1904)

Andrew Cecil Bradley (1851–1935) was born in Cheltenham, England. He is best known for *Shakespearean Tragedy* (1904), considered the preeminent work on Shakespeare throughout the first half of the twentieth century. Bradley worked as a professor at Liverpool, Glasgow, and finally Oxford, becoming chairperson of poetry there in 1901. Many of his lectures were later published, including *Poetry for Poetry's Sake* (1901) and *Oxford Lectures on Poetry* (1909).

The ideas of Wordsworth and of Coleridge about poetry have often been discussed and are familiar. Those of Shelley are much less so, and in his eloquent exposition of them there is a radiance which almost conceals them from many readers. I wish, at the cost of all the radiance, to try to see them and show them rather more distinctly. Even if they had little value for the theory of poetry, they would still have much as material for it, since they allow us to look into a poet's experience in conceiving and composing. And, in addition, they throw light on some of the chief characteristics of Shelley's own poetry.

His poems in their turn form one of the sources from which his ideas on the subject may be gathered. We have also some remarks in his letters and in prose pieces dealing with other topics. We have the prefaces to those of his works which he himself published. And, lastly, there is the *Defence of Poetry*. This essay was written in reply to an attack made on contemporary verse by Shelley's friend Peacock,—not a favourable specimen of Peacock's writing. The *Defence*, we can see, was hurriedly composed, and it remains a fragment, being only the first of three projected parts. It contains a good deal of historical matter, highly interesting, but too extensive to be made use of here. Being polemical, it no doubt exaggerates such of Shelley's views as collided with those of his antagonist. But, besides being the only full expression of these views, it is the most mature, for it was written within eighteen months of his death. It appears to owe very little either to Wordsworth's Prefaces or to Coleridge's *Biographia Literaria;* but there are a few reminiscences of Sidney's *Apology*, which Shelley had read just before he wrote his own *Defence;* and it shows, like much of his mature poetry, how deeply he was influenced by the more imaginative dialogues of Plato.

Anyone familiar with the manner in which Shelley in his verse habitually represents the world could guess at his general view of poetry. The world to him is a melancholy place, a 'dim vast vale of tears,' illuminated in flashes by the light of a hidden but glorious power. Nor is this power, as that favourite metaphor would imply, wholly outside the world. It works within it as a soul contending with obstruction and striving to penetrate and transform the whole mass. And though the fulness of its glory is concealed, its nature is known in outline. It is the realised perfection of everything good and beautiful on earth; or, in other words, all such goodness and beauty is its partial manifestation. 'All,' I say: for the splendour of nature, the love of lovers, every affection and virtue, any good action or just law, the wisdom of philosophy, the creations of art, the truths deformed by superstitious religion,—all are equally operations or appearances of the hidden power. It is of the first importance for the understanding of Shelley to realise how strong in him is the sense and conviction of this unity in life: it is one of his Platonic traits. The intellectual Beauty of his 'Hymn' is absolutely the same thing as the Liberty of his 'Ode,' the 'Great Spirit' of Love that he invokes to bring freedom to Naples, the One which in *Adonais* he contrasts with the Many, the Spirit of Nature of *Queen Mab*, and the Vision of *Alastor* and *Epipsychidion*. The skylark of the famous stanzas is free from our sorrows, not because it is below them, but because, as an embodiment of that perfection, it knows the rapture of love without its satiety, and understands death as we cannot. The voice of the mountain, if a whole nation could hear it with the poet's ear, would 'repeal large codes of fraud and woe'; it is the same voice as the reformer's and the martyr's. And in the far-off day when the 'plastic stress' of this power has mastered the last resistance and is all in all, outward nature, which now suffers with man, will be redeemed with him, and man, in becoming politically free, will become also the perfect lover. Evidently, then, poetry, as the world now is, must be one of the voices of this power, or one tone of its voice. To use the language so dear to Shelley, it is the revelation of those eternal ideas which lie behind the many-coloured, ever-shifting veil that we call reality or life. Or rather, it is one such revelation among many.

When we turn to the *Defence of Poetry* we meet substantially the same view. There is indeed a certain change; for Shelley is now philosophising and writing prose, and he wishes not to sing from the mid-sky, but, for a while at least, to argue with his friend on the earth. Hence at first we hear nothing of that perfect power at the heart of things, and poetry is considered as a creation rather than a revelation. But for Shelley, we soon discover, this would be a false antithesis. The poet creates, but this creation is no mere fancy of his; it represents 'those forms which are common to universal nature and existence,' and 'a poem is the very image of life expressed in its eternal truth.' We notice, further, that the more voluntary and conscious work of invention and execution is regarded as quite subordinate in the creative process. In that process the mind, obedient to an influence which it does not understand and cannot control, is driven to produce images of perfection which rather form themselves in it than are formed by it. The greatest stress is laid on this influence or inspiration; and in the end we learn that the origin of the whole process lies in certain exceptional moments when visitations of thought and feeling, elevating and delightful beyond all expression, but always arising unforeseen and departing unbidden, reach the soul; that these are, as it were, the interpenetration of a diviner nature through our own; and that the province of the poet is to arrest these apparitions, to veil them in language, to colour every other form he touches with their evanescent hues, and so to 'redeem from decay the visitations of the divinity in man.'

Even more decided is the emphasis laid on the unity of all the forms in which the 'divinity' or ideal power thus attests its presence. Indeed, throughout a large part of the essay, that 'Poetry' which Shelley is defending is something very much wider than poetry in the usual sense. The enemy he has to meet is the contention that poetry and its influence steadily decline as civilisation advances, and that they are giving place, and ought to give place, to reasoning and the pursuit of utility. His answer is that, on the contrary, imagination has been, is, and always will be, the prime source of everything that has intrinsic value in life. Reasoning, he declares, cannot create, it can only operate upon the products of imagination. Further, he holds that the predominance of mere reasoning and mere utility has become in great part an evil; for while it has accumulated masses of material goods and moral truths, we distribute the goods iniquitously and fail to apply the truths, because, for want of imagination, we have not sympathy in our hearts and do not feel what we know. The 'Poetry' which he defends, therefore, is the whole creative imagination with all its products. And these include not merely literature in verse, but, first, whatever prose writing is allied to that literature; and, next, all the other fine arts; and, finally, all actions, inventions, institutions, and even ideas and moral dispositions, which imagination brings into being in its effort to satisfy the longing for perfection. Painters and musicians are poets. Plato and Bacon, even Herodotus and Livy, were poets, though there is much in their works which is not poetry. So were the men who invented the arts of life, constructed laws for tribes or cities, disclosed, as sages or founders of religion, the excellence of justice and love. And every one, Shelley would say, who, perceiving the beauty of an imagined virtue or deed, translates the image into a fact, is so far a poet. For all these things come from imagination.

Shelley's exposition of this, which is probably the most original part of his theory, is not very clear; but, if I understand his meaning, that which he takes to happen in all these cases might be thus described. The imagination—that is to say, the soul imagining—has before it, or feels within it, something which, answering perfectly to its nature, fills it with delight and with a desire to realise what delights it. This something, for the sake of brevity, we may call an idea, so long as we remember that it need not be distinctly imagined and that it is always accompanied by emotion. The reason why such ideas delight the imagining soul is that they are, in fact, images or forebodings of its own perfection—of itself become perfect—in one aspect or another. These aspects are as various as the elements and forms of its own inner life and outward existence; and so the idea may be that of the perfect harmony of will and feeling (a virtue), or of the perfect union of soul with soul (love), or of the perfect order of certain social relations or forces (a law or institution), or of the perfect adjustment of intellectual elements (a truth); and so on. The formation and expression of any such idea is thus the work of Poetry in the widest sense; while at the same time (as we must add, to complete Shelley's thought) any such idea is a gleam or apparition of the perfect Intellectual Beauty.

I choose this particular title of the hidden power or divinity in order to point out (what the reader is left to observe for himself) that the imaginative idea is always regarded by Shelley as beautiful. It is, for example, desirable for itself and not merely as a means to a further result; and it has the formal characters of beauty. For, as will have been noticed in the instances given, it is always the image of an order, or harmony, or unity in variety, of the elements concerned. Shelley sometimes even speaks of their 'rhythm.' For example, he uses this word in reference to an action; and I quote the passage because, though it occurs at some distance from the exposition of his main view, it illustrates it well. He is saying that the true poetry of Rome, unlike that of Greece, did not fully express itself in poems. 'The true poetry of Rome lived in its institutions: for whatever of beautiful, true and majestic they contained, could have sprung only from the faculty which creates the order in which they consist. The life of Camillus; the death of Regulus; the expectation of the senators, in their god-like state, of the victorious Gauls; the refusal of the Republic to make peace with Hannibal after the battle of Cannae'—these he describes as 'a rhythm and order in the shows of life,' an order not arranged with a view to utility or outward result, but due to the imagination, which, 'beholding the beauty of this order, created it out of itself according to its own idea.'

II

If this, then, is the nature of Poetry in the widest sense, how does the poet, in the special sense, differ from other unusually creative souls? Not essentially in the inspiration and general substance of his poetry, but in the kind of expression he gives to them. In so far as he is a poet, his medium of expression, of course, is not virtue, or action, or law; poetry is one of the arts. And, again, it differs from the rest, because its particular vehicle is language. We have now to see, therefore, what Shelley has to say of the form of poetry, and especially of poetic language.

First, he claims for language the highest place among the vehicles of artistic expression, on the ground that it is the most direct and also the most plastic. It is itself produced by imagination instead of being simply encountered by it, and it has no relation except to imagination; whereas any more material medium has a nature of its own, and relations to other things in the material world, and this nature and these relations intervene between the artist's conception and his expression of it in the medium. It is to the superiority of its vehicle that Shelley attributes the greater fame which poetry has always enjoyed as compared with other arts. He forgets (if I may interpose a word of criticism) that the media of the other arts have, on their side, certain advantages over language, and that these perhaps counterbalance the inferiority which he notices. He would also have found it difficult to show that language, on its physical side, is any more a product of imagination than stone or pigments. And his idea that the medium in the other arts is an obstacle intervening between conception and expression is, to say the least, one-sided. A sculptor, painter, or musician, would probably reply that it is only the qualities of his medium that enable him to express at all; that what he expresses is inseparable from the vehicle of expression; and that he has no conceptions which are not from the beginning sculpturesque, pictorial, or musical. It is true, no doubt, that his medium is an obstacle as well as a medium; but this is also true of language.

But to resume. Language, Shelley goes on to say, receives in poetry a peculiar form. As it represents in its meaning a perfection which is always an order, harmony, or rhythm, so it itself, as so much sound, *is* an order, harmony, or rhythm. It is measured language, which is not the proper vehicle for the mere recital of facts or for mere reasoning. For Shelley, however, this measured language is not of necessity metrical. The order or measure may remain at the stage which it reaches in beautiful prose, like that of Plato, the melody of whose language, Shelley declares, is the most intense it is possible to conceive. It may again advance to metre; and he admits that metrical form is convenient, popular, and preferable, especially in poetry containing much action. But he will not have any new great poet tied down to it. It is not essential, while measure is absolutely so. For it is no mere accident of poetry that its language is measured, nor does a delight in this measure mean little. As sensitiveness to the order of the relations of sounds is always connected with sensitiveness to the order of the relations of thoughts, so also the harmony of the words is scarcely less indispensable than their meaning to the communication of the influence of poetry. 'Hence,' says Shelley, 'the vanity of translation: it were as wise to cast a violet into a crucible that you might discover the formal principle of its colour and odour, as seek to transfuse from one language into another the creations of a poet.' Strong words to come from the translator of the *Hymn to Mercury* and of Agathon's speech in the *Symposium!* And is not all that Shelley says of the difference between measured and unrhythmical language applicable, at least in some degree, to the difference between metrical and merely measured language? Could he really have supposed that metre is no more than a 'convenience,' which contributes nothing of any account to the influence of poetry? But I will not criticise. Let me rather point out how surprising, at first sight, and how significant, is Shelley's insistence on the importance of measure or rhythm. No one could assert more absolutely than he the identity of the general substance of poetry with that of moral life and action, of the other arts, and of the higher kinds of philosophy. And yet it would be difficult to go beyond the emphasis of his statement that the formal element (as he understood it) is indispensable to the effect of poetry.

Shelley, however, nowhere considers this element more at length. He has no discussions, like those of Wordsworth and Coleridge, on diction. He never says, with Keats, that he looks on fine phrases like a lover. We hear of his deep-drawn sigh of satisfaction as he finished reading a passage of Homer, but not of his shouting his delight, as he ramped through the meadows of Spenser, at some marvellous flower. When in his letters he refers to any poem he is reading, he scarcely ever mentions particular lines or expressions; and we have no evidence that, like Coleridge and Keats, he was a curious student of metrical effects or the relations of vowel-sounds. I doubt if all this is wholly accidental. Poetry was to him so essentially an effusion of aspiration, love and worship, that we can imagine his feeling it almost an impiety to break up its unity even for purposes of study, and to give a separate attention to its means of utterance. And what he does say on the subject confirms this impression. In the first place, as we have seen, he lays great stress on inspiration; and his statements, if exaggerated and misleading, must still reflect in some degree his own experience. No poem, he asserts, however inspired it may be, is more than a feeble shadow of the original conception; for when composition begins, inspiration is already on the decline. And so in a letter he speaks of the detail of execution destroying all wild and beautiful visions. Still, inspiration, if diminished by composition, is not wholly dispelled; and he appeals to the greatest poets of his day whether it is not an error to assert that the finest passages of poetry are produced by labour and study. Such toil he would restrict to those parts which connect the inspired passages, and he speaks with contempt of the fifty-six various readings of the first line of the *Orlando Furioso*. He seems to exaggerate on this matter because in the *Defence* his foe is cold reason and calculation. Elsewhere he writes more truly of the original conception as being obscure as well as intense; from which it would seem to follow that the feeble shadow, if darker, is at least more distinct than the original. He forgets, too, what is certainly the fact, that the poet in reshaping and correcting is able to revive in some degree the fire of the first impulse. And we know from himself that his greatest works cost him a severe labour not confined to the execution, while his manuscripts show plenty of various readings, if never so many as fifty-six in one line.

Still, what he says is highly characteristic of his own practice in composition. He allowed the rush of his ideas to have its way, without pausing to complete a troublesome line or to find a word that did not come; and the next day (if ever) he filled up the gaps and smoothed the ragged edges. And the result answers to his theory. Keats was right in telling him that he might be more of an artist. His language, indeed, unlike Wordsworth's or Byron's, is, in his mature work, always that of a poet; we never hear his mere speaking voice; but he is frequently diffuse and obscure, and even in fine passages his constructions are sometimes trailing and amorphous. The glowing metal rushes into the mould so vehemently that it overleaps the bounds and fails to find its way into all the little crevices. But no poetry is more manifestly inspired, and even when it is plainly imperfect it is sometimes so inspired that it is impossible to wish it changed. It has the rapture of the mystic, and that is too rare to lose. Tennyson quaintly said of the hymn 'Life of Life': 'He seems to go up into the air and burst.' It is true: and, if we are to speak of poems as fireworks, I would not compare 'Life of Life' with a great set piece of Homer or Shakespeare that illumines the whole sky; but, all the same, there is no more thrilling sight than the heavenward rush of a rocket, and it bursts at a height no other fire can reach.

In addition to his praise of inspiration Shelley has some scattered remarks on another point which show the same spirit. He could not bear in poetic language any approach to artifice, or any sign that the writer had a theory or system of style. He thought Keats's earlier poems faulty in this respect, and there is perhaps a reference to Wordsworth in the following sentence from the Preface to the *Revolt of Islam:* 'Nor have I permitted any system relating to mere words to divert the attention of the reader, from whatever interest I may have succeeded in creating, to my own ingenuity in contriving,—to disgust him according to the rules of criticism. I have simply clothed my thoughts in what appeared to me the most obvious and appropriate language. A person familiar with nature, and with the most celebrated productions of the human mind, can scarcely err in following the instinct, with respect to selection of language, produced by that familiarity.' His own poetic style certainly corresponds with his intention. It cannot give the kind of pleasure afforded by what may be called without disparagement a learned and artful style, such as Virgil's or Milton's; but, like the best writing of Shakespeare and Goethe, it is, with all its individuality, almost entirely free from mannerism and the other vices of self-consciousness, and appears to flow so directly from the thought that one is ashamed to admire it for itself. This is equally so whether the appropriate style is impassioned and highly figurative, or simple and even plain. It is indeed in the latter case that Shelley wins his greatest, because most difficult, triumph. In the dialogue part of *Julian and Maddalo* he has succeeded remarkably in keeping the style quite close to that of familiar though serious conversation, while making it nevertheless unmistakably poetic. And the *Cenci* is an example of a success less complete only because the problem was even harder. The ideal of the style of tragic drama in the nineteenth or twentieth century should surely be, not to reproduce with modifications the style of Shakespeare, but to do what Shakespeare did—to idealise, without deserting, the language of contemporary speech. Shelley in the *Cenci* seems to me to have come nearest to this ideal.

III

So much for general exposition. If now we consider more closely what Shelley says of the substance of poetry, a question at once arises. He may seem to think of poetry solely as the direct expression of perfection in some form, and accordingly to imagine its effect as simply joy or delighted aspiration. Much of his own poetry, too, is such an expression; and we understand when we find him saying that Homer embodied the ideal perfection of his age in human character, and unveiled in Achilles, Hector, and Ulysses 'the truth and beauty of friendship, patriotism, and persevering devotion to an object.' But poetry, it is obvious, is not wholly, perhaps not even mainly, of this kind. What is to be said, on Shelley's theory, of his own melancholy lyrics, those 'sweetest songs' that 'tell of saddest thought'? What of satire, of the epic of conflict and war, or of tragic exhibitions of violent and destructive passion? Does not his theory reflect the weakness of his own practice, his tendency to portray a thin and abstract ideal instead of interpreting the concrete detail of nature and life; and ought we not to oppose to it a theory which would consider poetry simply as a representation of fact?

To this last question I should answer No. Shelley's theory, rightly understood, will take in, I think, everything really poetic. And to a considerable extent he himself shows the way to meet these doubts. He did not mean that the *immediate* subject of poetry must be perfection in some form. The poet, he says, can colour with the hues of the ideal everything he touches. If so, he may write of absolutely anything so long as he *can* so colour it, and nothing would be excluded from his province except those things (if any such exist) in which no positive relation to the ideal, however indirect, can be shown or intimated. Thus to take the instance of Shelley's melancholy lyrics, clearly the lament which arises from loss of the ideal, and mourns the evanescence of its visitations or the desolation of its absence, is indirectly an expression of the ideal; and so on his theory is the simplest song of unhappy love or the simplest dirge. Further, he himself observes that, though the joy of poetry is often unalloyed, yet the pleasure of the 'highest portions of our being is frequently connected with the pain of the inferior,' that 'the pleasure that is in sorrow is sweeter than the pleasure of pleasure itself,' and that not sorrow only, but 'terror, anguish, despair itself, are often the chosen expressions of an approximation to the highest good.' That, then, which appeals poetically to such painful emotions will again be an indirect portrayal of the ideal; and it is clear, I think, that this was how Shelley in the *Defence* regarded heroic and tragic poetry, whether narrative or dramatic, with its manifestly imperfect characters and its exhibition of conflict and wild passion. He had, it is true, another and an unsatisfactory way of explaining the presence of these things in poetry; and I will refer to this in a moment. But he tells us that the Athenian tragedies represent the highest idealisms (his name for ideals) of passion and of power (not merely of virtue); and that in them we behold ourselves, 'under a thin disguise of circumstance, stripped of all but that ideal perfection and energy which every one feels to be the internal type of all that he loves, admires, and would become.' He writes of Milton's Satan in somewhat the same strain. The Shakespearean tragedy from which he most often quotes is one in which evil holds the stage, *Macbeth;* and he was inclined to think *King Lear,* which certainly is no direct portrait of perfection, the greatest drama in the world. Lastly, in the Preface to his own *Cenci* he truly says that, while the story is fearful and monstrous, 'the poetry which exists in these tempestuous sufferings and crimes,' if duly brought out, 'mitigates the pain of the contemplation of moral deformity': so that he regards Count Cenci himself as a *poetic* character, and therefore as in *some* sense an expression of the ideal. He does not further explain his meaning. Perhaps it was that the perfection which poetry is to exhibit includes, together with those qualities which win our immediate and entire approval or sympathy, others which are capable of becoming the instruments of evil. For these, the energy, power and passion of the soul though they may be perverted, are in themselves elements of perfection; and so, even in their perversion or their combination with moral deformity, they retain their value, they are not simply ugly or horrible, but appeal through emotions predominantly painful to the same love of the ideal which is directly satisfied by pictures of goodness and beauty. Now to these various considerations we shall wish to add others; but if we bear these in mind, I believe we shall find Shelley's theory wide enough, and must hold that the substance of poetry is never mere fact, but is always ideal, though its method of representation is sometimes more direct, sometimes more indirect.

Nevertheless, he does not seem to have made his view quite clear to himself, or to hold to it consistently. We are left with the impression, not merely that he personally preferred the direct method (as he was, of course, entitled to do), but that his use of it shows a certain weakness, and also that even in theory he unconsciously tends to regard it as the primary and proper method, and to admit only by a reluctant after-thought the representation of imperfection. Let me point out some signs of this. He considered his own *Cenci* as a poem inferior in kind to his other main works, even as a sort of accommodation to the public. With all his modesty he knew what to think of the neglected *Prometheus* and *Adonaïs,* but there is no sign that he, any more than the world, was aware that the character of Cenci was a creation without a parallel in our poetry since the seventeenth century. His enthusiasm for some second-rate and third-rate Italian paintings, and his failure to understand Michael Angelo, seem to show the same tendency. He could not enjoy comedy: it seemed to him simply cruel: he did not perceive that to show the absurdity of the imperfect is to glorify the perfect. And, as I mentioned just now, he wavers in his view of the representation of heroic and tragic imperfection. We find in the Preface to *Prometheus Unbound* the strange notion that Prometheus is a more poetic character than Milton's Satan because he is free from Satan's imperfections, which are said to interfere with the interest. And in the *Defence* a similar error appears. Achilles, Hector, Ulysses, though they exhibit ideal virtues, are, he admits, imperfect. Why, then, did Homer make them so? Because, he seems to reply, Homer's contemporaries regarded their vices (e.g. revengefulness and deceitfulness) as virtues. Homer accordingly had to conceal in the costume of these vices the unspotted beauty that he himself imagined; and, like Homer, 'few poets of the highest class have chosen to exhibit the beauty of their conceptions in its naked truth and splendour.' Now, this idea, to say nothing of its grotesque improbability in reference to Homer, and its probable baselessness in reference to most other poets, is quite inconsistent with that truer view of heroic and tragic character which was explained just now. It is an example of Shelley's tendency to abstract idealism or spurious Platonism. He is haunted by the fancy that if he could only get at the One, the eternal Idea, in complete aloofness from the Many, from life with all its change, decay, struggle, sorrow and evil, he would have reached the true object of poetry: as if the whole finite world were a mere mistake or illusion, the sheer opposite of the infinite One, and in no way or degree its manifestation. Life, he says—

> Life, like a dome of many-coloured glass,
> Stains the white radiance of eternity;

but the other side, the fact that the many colours *are* the white light broken, he tends to forget, by no means always, but in one, and that not the least inspired, of his moods. This is the source of that thinness and shallowness of which his view of the world and of history is justly accused, a view in which all imperfect being is apt to figure as absolutely gratuitous, and everything and everybody as pure white or pitch black. Hence also his ideals of good, whether as a character or as a mode of life, resting as they do on abstraction from the mass of real existence, tend to lack body and individuality; and indeed, if the existence of the many is a mere calamity, clearly the next best thing to their disappearance is that they should all be exactly alike and have as little character as possible. But we must remember that Shelley's strength and weakness are closely allied, and it may be that the very abstractness of his ideal was a condition of that quivering intensity of aspiration towards it in which his poetry is unequalled. We must not go for this to Homer and Shakespeare and Goethe; and if we go for it to Dante, we shall find, indeed, a mind far vaster than Shelley's, but also that dualism of which we complain in him, and the description of a heaven which, equally with Shelley's regenerated earth, is no place for mere mortality. In any case, as we have seen, the weakness in his poetical practice, though it occasionally appears also as a defect in his poetical theory, forms no necessary part of it.

IV

I pass to his views on a last point. If the business of poetry is somehow to express ideal perfection, it may seem to follow that the poet should embody in his poems his beliefs about this perfection and the way to approach it, and should thus have a moral purpose and aim to be a teacher. And in regard to Shelley this conclusion seems the more natural because his own poetry allows us to see clearly some of his beliefs about morality and moral progress. Yet alike in his Prefaces and in the *Defence* he takes up most decidedly the position that the poet ought neither to affect a moral aim nor to express his own conceptions of right and wrong. 'Didactic poetry,' he declares, 'is my abhorrence: nothing can be equally well expressed in prose that is not tedious and supererogatory in verse.' 'There was little danger,' he tells us in the *Defence,* 'that Homer or any of the eternal poets' should make a mistake in this matter; but 'those in whom the poetical faculty, though great, is less intense, as Euripides, Lucan, Tasso, Spenser, have frequently affected a moral aim, and the effect of their poetry is diminished in exact proportion to the degree in which they compel us to advert to this purpose.' These statements may appeal to us, but are they consistent with Shelley's main views of poetry? To answer this question we must observe what exactly it is that he means to condemn.

Shelley was one of the few persons who can literally be said to *love* their kind. He held most strongly, too, that poetry does benefit men, and benefits them morally. The moral purpose, then, to which he objects cannot well be a poet's general purpose of doing moral as well as other good through his poetry—such a purpose, I mean, as he may cherish when he contemplates his life and his life's work. And, indeed, it seems obvious that nobody with any humanity or any sense can object to that, except through some intellectual confusion. Nor, secondly, does Shelley mean, I think, to condemn even the writing of a particular poem with a view to a particular moral or practical effect; certainly, at least, if this was his meaning he was condemning some of his own poetry. Nor, thirdly, can he be referring to the portrayal of moral ideals; for that he regarded as one of the main functions of poetry, and in the very place where he says that didactic poetry is his abhorrence he also says, by way of contrast, that he has tried to familiarise the minds of his readers with beautiful idealisms of moral excellence. It appears, therefore, that what he is really attacking is the attempt to give, in the strict sense, moral *instruction*, to communicate doctrines, to offer argumentative statements of opinion on right and wrong, and more especially, I think, on controversial questions of the day. An example would be Wordsworth's discourse on education at the end of the *Excursion*, a discourse of which Shelley, we know, had a very low opinion. In short, his enemy is not the purpose of producing a moral effect, it is the appeal made for this purpose to the reasoning intellect. He says to the poet: By all means aim at bettering men; you are a man, and are bound to do so; but you are also a poet, and therefore your proper way of doing so is not by reasoning and preaching. His idea is of a piece with his general championship of imagination, and it is quite consistent with his main view of poetry.

What, then, are the *grounds* of this position? They are not clearly set out, but we can trace several, and they are all solid. Reasoning on moral subjects, moral philosophy, was by no means 'tedious' to Shelley; it seldom is to real poets. He loved it, and (outside his *Defence*) he rated its value very high. But he thought it tedious and out of place in poetry, because it can be equally well expressed in 'unmeasured' language—much better expressed, one may venture to add. You invent an art in order to effect by it a particular purpose which nothing else can effect as well. How foolish, then, to use this art for a purpose better served by something else! I know no answer to this argument, and its application is far wider than that given to it by Shelley. Secondly, Shelley remarks that a poet's own conceptions on moral subjects are usually those of his place and time, while the matter of his poem ought to be eternal, or, as we say, of permanent and universal interest. This, again, seems true, and has a wide application; and it holds good even when the poet, like Shelley himself, is in rebellion against orthodox moral opinion; for his heterodox opinions will equally show the marks of his place and time, and constitute a perishable element in his work. Doubtless no poetry can be without a perishable element; but that poetry has least of it which interprets life least through the medium of systematic and doctrinal ideas. The veil which time and place have hung between Homer or Shakespeare and the general reader of to-day is almost transparent, while even a poetry so intense as that of Dante and Milton is impeded in its passage to him by systems which may be unfamiliar, and, if familiar, may be distasteful.

Lastly—and this is Shelley's central argument—as poetry itself is directly due to imaginative inspiration and not to reasoning, so its true moral effect is produced through imagination and not through doctrine. Imagination is, for Shelley, 'the great instrument of moral good.' The 'secret of morals is love.' It is not 'for want of admirable doctrines that men hate and despise and censure and deceive and subjugate one another': it is for want of love. And love is 'a going out of our own nature, and an identification of ourselves with the beautiful which exists in thought, action or person not our own.' 'A man,' therefore, 'to be greatly good must imagine intensely and comprehensively.' And poetry ministers to moral good, the effect, by acting on its cause, imagination. It strengthens imagination as exercise strengthens a limb, and so it indirectly promotes morality. It also fills the imagination with beautiful impersonations of all that we should wish to be. But moral reasoning does not act upon the cause, it only analyses the effect; and the poet has no right to be content to analyse what he ought indirectly to create. Here, again, in his eagerness, Shelley cuts his antitheses too clean, but the defect is easily made good, and the main argument is sound.

Limits of time will compel me to be guilty of the same fault in adding a consideration which is in the spirit of Shelley's. The chief moral effect claimed for poetry by Shelley is exerted, primarily, by imagination on the emotions; but there is another influence, exerted primarily through imagination on the understanding. Poetry is largely an interpretation of life; and, considering what life is, that must mean a moral interpretation. This, to have poetic value, must satisfy imagination; but we value it also because it gives us knowledge, a wider comprehension, a new insight into ourselves and the world. Now, it may be held—and this view answers to a very general feeling among lovers of poetry now—that the most deep and original moral interpretation is not likely to be that which most shows a moral purpose or is most governed by reflective beliefs and opinions, and that as a rule we learn most from those who do not try to teach us, and whose opinions may even remain unknown to us: so that there is this weighty objection to the appearance of such purpose and opinions, that it tends to defeat its own intention. And the reason that I wish to suggest is this, that always we get most from the *genius* in a man of genius and not from the rest of him. Now, although poets often have unusual powers of reflective thought, the specific genius of a poet does not lie there, but in imagination. Therefore his deepest and most original interpretation is likely to come by the way of imagination. And the specific way of imagination is not to clothe in imagery consciously held ideas; it is to produce half-consciously a matter from which, when produced, the reader may, if he chooses, extract ideas. Poetry (I must exaggerate to be clear), psychologically considered, is not the *expression* of ideas or of a view of life; it is their discovery or creation, or rather both discovery and creation in one. The interpretation contained in *Hamlet* or *King Lear* was not brought ready-made to the old stories. What was brought to them was the huge substance of Shakespeare's imagination, in which all his experience and thought was latent; and this, dwelling and working on the stories with nothing but a dramatic purpose, and kindling into heat and motion, gradually discovered or created in them a meaning and a mass of truth about life, which was brought to birth by the process of composition, but never preceded it in the shape of ideas, and probably never, even after it, took that shape to the poet's mind. And *this* is the interpretation which we find inexhaustibly instructive, because Shakespeare's *genius* is in it. On the other hand, however much from curiosity and personal feeling towards him we may wish to know his opinions and beliefs about morals or religion or his own poems or Queen Elizabeth, we have not really any reason to suppose that their value would prove extraordinary. And so, to apply this generally, the opinions, reasonings and beliefs of poets are seldom of the same quality as their purely imaginative product. Occasionally, as with Goethe, they are not far off it; but sometimes they are intense without being profound, and more eccentric than original; and often they are very sane and sound, but not very different from those of wise men without genius. And therefore poetry is not the place for them. For we want in poetry a moral interpretation, but not the interpretation we have already. As a rule the genuine artist's quarrel with 'morality' in art is not really with morality, it is with a stereotyped or narrow morality; and when he refuses in his art to consider things from what he calls the moral point of view, his reasons are usually wrong, but his instinct is right.

Poetry itself confirms on the whole this contention, though doubtless in these last centuries a great poet's work will usually reveal more of conscious reflection than once it did. Homer and Shakespeare show no moral aim and no system of opinion. Milton was far from justifying the ways of God to men by the argumentation he put into divine and angelic lips; his truer moral insight is in the creations of his genius; for instance, in the character of Satan or the picture of the glorious humanity of Adam and Eve. Goethe himself could never have told the world what he was going to express in the

First Part of *Faust:* the poem told *him,* and it is one of the world's greatest. He knew too well what he was going to express in the Second Part, and with all its wisdom and beauty it is scarcely a great poem. Wordsworth's original message was delivered, not when he was a Godwinian semi-atheist, nor when he had subsided upon orthodoxy, but when his imagination, with a few hints from Coleridge, was creating a kind of natural religion; and this religion itself is more profoundly expressed in his descriptions of his experience than in his attempts to formulate it. The moral virtue of Tennyson is in poems like *Ulysses* and parts of *In Memoriam,* where sorrow and the consciousness of a deathless affection or an unquenchable desire for experience forced an utterance; but when in the *Idylls* he tried to found a great poem on explicit ideas about the soul and the ravages wrought in it by lawless passion, he succeeded but partially, because these ideas, however sound, were no product of his genius. And so the moral virtue of Shelley's poetry lay, not in his doctrines about the past and future of man, but in an intuition, which was the substance of his soul, of the unique value of love. In the end, for him, the truest name of that perfection called Intellectual Beauty, Liberty, Spirit of Nature, is Love. Whatever in the world has any worth is an expression of Love. Love sometimes talks. Love talking musically is Poetry.

—A.C. BRADLEY, "Shelley's View of Poetry," 1904,
Oxford Lectures on Poetry, 1909, pp. 151–174

Chronology

1792 Percy Bysshe Shelley is born on August 4 at Field Place, Horsham, Sussex, the son of a prosperous landowner and Whig Member of Parliament.

1802–04 Attends Sion Academy.

1804–10 Studied at Eton.

1810 Publishes *Zastrozzi,* a gothic novel, in March, followed by *Original Poetry by Voctor and Cazire* (Shelley and his sister) in September, and the publication of "Posthumous Fragments of Margaret Nicholson" in November. Takes up residence at University College, Oxford.

1811 Expelled from Oxford on March 25 for writing pamphlet *The Necessity of Atheism,* with T.J. Hogg, who is also expelled. Elopes in August with Harriet Westbrook, and they marry in Edinburgh.

1812 From February to April, Shelley goes to Ireland to take part in political agitation. In October, he meets William Godwin, whose *Poetical Justice* he had read at Oxford, and with whom he had corresponded.

1813 *Queen Mab* is published in May, and a daughter, Ianthe, is born in June.

1814 On July 28 Shelley elopes to continent with Mary Godwin and returns in September.

1815 In January, Shelley's grandfather, Sir Bysshe Shelley, dies leaving Shelley an income.

1816 William is born to Mary Godwin in January, and "Alastor" is published during that summer. Shelley, Mary, and Mary's half sister Claire join Byron in Geneva (Claire had seen Byron in London), and they leave for England on August 29. In December the body of Harriet Shelley is found in a pond in Hyde Park. Shelley marries Mary Godwin.

1817 Clara is born in September, and "Laon and Cythna" is printed in December.

1818 *The Revolt of Islam* is published in January, and Shelley moves to Italy in March. In September Clara dies, and Shelley begins *Prometheus Unbound* that autumn.

1819 William dies in June, and Shelley writes "Ode to the West Wind" in October. He begins *A Philosophical View of Reform* (published in 1920) in November. Percy Florence is born.

1820 *The Cenci* is published in the spring, and Shelley writes "To a Skylark" in June. *Prometheus Unbound* is published during the summer. From August 14 to 16, Shelley writes "Witch of Atlas."

1821 "A Defense of Poetry" is written in February and March. *Adonais* and *Epipsychidion* are published that summer.

1822 From May to June, Shelley works on *Triumph of Life*. On July 8, Shelley drowns at sea.

Index

abstract, the, 102, 106, 132, 133
Addison, Joseph, 53
"Address to the Irish People," 162–163
Address to the People on the Death of the Princess Charlotte, An, 2
Adonais: An Elegy on the Death of John Keats, 2, 14, 101, 135, 146, 172, 181
adversaries, 104
Aeschylus, 83, 129
The Persians, 2
affectations, 48, 50
Alastor: or, The Spirit of Solitude and Other Poems, 2, 54, 95, 140, 143, 150, 151, 153, 172
An Address to the Irish People, 1
Ancient Mariner (Coleridge), 56
Ariel, 47
"Ariel to Miranda," 156
aristocratic class, 124–125
Aristotle, 164
Arnold, Matthew, 70, 124–125, 126
art, 48, 148
    morality and, 87, 185
    nature and, 143, 145
    poetry as, 144, 175–176
    two kinds of artists, 114
"art for art's sake" movement, 62
asceticism, 144, 161

atheism, 23, 29, 71, 72, 73–74, 97, 104–105, 118, 136–137, 145. See also *Necessity of Atheism, The*
autobiographical poem. See *Epipsychidion*

Bacon, Lord, 91, 130, 160, 174
Bagehot, William, 41
    "Percy Bysshe Shelley," 51–53
"Barbara" (Wordsworth), 155
bard, 45
beauty, 45, 102, 116, 123, 184
    absence of, love poems, 156
    blasphemy and, 98–99
    image/vison of, 69, 138, 140, 149
    "Intellectual Beauty," 91, 106, 137, 167, 169, 174, 186
    "realm of true beauty," 136
    strangeness and, 48
    See also imagery
Beddoes, Thomas Lovell, 43–44
Behrendt, Stephen C., 8
Bible/biblical subjects, 52
Bieri, James, 7
Biographica Literaria (Coleridge), 172
biographies, 7, 111–112, 114, 115, 118
biography, 1–3
    birth, 1, 58, 101
    children, 1, 2

death/cremation/burial, 2, 12, 19, 26, 41, 102, 103, 122, 125, 142, 155
health problems/depression, 14, 68, 102, 122, 154
Life of Percy Bysshe Shelley, 67
marriage/wives, 1, 10, 12, 54, 86
See also education
Blackmur, R.P., xi
Blackwood, John, 49
Blackwood, William, 80
Blackwood's (magazine), 16, 44, 49, 80
Blake, William, xii, 70, 170
blasphemy, 71, 82, 97
Blessington, Countess of, 9, 12–15
Bloom, Harold, xi–xii
boating, 55, 101, 121
Bradley, A.C.
"Shelley's View of Poetry," 171–186
"Bridal Song," 156
broadsides, 1
Brooke, Stopford A.
"Some Thoughts on Shelley," 141–157
Brooks, Cleanth, xi
Brown, Armitage, 21
Browning, Robert, xi, xii, 7, 8, 22, 45–46, 70
"Introduction," 109–124
Pauline, 110
Burns, Robert, 50, 52, 104, 155, 156
Byron, Lord, xii, 2, 9, 12, 15–16, 22, 23, 28, 30, 50, 56, 70, 84, 93–94, 95, 104, 105, 122, 155, 178
compared to Shelley, 124–126
revolution, sympathy for, 159
Vision of Judgment, 89
Byron-Shelley *clique*, 37

Calvert, George Henry, 67
Calvinism, 21
Cambridge Review, 79
Cameron, Kenneth Neill, 7

Cenci, The (verse tragedy), 2, 46, 54, 85–89, 98, 108, 156, 179
Beatrice in, 17, 86, 87, 88, 146
Cenci in, 17, 88, 181
moral ugliness/artistic beauty in, 88
murder in, 87
passage from, 146
Preface to, 180
as unrivalled, 123
censorship, 67–68, 97
central argument, 184
changefulness, love of, 147–149, 151, 153
character, 19, 30, 92, 101, 113, 116, 118
appearance and, 10, 11, 23, 24, 29, 90
benevolence/virtue, 30, 58
dining habits, 30
gentleman, 23, 37, 81, 82, 107
judging of Shelley's, 117, 131
"madness," 41, 50, 69, 107, 122
nonconformist, 162
self-preservation and, 28
sincerity/honesty, 48, 69, 91, 105, 119
temperament, 25, 69, 90, 131–132, 147, 154
thirst for knowledge, 29
Charles I, 2, 156
Chatterton, Thomas, 122
Chaucer, xii
choral dramas, 56
Christabel (Coleridge), 56
Christianity/Christians, 17, 20–21, 109, 170
capital dogmas of, 121
censorship by, 71
Christian virtues, 105
Churchdom and, 120
delusions of, genius and, 29
following Christ, 121, 145
fragmentary Essay on, 33
Clairmont, Claire, 2, 26, 36, 37

Clairmont, Mary Jane, 85
classes of poetry, 101
Clermont, Jane, 37
"Cloud, The," 101, 149, 152
Cockney School, 18, 80
Coleridge, Edward, 35
Coleridge, Samuel T. , xii, 9, 16, 46,
    47, 87, 121, 125, 134, 171, 177, 186
    Ancient Mariner, 56
    Biographica Literaria, 172
    Christabel, 56
    Kubla Khan, 56
    Poems on Various Subjects, 42
Coleridge, Sir John, 35
"Coliseum, A Fragment, The," 146
Collins, William, 50
composition, style of, 177–178. *See also* language/style
Congreve, William, 58
Coriolanus (Shakespeare), 156
Cory, William
    "Shelley at Eton," 34–36
Crane, Hart, xi, xii
cruelty, hatred of, 106, 107

Dante Alighieri, 182, 184
Vita Nuova, 132
dawn, poem on, 151–152
De Quincey, Thomas, 9, 86–87
    "Notes on Gilfillan's Literary Portraits," 16–18
death and life, 136, 140
Declaration of Rights, 71, 74
Defence of Poetry, A, 2, 33, 172, 173, 177, 180, 181, 182
desire, 163, 165
D'Holbach (baron), 106
Dickinson, Emily, xii
'didactic poetry,' 182, 183
discredit/harsh criticism, 76, 77, 97–98, 99, 105, 108, 117
disillusionment, 165
Donne, John, xii, 58
Dowden, Edward, 7
    "Last Words on Shelley," 157–166

dreams, 130, 133, 134, 136, 137, 154
    love dying into, 155
    realities and, 138, 139
    stars and, 168
    visions/ideals and, 162, 163
Dryden, John, 50
dualism, 182

Edinburgh Review, 17, 44
education, 105, 138. *See also* Eton College/Etonian; Oxford University
Egham Hill, 33
Eliot, George
    "Stradivarius," 165
Eliot, T.S., xi
Ellis, A., 67
Engelberg, Karsten Klejs, 8
England, 30, 46, 54, 61, 108, 142, 143
    abuses in, attacks on, 144
    literary movement in, 159
    poetic conventions in, 143
"English Poets of the Nineteenth Century" (Whipple), 104–109
enigmatical conduct, 105
Epipsychidion, 2, 52, 79, 140, 156, 172
    symbolism in, 167, 169
    vulgar/beauties of, 132
errors, 102, 105
    of excess, 97–98
    excused for, 63, 97, 161
    opposing views, 105–106
Essays, Letters from Abroad, Translations and Fragments, 3, 13
eternity, 136, 137, 154
Eton College/Etonian, 1, 53–54, 77–78
    "Shelley at Eton," 34–36
Euganean Hills, 150. *See also* "Lines written among the Euganean Hills"
evil and good, 119, 120, 143, 145
    attack on evil, poetry and, 144
    death as last evil, 136
    evil of matter, 144

evils poets suffer under, 141
returning good for evil, 161
Examiner, 18, 19, 81, 94
Excursion (Wordsworth), 149, 183
exile. *See under* Italy

Faery Queen (Spencer), 134
Falkner (Mary Shelley), 13
fantasies, 158, 162
Faust (Goethe), 166, 186
Fichte, Johann Gottlieb, 136–137
    Vocation of Man, 135–136
"Five English Poets" (Rossetti), 56
Fonblanque, Albany
    "Literary Notices," 94–96
Forman, H. Buxton, 8, 78
Forster, Joseph, 67
Fortunes of Perkin Warbeck (M. Shelley), 13
Four Ages of Poetry, The (Peacock), 2, 33
fragments, 33, 117, 121, 123, 143
Frankenstein (M. Shelley), 12
freedom, 56, 132, 186
    love of, 103, 107
    political freedom, 100, 159
French Revolution, 100, 159, 165
    Reign of Terror and, 144
Frost, Roberts, xii

Gardner, Margaret, 15–16
Garnett, Richard, 7, 8
genius, 10, 48, 50, 60, 90, 92, 96, 106, 109, 116, 118, 143, 167–168
    Christianity and, 29
    in composition, 108, 123
    creative mind/poetic genius, 117, 169
    ignored/abused in own time, 142
    infirmities of, 104
    irradiations of, 103
    pliancy of, 98
    of poet, imagination and, 184–185
    subjective/objective poet, 112

symbolism and, 70
    William Hazlitt on, 69
Gisborne, Maria, 37
Godwin, Mary Wollstonecraft, 2, 12, 23, 85. *See also* Shelley, Mary
Godwin, William, 1, 9, 12, 79, 80, 129, 134, 135, 159, 163, 186
    on *The Cenci*, 85–86
    perfectibility and, 137–138
    prosaic system of, 139
    "rights of man" and, 166
"Godwin and Shelley" (Stephen), 127–141
Godwinism, 127
Goethe, Johann Wolfgang von, 9, 126, 158, 178, 182, 185
    Faust, 166
    Werther, 166
Gordon, George. *See* Byron, Lord
Gosse, Edmund, 42
    "Shelley in 1892," 58–61
government, 82, 104, 107, 159
Graham, William, 36–37
Gray, Thomas, 50, 58
Greece, 30, 62, 165, 175
Greek-style poetry, 19, 47, 55, 151–152
    English poetry and, 61
gunpowder, use of, 35

Harden, W. Tyas, 8, 110
Hardy, Thomas, xi
"hate of hate," 121
Hazlitt, William, 8, 43, 80, 98
    conflicted view of, 69
    "On Paradox and Commonplace," 10–12
    on *Posthumous Poems*, 89–93
Hellas, 2, 14, 122, 133, 149, 167
Hermetists, 168
heroism, 87, 101, 106, 158–159, 163, 180
history, 129
Hogg, Thomas Jefferson, 1, 8, 9, 23–26, 31

The Life of Percy Bysshe Shelley, 7
Homer, 177, 178, 179, 181, 182, 184, 185
   Illiad, 46
humanity
   infinite capability of man, 108
   progress of/future state, 154, 162
   social improvement, 91, 93
   "subject to law," 159
Hume, David, 106
humourist, 56
Hunt, Leigh, 2, 8, 11, 14, 18–21, 23, 41, 46–47, 68, 80, 98, 163
   Examiner, 94
   Reflector, 43
   on Shelley, 109
"Hymn to Intellectual Beauty," 19, 172

"I fear thy kisses gentle maiden," 68
Ideal Pantheist, 145
idealism, 53, 133, 157, 166, 179–180
   abstractions and, 132, 181
   false ideals, 162
   ideal forest, 151
   ideal regret, 155
   moral excellence, 120, 183
   of reality, 102, 108
   transcendental/eternal world, 137
ideality, 56, 143, 155
Idylls of the King (Tennyson), 186
imagery, 41, 47, 59, 81, 108, 155
   celestial, 37, 52–53, 121, 140
   changefulness of, 147–148, 151, 153
   of landscapes, 134, 150–151
   sky/sunrise/sunset, 50, 116, 148, 149, 150, 156
   See also symbolism
imagination, 59, 74, 107, 137, 147, 160
   abstractions and, 106
   genius and, 184–185
   as instrument of moral good, 184
   language and, 175–176
   like minds and, 102
   passion of love and, 164
   scenery and, 151
   soul and, 174
imaginative poetry, 101
immortality, 29, 134
impiety, 72, 73
In Memoriam (Tennyson), 186
incest, 87
incomprehensibility, 130
   of ambiguous passages, 118
   of ideal world, 137
   metaphysics and, 50
indefinite, the, 146, 148–149, 155
individuality, 59, 143, 151–153
inspiration, 101
intellect, 97, 144, 140, 147, 166
   depth/fineness of, 106, 116
   "Intellectual Beauty," 91, 105, 137, 167, 169, 174, 186
   transcendent powers of, 8–9
"Invitation, The," 156
Ion, 102
Ireland, 1, 14, 165
irreligion, 109
Italy, 12, 20, 62, 101, 102, 143
   burial in, 142
   in exile in, 142, 145
   Leghorn (Livorno), 2, 14, 26, 29
   Lerici, 22, 148, 156
   Naples, 2, 68, 122, 172
   Pisa, 2, 14, 16, 22, 23, 122, 151, 156
   See also Rome, Italy

Jeaffreson, John Cordy, 76–79
"John Bunyan" (essay), 44–45
Jonson, Ben, xii, 55
joy, 18, 163, 165
Julian and Maddalo, 2, 122, 123, 149, 178
   famous saying in, 141

Junius, 122

Keats, John, xii, 44, 56, 70, 80, 81, 87, 125, 135, 143, 169, 170–171, 177, 178
    "Ode to Autumn," 126
    "realm of true beauty," 136
kingcraft, 144
Kipling, Rudyard, 36
Knight's Quarterly Magazine, 44
Kubla Khan (Coleridge), 56

La Fayette, Marquis de, 58
Lake Poets, 44
Lamb, Charles, 9, 19, 42–43
Lamb, John and Elizabeth, 42
Landor, Walter Savage, 125, 127
    "To Shelley," 21–22
language/style, 56, 99, 134, 136–137, 138
    allegory, 82, 89
    diction, 116, 177
    fragmentary style, 148
    imagination and, 175–176
    measured/metrical, 176
    metaphor, 172
    otherworldly qualities, 41
    poetic language, 81, 178, 179
    sanitation of, 67
"Laodameia" (Wordsworth), 155
Laon and Cyntha, 2, 167. See also *Revolt of Islam, The*
Last Man, The (Mary Shelley), 13
law, destruction of, 130
letters of Shelley, 7, 109–124
    forgeries, 32, 110–111
    supplementary letters, 118, 124
Liberal, 18
libertinism, 104
life and death, 136
"Life of Life" (hymn), 178
Life of Percy Bysshe Shelley, 23, 31
Life of the Poet (Medwin), 31
"Lines written among the Euganean Hills," 101, 126, 140

"Lines written in Dejection near Naples," 14
Lockhart, John Gibson, 44, 68–69
    "Observations on *The Revolt of Islam*," 80–81
Lodore (M. Shelley), 13
love, 9, 28, 101, 103, 113, 122–123, 182, 186
    human love/passion, 147, 155, 156, 164
    lovers, 117–118
    as ultimate reality, 137, 154, 184
    whom the gods love, 166
lyrical poems, 19, 51, 56, 143, 155, 164

Macaulay, Thomas Babington, 41
    "John Bunyan," 44–45
Maginn, William, 49
Marlowe, Christopher, xii
marriage, 129, 130–131, 132
Marvell, Andrew, xii
Marx, Karl, xi
"Mask of Anarchy, The," 2, 163
materialist, 134, 145, 158
Matilda (Mary Shelley), 12
"May-day Night," 96
Medwin, Thomas, 7, 31, 67
melancholy doctrine, 140–141
memoirs, 7, 117
Meredith, George
    "The Poetry of Shelley," 49
metaphysics, 13, 45, 46, 50, 89, 102, 134, 137, 143
metrical form/rhythm, 176. *See also* "singer"/songs
Michael Angelo, 55, 122, 181
Middleton, Charles S., 7
millennium, conception of, 128
Milton, John, xii, 46, 47, 50, 84, 85, 98, 126, 178, 180, 181, 184, 185
misanthropy, 53, 129
modernity, 61
Moir, David Macbeth, 41, 49–51
monomania, 17

Monroe, Harriet
    "With a Copy of Shelley," 57–58
mood, exalted poetic, 117, 118–119, 140
Moore, Thomas, 105
morality/moral good, 12, 50, 51, 105, 148, 157
    art and, 87, 185
    breaking moral laws, 131
    of the heart, 154
    misconstruction of nature, 117
    moral character of writing, 98
    moral nature/aim, 9, 17, 115, 116
    moral ugliness, 87
    moral weakness, 131
    reasoning and, 183–184
Morning and Evening Star, 70, 167–168
Moxon, Edward, 7, 32, 111

nature, 50, 126–127, 140, 151–153
    ephemeral scenery, 132–133
    evil outside of, 128
    love of, 56, 91
    observation of, 54, 55, 60
    representation of, 143, 145
    source of, 146–147
    "state of nature," 129
    See also imagery
Necessity of Atheism, The, 1, 23, 70–74
    Deity negation in, 71, 72, 73
New Critical disciples, xi

objective/subjective poet, 111–114
"Ode to Autumn" (Keats), 126
"Ode to Heaven," 146
"Ode to Liberty," 2, 144, 153–154, 167, 172
"Ode to Naples," 2, 123, 126
"Ode to the West Wind," 2, 150
Ollier, Edmund, 32
"On a poet's lips I slept," 154
oppression, hatred of, 54, 106, 107, 119
optimism/pessimism, 129, 159

Original Poetry by Victor and Cazire, 1
originality, 29, 41, 48, 84
Orlando Furioso, 177
Orphic style, 47
Ossians, 61
Ovid's fables, 90–91
Oxford University, 24, 25, 61–62, 79
    explusion from, 1, 23, 54, 71, 106

pamphlets, 1, 119–120, 139
panegyrists, 97
pantheism, 55, 145, 147, 151, 153, 155
paradox, 90, 92
Parkes, Kineton, 8
Parliament, 1, 20, 44, 45, 60, 94
Pauline (Browning), 110
Peacock, Thomas Love, 1, 2, 9, 23, 68, 172
    "Memoirs of Percy Bysshe Shelley," 33–34
Pearson, Howard S., 8
Perfection, 179–180, 181, 182
Persians, The (Aeschylus), 2
perversion in thinking, 81
pessimism/optimism, 129, 159
Peter Bell the Third, 2, 89
Petrarch, xi
philanthropy, 108, 131, 139
"Philosophical View of Reform, A," 2
philosophies, 12
    comprehension of, 43
    imperfection of, 140
    as radical, 11
    See also theories
"Philosophy of Shelley's Poetry, The" (Yeats), 167–171
Pilford, Shoreham and Elizabeth, 1
Pindar, xi
Plato, 106, 112, 137, 163, 170, 172, 174
    Praise of Love, 102, 148
Platonism, 181
Poe, Edgar Allan, 41
    "Elizabeth Barrett Browning," 47–48

poet
   described, 106
   double faculty of, 111
   poet's virtue, 116
poetasters/rhyme-stringers, 105
poetic conventions, 143
poetic faculty, two modes, 113–114
poetic form, attention to, 61
Poetical Works of Percy Bysshe Shelley, 3, 13, 68
poetry
   created from nothing, 89
   as interpretation of life, 184, 185
   subjects for, 51, 69
   true object of, 181
Poet's Corner, Westminster Abbey, 45
point of view
   appraisal of poetry and, 123
   moral point of view, 185
   view of life, 157
   youth and, 59, 102
polemical works, 56, 144, 145
political causes, 1, 97, 163
   communist, 54
   political freedom, 100
   republican, 54
   political philosophies, 8, 122
   political poems, 68, 144
Pope, Alexander, xii, 50
popular poetry, 101
Posthumoous Fragments of Margaret Nicholson (Shelley and Hogg), 1
Posthumous Poems of Percy Bysshe Shelley, 3, 43, 94–95, 96–98
   publication of, 13, 67, 69
power, 49, 87, 115, 121, 123, 172
   variety of, 98, 116
   weakness and, 153
   worship of, 137
Praise of Love (Plato), 102
"Preface" to Shelley's works, 8, 12–15, 69, 99, 100–103. See also *Posthumous Poems of Percy Bysshe Shelley*
Pre-Raphaelite Brotherhood, 56
priestcraft, 144, 145
Prince Athanase, 167, 168
Prometheus Unbound, 2, 11, 14, 54, 82–83, 85, 105, 108, 126, 128, 145, 151, 153
   conclusion of, 128
   Demogorgon in, 82, 130, 146
   doctrine/philosophy behind, 161
   ideals and, 162
   imagery in, 149
   Jupiter in, 82, 83, 129–130
   magnificent poetry of, 144
   mixed reviews for, 68, 70
   nymphs in, 140
   Panthea and Asia in, 133–134
   Preface to, 181
Proposal for Putting Reform to the Vote Throughout the Kingdom, A, 2
Providence, 104, 114

Quarterly Review, 35, 44, 81, 135
Queen Mab, 1, 54, 106, 121, 138, 149, 172
   crude poetry of, 127, 128
   incoherence in, 130

"radical" writing/views, 67, 69
Raleigh, Walter, 79–80
rapture, 164, 173
reality, 138, 173
   clear perception of, 163
   deepest realities, 154
   dreams and, 139
reason/reasoning, 138, 143, 183–184
"Recollection, The," 151, 156
reconstruction of society, 159
Redding, Cyrus, 7
Redpath, Theodore, 8
Reflector, 43
Reform Bill, 165
reformist theories, 12

Reiman, Donald H., 7, 8
religion, 51, 82, 105, 186
    antagonism to, 17, 56, 107
    boyish doubts about, 106
    opinions/philosophies on, 8, 52, 105
    as "prolific fiend," 128
    realm of true Beauty and, 136
    right/will of God, 145
Revolt of Islam, The, 2, 54, 80–81, 105, 138, 144, 162
    cruelties of schools in, 35
    mixed reviews for, 68, 69, 70
    polemical element in, 144, 145
    Preface to, 178
    scenery in, 143, 150
revolution/revolutionary instinct, 61, 71, 125, 138, 144, 158, 159, 162. *See also* French Revolution; government
Robinson, Henry Crabb, 9
Rogers, Samuel, 9, 22–23
romantic poets, 8, 67
Rome, Italy, 14, 25, 88, 123, 175
    disposition of ashes in, 2–3, 15
    Protestant Cemetery in, 26–27
    true poetry of, 175
Rosalind and Helen, 2, 144, 151
Rossetti, Dante Gabriel, 62
"Percy Bysshe Shelley," 56–57
Rossetti, William Michael, 26
Rousseau, Jean-Jacques, 129, 162, 163, 167, 168, 170
Rowley poetry, 122

Saintsbury, George, 42, 62–63
"satanic school, the," 72
satirist/satire, 56, 89, 129
Saturday Review, 35, 62
scandalous material, 68
Scott, R. Pickett, 50, 86–89, 157
Scott, Walter, 80
senses, the, 71
"Sensitive Plant, The," 2, 48, 46, 152
    passage from, 135

    stanzas deleted from, 68
sexual oppression, 120
Shakespeare, William, xi, xii, 23, 50, 55, 85, 109, 126, 133, 166, 178, 179, 182, 184
    Coriolanus, 156
    familiar text of, 133
    Hamlet, 185
    King Lear, 180, 185
    Macbeth, 180
    Merchant of Venice, 46
    Othello, 111
    Richard the Second, 19
    The Tempest, 133
Shelley, Lady Jane, 157
"Preface by the Editor," 30–33
Shelley, Mary, 2, 3, 8, 26, 169
    Frankenstein, 12
    Journal, 3
    passage by, 95
    publication of husband's works, 67, 70
    See also Godwin, Mary Wollstoncraft; "Preface" to Shelley's works
Shelley, Percy Florence, 12, 13, 32, 68, 157
Shelley, Sir Bysshe, 1, 2
Shelley, Sir Timothy, 1, 13, 33, 67, 68, 69
Shelley, William, 12
Shelley Society, 8, 70, 110, 141–142
Shelleyans, 78, 84
Shelleyolaters, 37
"Shelley's View of Poetry" (Bradley), 171–186
Sheridan, Richard Brinsley, 22, 23
"shocking" the audience, 8
"Similes for Two Political Characters of 1819," 68
"singer"/songs, 70, 154, 157, 164
    musical verse, 89
    singing-god, 125
Skipsey, Joseph, 7
skylark, 49, 90. *See also To a Skylark*

"Skylark" (Wordsworth), 152
Smith, George Barnett, 7, 8
Socrates, 102, 163, 170
solitude, 13
somnambulism, 123
"Sonnet: England in 1819," 2
sorrow, 105–106, 179–180
Sotheran, Charles, 8
Southey, Robert, 1, 7, 9, 22, 125, 127, 134
Spenser, Edmund, xi, xii, 42, 55, 63, 177, 182
  Faery Queen, 134
spirituality/soul, 101, 104, 108–109, 122, 158, 186
  attributes of soul, 115, 163
  'divinity' in man, 173
  imagining soul, 174
  spiritual audacity, 56
  spiritual comprehension, 113
  spiritual transparency, 116
St. Irvyn; or The Rosicrucian, 1, 74–76, 79
St. John, Jane Gibson, 13
Stacey, Sophia, 68
Stephen, Sir Leslie
  "Godwin and Shelley," 127–141
Stevens, Wallace, xii
"Stradivarius" (Eliot), 165
sunrise/sunset, 133
superstition, 139
Swift, Jonathan, 53, 129
Swinburne, Algernon Charles xi, 42, 58, 59, 70
  "Notes on the Text of Shelley," 124–127
  "The Centenary of Shelley," 61–62
Sydney, Sir Phillip, 55
Sykes, Frederick Henry, 67
symbolism, 130, 133, 166
  fantastic figures, 54
  moon, 169–170, 171
  stars, 70, 167, 168, 170, 171
  sun, 170

See also imagery
Symonds, John Addington, 7, 41, 42, 55–56
sympathies, 12, 91, 103, 112, 118, 119, 139
  critical sympathy, absence of, 143
  for nature and mankind, 104, 154
  for the oppressed, 106, 107, 119–120
  poetic sensibility and, 132
  sheltered from, 102
  universality of, 108
Symposium, 2, 102

Taine, Hippolyte, 41, 53–55
Tate, Allen, xi
Taylor, Jeremy, 98, 170
Tennyson, Alfred, xii, 178
  Idylls of the King, 186
  In Memoriam, 186
  Ulysses, 186
Thames River, 55, 101, 143
theists, 72, 134
theology, 144–145, 151. *See also* religion
theories, 120, 138, 157
  Godwin's theories, 128
  of government, 104
  ideal perfection, 179–180
  metaphysical/ethical, 45
  ultimate reality, 137
"To a Cloud," 2
"To a Lady, with a Guitar," 47
To a Skylark, 2, 46, 101, 140, 152, 153, 172–173
"To Night," 156
Todhunter, John, 7
Tories, 18, 80
transcendental/eternal world, 137, 139, 140
translations, 96
  Hymn to Mercury, 176
  Praise of Love, 102

Trelawny, Edward John, 8, 9, 23, 26–30
Trent, William P., 68
    "Apropos of Shelley," 84–85
Triumph of Life, The, xi, 2, 96, 101, 151, 156–157, 167, 168–169
    as last production, 14, 163
    symbolism in, 170
Trotsky, Leon, xi
truth, 50, 61, 91, 117, 121
    fact and, 137
    interests of mankind and, 72, 73
    poetic expression of, 173
    purity of, 119
Turner, J.M.W., 134, 150
tyrants/tyranny, 139
    antagonism to, 56, 107

Ulysses (Tennyson), 186
universality of poetry, 154
University College, Oxford, 1
unpopularity, 104
    haters of Shelley, 118
unreality, 108–109
Utopia, 52, 53, 153

Valperga (Mary Shelley), 13
Verocchio, Andrea, 124
Vision of Judgment (Byron), 89
"Vision of the Sea," 150
Vita Nuova (Dante), 132
Viviani, Emilia, 2, 170
Vocation of Man (Ficte), 135

Wallenstein (Schiller), 17
Werther (Goethe), 166
Westbrook, Harriet, 1, 12, 23, 157
Wheatley, Kim, 8
Whipple, Edwin P., 70, 104–109
White, Newman Ivey, 7, 8

Whitman, Walt, xii, 61
Williams, Edward, 2, 14, 23, 26, 27
Williams, Jane, 2, 23, 27, 37
Wilson, John, 41, 49
    "Preface," 44
wisdom, 159–160
Wise, Thomas James, 7
Witch of Atlas, The, 2, 14, 46, 68, 99, 101
"With a Copy of Shelley" (Monroe), 57–58
Wollstonecraft, Mary, 12, 85
Woodberry, George Edward, 67
Wordsworth, William, xii, 9, 16, 22, 41–42, 53, 125, 145–146, 147, 158, 163, 164–165, 171, 177, 178, 186
    antithesis to Shelley, 56
    "Barbara," 155
    descriptive power of, 151
    Excursion, 149, 183
    "Laodameia," 155
    nature and, 152, 153
    passage by, 52
    Prefaces of, 172
    "Skylark," 152
    universality of, 154
worshippers of Shelley, 132

Yeats, William Butler xi, xii, 70,
    "The Philosophy of Shelley's Poetry," 167–171
youthfulness, 59, 102, 118, 120, 122

Zastrozzi, 1, 74–76, 77–79, 122
    Adonis Verezzi in, 74, 75–76
    dagger scene in, 77
    Matilda in, 75–76
    negative commentary on, 68
    Zastrozzi in, 74, 75–76

PR
5438
.P435
2009